Aaron Philippe Toll (Ed.)

Dragon Ball Z: Ultimate Tenkaichi

Aaron Philippe Toll (Ed.)

Dragon Ball Z: Ultimate Tenkaichi

Dragon Ball, Manga, Anime, Spike (company), Namco Bandai Games, PlayStation 3

Ceed Publishing

Imprint

Publisher:
Ceed Publishing is a trademark of
International Book Market Service Ltd., 17 Rue Meldrum, Beau Bassin, 1713-01 Mauritius
Email: info@bookmarketservice.com
Website: www.bookmarketservice.com

Published in 2011

Printed in: U.S.A., U.K., Germany. This book was not produced in Mauritius.

ISBN: 978-613-9-99727-5

Contents

References

Dragon Ball Z: Ultimate Tenkaichi

Dragon Ball Z: Ultimate Tenkaichi	
Developer(s)	Spike
Publisher(s)	Namco Bandai
Composer(s)	Hiroshi Takaki
Platform(s)	PlayStation 3 Xbox 360
Release date(s)	• JP December 8, 2011 • NA October 25, 2011 • EU October 28, 2011
Genre(s)	Versus fighting
Mode(s)	Single player Multiplayer Online multiplayer
Rating(s)	• CERO: A • ESRB: T • PEGI: 12
Media/distribution	Blu-ray Disc DVD

Dragon Ball Z: Ultimate Tenkaichi, known in Japan as ***Dragon Ball: Ultimate Blast*** (*Doragon Bōru Arutimetto Burasuto*), is a video game based of the manga and anime franchise *Dragon Ball*. It is developed by Spike and will be distributed by Namco Bandai beginning in late October of 2011 for the PlayStation 3 and Xbox 360 gaming consoles.

The game is a 3D fighter that allows players to take controls of various characters from the *Dragon Ball* franchise or created by the player to either fight against the AI, or with another player locally or online.

Development

The game was first announced under the working title ***Dragon Ball Game Project Age 2011*** in the early May issue of *Weekly Shonen Jump* magazine. The announcement featured several screenshots with Son Goku and Vegeta, in both normal and Super Saiyan forms, highlighting the game's mechanics. The announcement stated that the game would feature superb visuals and would be released in Japan sometime in autumn of 2011 for both PlayStation 3 and Xbox 360 gaming consoles.[1] A few days later the North American branch of Namco Bandai released a press statement announcing that they would be releasing the game throughout North America. They also confirmed that Spike would return to work on the game.[2] The June issue of *V Jump* magazine showcased several screenshots of

Goku and Vegeta, in his Oozaru form, demonstrating a type of timed button action mechanic. It also confirmed the return of the destructible environments mechanics from previous *Budokai Tenkaichi* and *Raging Blast* games.[3]

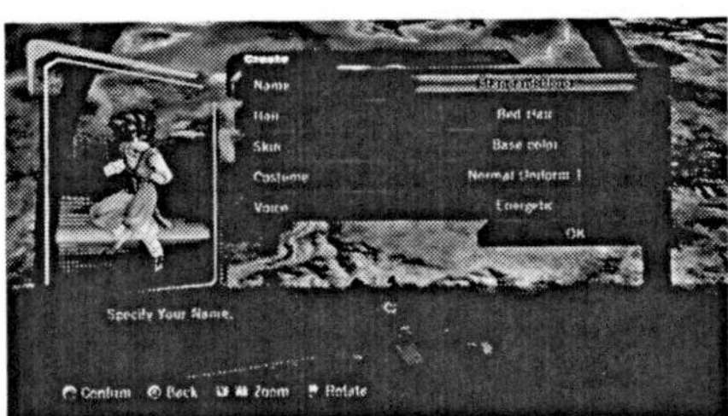
Hero Mode allows players to create a new character using new and existing attributes for hair, complexion, clothing, voice, and attack moves.

In the following Namco Bandai press release, it was announced that the game's official English title would be *Ultimate Tenkaichi* and that its release date would be October 25,[4] while its European release date was revealed to be October 28. [5] The title was chosen by fans through a survey conducted by Namco Bandai along with four other possible choices.[6] In response, Namco Bandai marketing president Carlson Choi stated their reasoning for the outreach by describing the fans as "the most passionate gamers and anime experts in the world."[4] Later in July, it was announced that the game would not only include the return of *GT* characters to the roster, but also include a type of character creation and customization mode.[7] [8] The August 8 issue of *Shonen Jump* featured screenshots of the story mode, showcasing a type of free-roam mechanic similar to that found in *Budokai 3* and the later *Budokai Tenkaichi* games, where the player could explore the entire planet from the sky and land in certain areas.[9] The following issue of *V Jump* revealed that the game will be released under the title of *Ultimate Blast* in Japan, with the release date for December 8.[10] In early September, more details were divulged on the character creator mode referred to as "Hero Mode". The mode gives players the option of customizing a character of their choosing with a variety of existing and original outfits and hairstyles, in addition to the ability to change and adjust their color schemes. [11] [12] The following issue of *Shonen Jump* revealed that Japanese release would include a new opening introduction in traditional 2D animation set to the song "Cha-La Head-Cha-La" by Hironobu Kageyama.[13]

Promotions

Preorders for the Japanese release will come with a life-sized replica of the *Sūshinchū* (*Four Star Ball*).[10] While in Europe, preorders will come a free capsule pen with their purchase.[5] The European version will also have a special collectors edition that will come boxed with a Son Gohan figure in Super Saiyan form dressed in his Great Saiyaman costume.[14]

Reception

IGN gave the game a 7.5 out of 10, praising the Graphics and Explosive Combat, but disliked the storytelling and the repetitive nature of the battles.[22] [23]

Gamespot gave the game a 4.0 out of 10, praising the Graphics, but detested the game's Combat System as there is no sense of originality; the Boss Fights due to the frustrating Quick Time Event Stages; and the Character Creation System due to the limited amount of items the game has given. [20] [21]

Destructoid gave the game a 7.5 saying "While combat can be a bit on the repetitive side, there's still that hint of satisfaction when you decimate your foe with an canyon-creating power move. There's a lot here for fans to enjoy, and while the story may be streamlined for new fans of the franchise, I can't really recommend it to anyone but the hardcore." [24]

References

[1] "Dragon Ball Game Project Age 2011!!! [Dragon Ball Game Project Age 2011 Dynamic Level!!!]" (in Japanese). *Shonen Jump* (101-8050 Chiyoda, Tokyo Hitotsubashi 2-5-10: Shueisha): 19. May 8, 2011.

[2] "Namco Bandai Games Announces New Dragon Ball Game Project For North America, Dragon Ball Game Project Age 2011 scheduled for release on Xbox 360 and PlayStation 3 for Fall 2011." (http://ps3.ign.com/articles/116/1167580p1.html). San Francisco, California: IGN. May 11, 2011. . Retrieved July 28, 2011.

[3] "!!! [Maximum impact!!!]" (in Japanese). *V Jump* (101-8050 Chiyoda, Tokyo Hitotsubashi 2-5-10: Shueisha): 81. June 2011.

[4] Rojas (June 30, 2011). "NAMCO BANDAI GAMES LAUNCHES DRAGON BALL Z® ULTIMATE TENKAICHI THIS OCTOBER" (http://www.funimation.com/rojas/blog/namco-bandai-games-launches-dragon-ball-z-ultimate-tenkaichi-this-october/221661). Fort Worth, Texas: Funimation. . Retrieved July 29, 2011.

[5] "The fans have decided! Namco Bandai Games launches DRAGON BALL Z ULTIMATE TENKAICHI on October 28th 2011" (http://www.uk.namcobandaigames.eu/news/2011/06/30/the-fans-have-decided-namco-bandai-games-launches-dragon-ball-z-ultimate-tenkaichi-on-october-28th-2011/1865). Namco Bandai EU. June 30, 2011. . Retrieved July 30, 2011.

[6] "Dragon Ball Games Survey" (http://www.surveymonkey.com/s.aspx?sm=dwYOalL5LucyUkJUYv0BVw==). Palo Alto, California: SurveyMonkey. May 11, 2011. . Retrieved July 28, 2011.

[7] Amanda L. Kondolojy. "Goku Comes Back for Another Round" (http://www.cheatcc.com/ps3/rev/dragonballzultimatetankaichipreview.html#.TjP2Y2GwXNY). Cheat Code Central. . Retrieved July 30, 2011.

[8] Sutton, Kye (July 14, 2011). "DragonBall Z: Ultimate Tenkaichi to feature create-a-character mode!" (http://www.mcmbuzz.com/2011/07/14/dragonball-z-ultimate-tenkaichi-to-feature-create-a-character-mode/). MCM Buzz. . Retrieved July 30, 2011.

[9] "!!D B A g e !!! [Story mode's long-awaited release!! "DB" AGE Giant Jump!!!]" (in Japanese). *Shonen Jump* (101-8050 Chiyoda, Tokyo Hitotsubashi 2-5-10: Shueisha): 19. August 8, 2011.

[10] " D B !! !!! ["DB" Experience the story of Bardock!! The Super Dimension Fly Away!!!]" (in Japanese). *V Jump* (101-8050 Chiyoda, Tokyo Hitotsubashi 2-5-10: Shueisha): 135 & 136. August 2011.

[11] Sutton, Kye (September 2, 2011). "Bandai Sama comes to the rescue in Ultimate Tenkaichi's Hero Mode!" (http://www.mcmbuzz.com/2011/09/02/bandai-sama-comes-to-the-rescue-in-ultimate-tenkaichis-hero-mode/). MCM Buzz. . Retrieved September 3, 2011.

[12] Velez, Joshua (September 1, 2011). "Dragon Ball Z Ultimate Tenkaichi's Character Creation Trailer" (http://www.dualshockers.com/2011/09/01/dragon-ball-z-ultimate-tenkaichis-character-creation-trailer/). Dualshockers. . Retrieved September 3, 2011.

[13] " !! !!! [Kimi public who battle!! Avatar thousand times over!!!]" (in Japanese). *Shonen Jump* (101-8050 Chiyoda, Tokyo Hitotsubashi 2-5-10: Shueisha): 135. September 12, 2011.

[14] Sutton, Kye (September 2, 2011). "DragonBall Z: Ultimate Tenkaichi limited collector's edition detailed" (http://www.mcmbuzz.com/2011/09/02/dragonball-z-ultimate-tenkaichi-limited-collectors-edition-detailed/#comment-560). MCM Buzz. . Retrieved September 3, 2011.

[15] "Dragon Ball Z: Ultimate Tenkaichi for PlayStation 3 - GameRankings" (http://www.gamerankings.com/ps3/630324-dragon-ball-z-ultimate-tenkaichi/index.html). Metacritic. . Retrieved 2011-10-29.

[16] "Dragon Ball Z: Ultimate Tenkaichi for Xbox 360 - GameRankings" (http://www.gamerankings.com/xbox360/630325-dragon-ball-z-ultimate-tenkaichi/index.html). Metacritic. . Retrieved 2011-10-29.

[17] "Dragon Ball Z: Ultimate Tenkaichi Critic Reviews for PlayStation 3 at Metacritic.com" (http://www.metacritic.com/game/playstation-3/dragon-ball-z-ultimate-tenkaichi/critic-reviews). Metacritic. . Retrieved 2011-10-29.

[18] "Dragon Ball Z: Ultimate Tenkaichi Critic Reviews for Xbox 360 at Metacritic.com" (http://www.metacritic.com/game/xbox-360/dragon-ball-z-ultimate-tenkaichi/critic-reviews). Metacritic. . Retrieved 2011-10-29.

[19] McKinley, Noble (2011-10-26). "Dragon Ball Z Ultimate Tenkaichi Review from GamePro" (http://www.webcitation.org/query?url=http://www.gamepro.com/article/reviews/224326/review-dragon-ball-z-ultimate-tenkaichi-360-ps3/&date=2011-11-30+22:31:06). GamePro. Archived from the original (http://www.gamepro.com/article/reviews/224326/review-dragon-ball-z-ultimate-tenkaichi-360-ps3/) on 2011-12-01. . Retrieved 2011-10-29.

[20] Petit, Carolyn (2011-10-28). "Dragon Ball Z: Ultimate Tenkaichi Review, Dragon Ball Z: Ultimate Tenkaichi Xbox 360 Review - GameSpot.com" (http://www.gamespot.com/xbox360/action/dragon-ball-game-project-age-2011/review.html). GameSpot. . Retrieved 2011-10-23.

[21] Petit, Carolyn (2011-10-28). "Dragon Ball Z: Ultimate Tenkaichi Review, Dragon Ball Z: Ultimate Tenkaichi PS3 Review - GameSpot.com" (http://www.gamespot.com/ps3/action/dragon-ball-game-project-age-2011/review.html). GameSpot. . Retrieved 2011-10-29.

[22] George, Richard (2011-10-25). "Dragon Ball Z: Ultimate Tenkaichi Review - PlayStation 3 Review at IGN" (http://ps3.ign.com/articles/121/1210625p1.html). IGN.com. . Retrieved 2011-10-29.

[23] George, Richard (2011-10-25). "Dragon Ball Z: Ultimate Tenkaichi Review - Xbox 360 Review at IGN" (http://xbox360.ign.com/articles/121/1210630p1.html). IGN.com. . Retrieved 2011-10-29.

[24] http://www.destructoid.com/review-dragon-ball-z-ultimate-tenkaichi-215488.phtml

External Links

- Official website (http://http://www.namcobandaigames.com/console/dragon-ball-z-ultimate-tenkaichi)
- Official website (http://http://b.bngi-channel.jp/dba/) (Japanese)

Dragon Ball

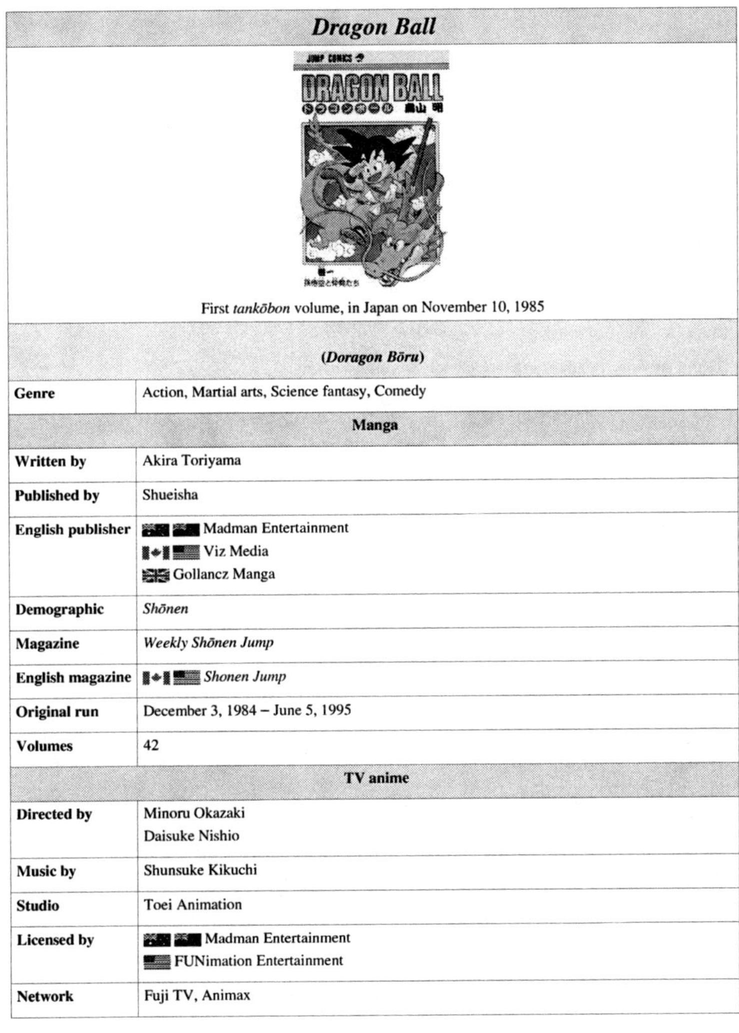

Dragon Ball	
First *tankōbon* volume, in Japan on November 10, 1985	
(Doragon Bōru)	
Genre	Action, Martial arts, Science fantasy, Comedy
Manga	
Written by	Akira Toriyama
Published by	Shueisha
English publisher	Madman Entertainment Viz Media Gollancz Manga
Demographic	*Shōnen*
Magazine	*Weekly Shōnen Jump*
English magazine	*Shonen Jump*
Original run	December 3, 1984 – June 5, 1995
Volumes	42
TV anime	
Directed by	Minoru Okazaki Daisuke Nishio
Music by	Shunsuke Kikuchi
Studio	Toei Animation
Licensed by	Madman Entertainment FUNimation Entertainment
Network	Fuji TV, Animax

English network	Cartoon Network (Toonami) YTV ANT1 Cartoon Network GMA-7, RPN-9 CNX, Toonami Cartoon Network (Toonami), Colours TV, Funimation Channel, Toonami Jetstream, KIKU 2x2
Original run	February 26, 1986 – April 12, 1989
Episodes	153
TV anime	
Dragon Ball Z	
Directed by	Daisuke Nishio
Music by	Shunsuke Kikuchi
Studio	Toei Animation
Licensed by	Madman Entertainment Manga Entertainment [1] FUNimation Entertainment
Network	Fuji TV, Animax, Tokyo MX
English network	Network Ten, Cartoon Network (Toonami) YTV ANT1 Cartoon Network, TV3 GMA-7, RPN-9 Cartoon Network, CNX, Toonami Cartoon Network (Toonami) 2x2
Original run	April 26, 1989 – January 31, 1996
Episodes	291
TV anime	
Dragon Ball GT	
Directed by	Osamu Kasai
Music by	Akihito Tokunaga
Studio	Toei Animation
Licensed by	Madman Entertainment FUNimation Entertainment
Network	Fuji TV, Animax
English network	Network Ten, Cartoon Network (Toonami) YTV Cartoon Network, TV3 GMA-7 Cartoon Network (Toonami) CNX, Toonami
Original run	February 7, 1996 – November 19, 1997
Episodes	64
TV anime	

Dragon Ball Kai / Dragon Ball Z Kai	
Directed by	Yasuhiro Nowatari
Music by	Kenji Yamamoto (1–95), Shunsuke Kikuchi (96–97, Extra)
Studio	Toei Animation
Licensed by	FUNimation Entertainment
Network	Fuji TV
English network	Nicktoons, The CW (Toonzai)
Original run	April 5, 2009 – March 27, 2011
Episodes	98
Related	

• Films, specials, and original video animations
• Video games
• Soundtracks
• Collectible card games
  • *Dragon Ball*
  • *Dragon Ball Z*	

Dragon Ball (*Doragon Bōru*) is a Japanese manga series written and illustrated by Akira Toriyama. It was originally serialized in *Weekly Shōnen Jump* from 1984 to 1995; later the 519 individual chapters were published into 42 *tankōbon* volumes by Shueisha. *Dragon Ball* was inspired by the classical Chinese novel *Journey to the West*. It follows the adventures of Goku from his childhood through adulthood as he trains in martial arts and explores the world in search of the seven mystical orbs known as the Dragon Balls, which can summon a wish-granting dragon when gathered. Along his journey, Goku makes several friends and battles a wide variety of villains, many of whom also seek the Dragon Balls for their own desires.

The 42 *tankōbon* have been adapted into two anime series produced by Toei Animation: *Dragon Ball* and *Dragon Ball Z*, which together were broadcast in Japan from 1986 to 1996. Additionally, Toei has developed seventeen animated feature films and three television specials, as well as an anime sequel titled *Dragon Ball GT*, which takes place after the events of the manga. From 2009 to 2011, Toei broadcast a revised, faster-paced version of *Dragon Ball Z* under the name of *Dragon Ball Kai*, in which most of the original version's footage not featured in the manga was removed. Several companies have developed various types of merchandising such as a collectible trading card game, and a large number of video games.

The manga series was licensed for an English-language release in North America by Viz Media, in the United Kingdom by Gollancz Manga, Australia and New Zealand by Chuang Yi and Malay-language release in Malaysia by Comics House. The entire anime series has been licensed by Funimation Entertainment for an English-language release in the United States, although the series has not always been dubbed by the same studio. In China, a live-action film adaptation was produced in 1989. In 2002, 20th Century Fox acquired the rights to produce an American-made live-action film that received a negative reception from critics and fans; the movie was released on April 10, 2009 in the United States.

Since its release, *Dragon Ball* has become one of the most successful manga and anime series of all time. The manga's 42 volumes have sold over 152 million copies in Japan and more than 200 million copies worldwide. Reviewers have praised the art, characterization, and humor of the story. It is widely regarded as one of the greatest manga series ever made, with many manga artists citing *Dragon Ball* as a source of inspiration for their own now popular works. The anime, particularly *Dragon Ball Z*, is also highly popular in various countries and was arguably one of the most influential in greatly boosting the popularity of Japanese animation in Western culture.

Plot summary

The series begins with a young monkey-tailed boy named Goku befriending a teenage girl named Bulma. Together, they go on a quest to find the seven Dragon Balls. Goku later undergoes rigorous training regimes and educational programs under the martial artist Master Roshi in order to fight in the World Martial Arts Tournament, a competition involving the most powerful fighters in the world. During his training time with Roshi, he meets a bald Shaolin monk named Krillin, who starts out as Goku's rival classmate; however, they soon become best friends. Outside the tournaments, Goku faces diverse villains such as Emperor Pilaf, the Red Ribbon Army, and the Namekian Piccolo Daimao. Several of the enemies and rivals Goku encounters eventually become his allies and close friends, including the desert bandit Yamcha, the assassin Tien Shinhan, the samurai Yajirobe, and Piccolo Daimao's offspring/reincarnation Piccolo Jr.

As a young adult, Goku meets his older brother Raditz, who reveals to him that they are members of a nearly extinct extraterrestrial race called the Saiyans. The Saiyans had sent Goku to Earth as an infant to conquer the planet for them, but he suffered a severe head injury soon after his arrival and lost all memory of his mission, as well as his blood-thirsty Saiyan nature. Goku refuses to help Raditz continue the mission, and afterwards begins to encounter other enemies from space, most notably the Saiyan prince Vegeta, who becomes his rival and eventually another ally. Goku also encounters Frieza, the galactic emperor responsible for the destruction of almost the entire Saiyan race, whose actions cause Goku to transform into a fabled Super Saiyan. After an epic battle on the planet Namek, Goku defeats Frieza, avenging the lives of billions across the galaxy.

Some time later, a group of androids from the former Red Ribbon Army appear, seeking revenge against Goku. During this time, an evil life form called Cell emerges and, after absorbing two of the androids to increase his power, holds his own fighting tournament to decide the fate of the Earth called the Cell Games. However, Cell is eventually defeated by Goku's son Gohan. Seven years later, Goku is drawn into another battle for the universe against a magical monster named Majin Buu. After many battles, Goku destroys Buu. Ten years later, at another World Martial Arts Tournament, Goku meets Buu's human reincarnation, Uub. Leaving the match between the two of them unfinished, Goku takes Uub away on a journey to train him as the Earth's next defender.

Themes

At its core, *Dragon Ball* maintains the central tenets of the *Weekly Shōnen Jump* philosophy of "friendship, struggle, and victory". As the series shifts from a "heart warming" story into a more action-oriented piece, the protagonists go through an unending cycle of fighting, winning, losing and improving. They continue this cycle by using miraculous devices to achieve life after death and continue to learn lessons as they defeat their challengers.[2] The series also follows the idea that people can better themselves and achieve their goals by constantly challenging themselves.[3]

Production

Wanting to break from the Western influences common in his other series, Akira Toriyama loosely modeled *Dragon Ball* on the classic Chinese novel *Journey to the West*.[4] [5] He also redeveloped one of his earlier one shot manga series, *Dragon Boy*, which was initially serialized in *Fresh Jump* and released in a single *tankōbon* volume in 1983.[5] This short work combined the comedic style of Toriyama's successful six-year series *Dr. Slump* with a more action-oriented plot and paid homage to famous martial art actor Jackie Chan.[5] [6] Toriyama notes that his goal for the series was to tell an "unconventional and contradictory" story.[7]

In the early concept of the series, Goku and Piccolo were from Earth. With the introduction of Kami, the idea of having fights from other planets was established and Goku and Piccolo were changed to alien species.[8] For the female characters, Toriyama felt it was not fun to draw "weak females" so he created women that he felt were not only "beautiful and sexy", but also "strong".[7] Going against the normal convention that the strongest characters should be the largest in terms of physical size, he designed many of *Dragon Ball*'s most powerful characters with small statures, including the protagonist, Goku.[7]

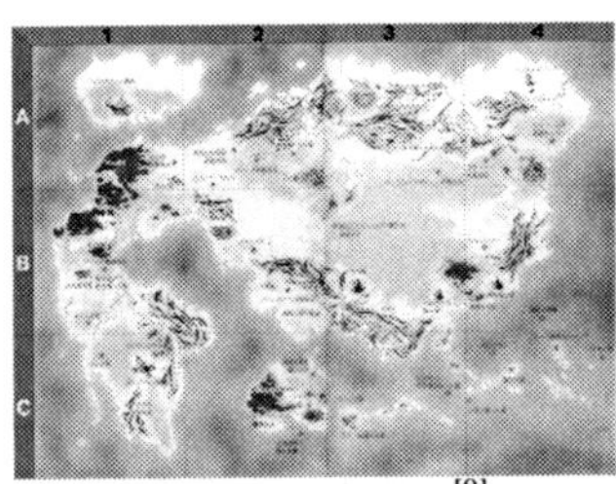
The Earth of *Dragon Ball*[9]

The fighting techniques were initially unnamed, but the series editor felt it would be better to name them all. Toriyama proceeded to create names for all of the techniques, except for the *Kamehameha* which his wife named when Toriyama was indecisive about what it should be called.[8] When creating the fictional world of the series, Toriyama decided to create it from his own imagination to avoid referencing popular culture. However the island where the World Martial Arts Tournament is held is modeled after Bali. When including fights in the manga, Toriyama had the characters go to uninhabited locations to avoid difficulties in drawing destroyed buildings. In order to advance the story quickly, he also gave most fighters the ability to fly so they could travel to other parts of the world without inconvenience. This was also the reasoning behind Goku learning to use *Instant Transmission* (thus allowing characters to teleport to any planet in a second).[8]

After the first chapters were released, readers commented that Goku seemed rather plain, so his appearance was changed. New characters (such as Master Roshi and Krillin) were added and martial arts tournaments were included to give the manga a greater emphasis on fighting. Anticipating that readers would expect Goku to win the tournaments, Toriyama had him lose the first two while continuing his initial goal of having Goku be the champion and hero. After Cell's death, he intended for Gohan to replace Goku as the series' protagonist, but then felt the character was not suited for the role and changed his mind.[10]

Toriyama based the Red Ribbon Army on a video game he had played named *Spartan X* in which enemies tended to appear very fast. After the second tournament concluded, Toriyama wanted to have a villain who would be a true "bad guy". After creating Piccolo as the new villain, he noted that it was one of the most interesting parts of the stories and that he and his son became the favorite characters of the series. With Goku established as the strongest fighter on Earth, Toriyama decided to increase the number of villains that came from outer space. Finding the escalating enemies to be a pain to work with, he created the Ginyu Force to add more balance to the series.[10] During this period of the series, Toriyama placed less emphasis on the series' art work, simplifying the lines and sometimes making things "too square." He found himself having problems determining the colors for characters and sometimes ended up changing them unintentionally mid-story.[6] In later accounts, Toriyama noted that he didn't plan out the details of the story, resulting in strange occurrences and discrepancies later in the series.[11]

Media

Manga

Dragon Ball

Written and illustrated by Akira Toriyama, *Dragon Ball* was initially serialized in the manga anthology *Weekly Shōnen Jump* starting on December 3, 1984.[5] The series ended on June 5, 1995 when Toriyama grew exhausted and felt he needed a break from drawing.[5] The 519 individual chapters were published into 42 *tankōbon* volumes by Shueisha from November 10, 1985 through August 4, 1995.[12] [13] [14] In 2004, the chapters were re-released in a collection of 34 *kanzenban* volumes, which included a slightly rewritten ending, new covers, and color artwork from its *Weekly Shōnen Jump* run. Toriyama also created a short series, *Neko Majin*, that became a self-parody of *Dragon Ball*. First appearing in *Weekly Shōnen Jump* in August 1999, the eight chapter series was released sporadically until it was completed in 2005. These chapters were compiled into a "kanzenban"-style package for release in Japan on April 4, 2005.[15]

The *Dragon Ball* manga was licensed for release in English in North America by Viz Media which has released all 42 volumes. Viz released volumes 17 through 42 under the title *Dragon Ball Z* to mimic the name of the anime series adaptated from those volumes, feeling it would reduce the potential for confusion by its readers. They initially released both series in a monthly comic book format starting in 1998, and later began collecting them in graphic novels.[16] [17] The first 10 volumes of both series were re-released from March to May 2003 under the "Shonen Jump" format, with *Dragon Ball* being completed on August 3, 2004 and *Dragon Ball Z* finishing on June 6, 2006.[18] [19] In June 2008, Viz began re-releasing the two series in a wideban format called "VIZBIG Edition," which collects three individual volumes into a single large volume.[20] [21]

In 2006, Toriyama and *One Piece* author Eiichiro Oda teamed up to create a single chapter crossover of their individual hit series. Entitled *Cross Epoch*, the chapter was published in the December 25, 2006 issue of *Weekly Shōnen Jump*.

Spin-offs

A manga adaptation of *Dragon Ball: Yo! Son Goku and His Friends Return!!* illustrated by Ooishi Naho, was published in the March 21, 2009 and April 21, 2009 issues of *V Jump*.[22]

A spinoff manga titled *Dragon Ball SD*, written by Ooishi Naho, has been published in Shueisha's *Super Strong Jump* magazine beginning in December 2010.[23] A second issue was released in April 2011.[24] This manga is a retelling of Goku's adventures as a child, with many details changed.[25]

Dragon Ball: Episode of Bardock is a special manga made by Ooishi Naho released in the July 2011 issue of V-Jump. As the title indicates the story revolves around Bardock, Goku's father, who in this 3-part special is featured in a "what-if" scenario in which he didn't die at the hands of Frieza and gets to fight his enemy as a Super Saiyan. The fact that Bardock will appear as a Super Saiyan is based on the *Dragon Ball Heroes* card featuring him as one.[26] This manga is a sequel to the TV special *Bardock - The Father of Goku* with some key details changed.

Anime series

Dragon Ball

Toei Animation produced an anime series based on the manga chapters, also titled *Dragon Ball*. The series premiered in Japan on Fuji Television on February 26, 1986 and ran until April 12, 1989, lasting 153 episodes.[5]

Harmony Gold USA licensed the series for an English-language release in the United States in 1989. In their voice dub of the series, Harmony renamed almost all of the characters, for example, Goku was renamed "Zero." This dub version was test-marketed in several cities, but was cancelled before it could be broadcast to the general public.

In 1995, Funimation Entertainment acquired the license for the distribution of *Dragon Ball* in the U.S., as well as its sequel series *Dragon Ball Z*. Funimation contracted BLT Productions to create an English voice track for the first anime at their Canadian-based dubbing studio and the dubbed episodes were edited for content.[27] Thirteen episodes aired in first-run syndication during the fall of 1995 before Funimation cancelled the project due to low ratings and decided to shift their focus on the more action-oriented *Dragon Ball Z*. Vidmark Entertainment (later known as Trimark Pictures) purchased the home video distribution rights for these dubbed episodes sometime after.[5] In March 2001, following the success of *Dragon Ball Z* on Cartoon Network, Funimation announced the return of *Dragon Ball* to American television, featuring a new English audio track produced at their own Texas-based dubbing studio, as well as slightly less editing, and left the original background music intact unlike their dubs of the two sequel series.[27] [28] The re-dubbed episodes aired on Cartoon Network from August 20, 2001 to December 1, 2003.[29] Funimation also broadcast the series on Colours TV and their own Funimation Channel starting in 2006.[30]

Funimation began releasing their in-house dub to Region 1 DVD box sets in March 2003. Each box set, spanning an entire saga of the series, included the English dub track and the original Japanese audio track with optional English subtitles. However, they were unable to release the first thirteen episodes at the time, due to Lionsgate Entertainment holding the distribution rights to their original dub of the same episodes, having acquired them from Trimark after the company became defunct. After Lionsgate's license to the first thirteen episodes expired in 2009, Funimation remastered and re-released the complete *Dragon Ball* series to DVD in five individual season box sets, with the first set released on September 15, 2009 and the final set released on July 27, 2010.

Dragon Ball Z

With the ending of *Dragon Ball*, Toei Animation quickly released a second anime series, *Dragon Ball Z* (Z ( ) *Doragon Bōru Zetto*, commonly abbreviated as *DBZ*). Picking up where the first left off, *Dragon Ball Z* is adapted from the final twenty-six volumes of the manga series, it premiered in Japan on Fuji Television on April 26, 1989, taking over its predecessor's time slot, and ran for 291 episodes until its conclusion on January 31, 1996.[5]

Following the short-lived dub of *Dragon Ball* in 1995, Funimation Entertainment began production on an English-language release of *Dragon Ball Z*. They collaborated with Saban Entertainment to finance and distribute the series to television, sub-licensed home video distribution to Pioneer Entertainment (later known as Geneon Universal Entertainment), contracted Ocean Productions to dub the anime into English, and hired Shuki Levy to compose an alternate musical score. This dub of *Dragon Ball Z* was heavily edited for content, as well as length; reducing the first 67 episodes into 53. The series premiered in the U.S. on September 13, 1996 in first-run syndication, but also struggled to find a substantial audience during its run and was ultimately cancelled after two seasons. On August 31, 1998, however, these cancelled dubbed episodes began airing on Cartoon Network's weekday-afternoon programming block, Toonami, where the series received much more popularity. With new success, Funimation continued production on the series by themselves, now with less editing due to fewer restrictions on cable programing. However, they could no longer afford the services of either the Ocean voice cast or Shuki Levy's music without Saban's financial assistance, resulting in the creation of their own in-house voice cast and a new musical score composed by Bruce Faulconer. *Dragon Ball Z* was now in full production in the U.S. and the new dub of the series aired on Cartoon Network from September 13, 1999 to April 7, 2003. In 2004, Geneon's distribution rights to the first 53/67 episodes of *Dragon Ball Z* expired, allowing Funimation to re-dub them with their in-house voice cast and restore the removed content. These re-dubbed episodes aired on Cartoon Network during the summer of 2005.[31] [32]

In 2006, Funimation remastered the episodes cropped to 16:9 widescreen format and then began re-releasing the series to Region 1 DVD in nine individual season box sets, with the first set released on February 6, 2007 and the final set released on May 19, 2009. These sets were notable for including the option of hearing Funimation's in-house dub alongside the original Japanese music, an option that had previously not been available. Other options included hearing the in-house dub with the American soundtrack composed by Bruce Faulconer and Nathan Johnson, and a third option included watching the original Japanese version, with the original Japanese soundtrack and English subtitles. In July 2009, Funimation announced that they would be re-releasing *Dragon Ball Z* in a new seven-volume DVD set called the "Dragon Boxes." Based on the original series masters with frame-by-frame restoration, the first set was released on November 10, 2009 and the final set was released on October 11, 2011.[33] Unlike the season box sets, Funimation's "Dragon Box" release is presented in the original 4:3 fullscreen format.[34]

Funimation and Toei released a statement in January 2011 confirming that they would stream *Dragon Ball Z* within 30 minutes before their simulcast of *One Piece*.[35] *Dragon Ball Z* is now being streamed on Hulu and Toonzaki, containing the English dub with the Japanese music and uncut footage, as well as subtitled Japanese episodes.

In July 2011, Funimation announced plans to release *Dragon Ball Z* in Blu-Ray format. *Dragon Ball Z Level 1.1*, containing the first 17 episodes, was released on November 8, 2011.[36] [37] [38]

Dragon Ball GT

Produced by Toei Animation, *Dragon Ball GT* (G T ( ) *Doragon Bōru Jī Tī, G(rand) T(our)*[5]) premiered on Fuji TV on February 2, 1996, and ran until November 19, 1997. Unlike the first two series, it was not based on the original *Dragon Ball* manga.[39] The series lasted 64 episodes.[5] In *Dragon Ball GT*, Goku is turned back into a child by the Black Star Dragon Balls and is forced to travel across the galaxy to retrieve them.

Following the success of *Dragon Ball* and *Dragon Ball Z* on Cartoon Network, Funimation Entertainment licensed *Dragon Ball GT* for distribution in the U.S. as well. Funimation's dub of the series aired on Cartoon Network from November 14, 2003 to April 16, 2005. The television broadcast initially skipped the first sixteen episodes of the series. Instead, Funimation created a composition episode entitled "A Grand Problem," which used scenes from the skipped episodes to summarize the story. The skipped episodes, advertised as "The Lost Episodes," were later aired after the remaining episodes of the series had been broadcast.

Funimation later released their dub to bilingual Region 1 DVD in two season box sets, with the first set released on December 9, 2008 and the final set released on February 10, 2009, which also featured the *Dragon Ball GT* TV special, *A Hero's Legacy*. In a similar fashion to their DVD releases for *Dragon Ball* Z, the DVD box sets have the option of hearing the English dub alongside the original Japanese music, and the rap song used for the TV airing of the show (nicknamed by fans "Step Into the Grand Tour") has been replaced by English-dubbed versions of the original Japanese opening and ending songs. Funimation later released a "Complete Series" box set of *Dragon Ball GT* (using the same discs as the two season sets, but with different packaging) on September 21, 2010.[40]

Dragon Ball Z Kai

In February 2009, Toei Animation announced that it would begin broadcasting a revised version of *Dragon Ball Z* as part of the series' 20th anniversary celebrations. The series premiered on Fuji TV in Japan on April 5, 2009, under the name *Dragon Ball Kai* (() *Doragon Bōru Kai*, lit. "Dragon Ball Revised"), with the episodes remastered for HDTV, featuring updated opening and ending sequences, and a rerecording of the vocal tracks by most of the original cast.[41] [42] The footage was also re-edited to more closely follow the manga, resulting in a faster moving story, and damaged frames removed.[43] As such, it is a new version of *Dragon Ball Z* created from the original footage.

On March 9, 2011, Toei announced that due to Kenji Yamamoto's score for *Dragon Ball Kai* infringing on the rights of an unknown third party, the score for remaining episodes and replays of previous episodes would be replaced.[44] Later reports from Toei claimed that with the exception of the series' opening and closing songs, as well as eyecatch music, Yamamoto's score was replaced with Shunsuke Kikuchi's original score from *Dragon Ball Z*. The series concluded with the finale of the Cell arc as opposed to including the Buu arc, which served as the ending of both the manga and original *Dragon Ball Z* series. It was originally planned to run 98 episodes, however due to the Tōhoku offshore earthquake and tsunami, the final episode of *Dragon Ball Kai* was not aired and the series ended on its 97th episode in Japan on March 27, 2011.

Like all other *Dragon Ball*-based anime, Funimation Entertainment licensed *Dragon Ball Kai* for an English-language release, under the title *Dragon Ball Z Kai*. The series made its U.S. premiere on Nicktoons on May 24, 2010.[45] [46] In addition to Nicktoons, the series also began airing on The CW's Saturday-morning programming block, Toonzai, on August 14, 2010.[47] Both the Nicktoons and Toonzai airings are edited for content, though the Toonzai version is censored even more so than Nicktoons', most likely due to The CW being a broadcast network. Both these versions are rated TV-Y7. However, in addition to the TV airings, Funimation is also releasing bilingual Region 1 DVD and Blu-Ray box sets of the show. These box sets contain the original Japanese audio track with English subtitles, as well as the uncut version of the English dub, which does not contain any of the edits made for the TV airings.[48] The uncut version is rated TV-PG due to violence and profanity.[49]

Anime films

Seventeen anime films based on the *Dragon Ball* series have been released in Japan. The first three films were based on the original *Dragon Ball* anime series. The remaining films included thirteen *Dragon Ball Z* films and one tenth anniversary special (also based on the first anime series). However, the films are generally either alternate re-tellings of certain story arcs or extra side-stories that don't fit in with the same timeline as the series. Funimation Entertainment has licensed and released all of the films to home video in North America.

Specials

Three television specials based on the metaseries were released in Japan. The first, *Bardock – The Father of Goku*, was released on October 17, 1990. It is a prequel to the series, set years before the start of the manga and details how Goku's father, Bardock, discovers that Frieza is planning to kill all the other Saiyans, and his efforts to stop him. The second special, *The History of Trunks* was released on March 24, 1993. Based on an extra chapter of the original manga, it is set in a parallel universe where most of the series characters are killed by the evil androids and focuses on Bulma and Vegeta's son Trunks. *A Hero's Legacy*, released on March 26, 1997, is set 100 years after the end of *Dragon Ball GT*. It features one of Goku's descendants who begins looking for the Dragon Balls in order to help his sick grandmother, Pan.

Two other specials were also released in Japan. A two-episode original video animation (OVA) series titled *Dragon Ball Z Side Story: Plan to Eradicate the Saiyans*, based on the Famicom video game of the same name, was released in 1993 and was set during *Dragon Ball Z*.[50] Another special, *Dragon Ball: Yo! Son Goku and His Friends Return!!*, premiered at the Jump Super Anime Tour on November 24, 2008. The special is set two years after the defeat of the evil Buu and has Goku and his friends facing against new enemies, Abo and Kado, and meeting Vegeta's younger brother, Tarble and his wife, Gure.

In November 2011, *V-Jump* announced an animated adaptation from Bardock's spinoff manga by Naho Ōishi to air in Japan in April 2012.[51]

Video games

The *Dragon Ball* franchise has spawned multiple video games across various genres and platforms. Earlier games of the series included a system of card battling and were released for the Nintendo Entertainment System following the storyline of the series.[52] Starting Super Nintendo Entertainment System, the Sega Saturn and the PlayStation most of the games were from the fighting genre including the series *Super Butoden*.[53] The first *Dragon Ball* game to be released in the United States was *Dragon Ball GT: Final Bout* for the PlayStation on July 31, 1997.[54] For the PlayStation 2 and PlayStation Portable games the characters were redone in 3D cel-shaded graphics. These games included the *Dragon Ball Z: Budokai* series and the *Dragon Ball Z: Budokai Tenkaichi* series.[55] [56] *Dragon Ball Z: Burst Limit* was the first game of the series developed for the PlayStation 3 and Xbox 360.[57] A massively multiplayer online role-playing game called *Dragon Ball Online* is currently playable. It has been stated that Akira Toriyama has been working on character designs for this project for several years, and the game is available in Japan and South Korea.[58]

Soundtracks

Myriad soundtracks were released to the anime, movies and the games. The music for the first two anime *Dragon Ball* and *Z* and its films was directed by Shunsuke Kikuchi, while the music from *GT* was directed by Akihito Tokunaga and the music from *Kai* was directed by Kenji Yamamoto. For the first anime, the soundtracks released were *Dragon Ball: Music Collection* in 1985 and *Dragon Ball: Complete Song Collection* in 1991 although they were reissued in 2007 and 2003, respectively.[59] For the second anime, the soundtrack series released were *Dragon Ball Z Hit Song Collection Series*. It was produced and released by Columbia Records of Japan from July 21, 1989 to March 20, 1996 the show's entire lifespan. On September 20, 2006 Columbia re-released the Hit Song Collection on their Animex 1300 series.[60] [61] Other CDs released are compilations, video games and films soundtracks as well as music from the English versions.[62]

Live-action films

A live-action Mandarin Chinese film adaptation of the series, *Dragon Ball: The Magic Begins*, was released in Taiwan in 1989.[5] Considered a "tacky" version of the story by critics,[5] the plot revolves around a rag-tag group of heroes, led by "Monkey Boy" (Goku) trying to stop King Horn from using the wish-granting "Dragon Pearls" (Dragon Balls) to rule the world.

In December of 1990, the unofficial live-action Korean film *Dragon Ball: Ssawora Son Goku, Igyeora Son Goku* was released. The movie follows the original *Dragon Ball* story, and does so more closely than *The Magic Begins*. This live action adaption from Korea is an adaptation of the events of the first *Dragon Ball* story arc and a little of the Saiyan arc of *Dragon Ball Z*, as Vegeta's partner Nappa makes an appearance in this film.

In March 2002, 20th Century Fox acquired feature film rights to the *Dragon Ball* franchise[63] and began production on an American live action film entitled *Dragonball Evolution*.[64] Ben Ramsey was tapped to create a screenplay based on *Dragon Ball Z*.[65] Directed by James Wong and produced by Stephen Chow,[64] the film was released in the United States on April 10, 2009.[66] The film was largely considered a failure by both critics and *Dragon Ball* fans,[67] and it only grossed $57 million at the box office.[68]

Art books

There are two companion books to the series, one called *Dragon Ball - The Complete Illustrations*, first published in Japan in 1995, which was then translated and printed in 2008 by VIZ Media. It contains all the 264 coloured illustrations Akira Toriyama done for the Weekly Jump magazines' covers, bonus giveaways and specials, and all the covers for the 42 tankōbon. It also includes an interview with Akira Toriyama on his work process. In Japan it was originally released as Volume 1 of the 7 part Daizenshuu series.

The other was *Dragon Ball GT Perfect Files*, released in May 1997 and December 1997 by Shueisha's *Jump Comics Selection* imprint. It include series information, illustration galleries, behind-the-scenes information, and more. They were out of print for many years, but were re-released in April 2006 and this edition is still in print.[69] [70]

Cover art of *Dragon Ball - The Complete Illustrations*.

Collectible cards

There have been collectible cards, based on the *Dragon Ball*, *Dragon Ball Z*, and *Dragon Ball GT* series, released under Bandai. They feature various scenes from the manga and anime stills, plus exclusive artwork from all three series. They were previously released in other countries, like Taiwan and Singapore, before making their debut in the United States in July 2008.

Reception

Manga

Dragon Ball is one of the most popular manga series of all time, and it continues to enjoy high readership today. By 2000, more than 126 million copies of its *tankōbon* volumes had been sold in Japan alone.[2] By 2007, this number had grown to pass 152 million in Japan and 200 million worldwide.[71] In *Little Boy: The Art of Japan's Exploding Subculture* Takashi Murakami notes that *Dragon Ball*'s "never-ending cyclical narrative moves forward plausibly, seamlessly, and with great finesse."[2] Goku's journey and his ever growing strength resulted in the character winning "the admiration of young boys everywhere".[4]

In a survey conducted by Oricon in 2007 between 1,000 people, Goku, the main character of the franchise, ranked first place as the "Strongest Manga character of all time."[72] Manga artists, such as *Naruto* creator Masashi Kishimoto and *One Piece* creator Eiichiro Oda, have stated that Goku inspired their series' main protagonists as well as series structure.[73] [74] *Dragon Ball* was also high inspiration to *Yaiba*, the manga written by Gosho Aoyama, best known as the creator of the manga series *Detective Conan*. Both *Yaiba* and *Dragon Ball* began as a kind of light-hearted gag manga, but towards the end of their run the tone became more serious and action-packed. When TV Asahi conducted an online poll for the top one hundred anime, the *Dragon Ball* series came in place twelve.[75]

Animerica felt the series had "worldwide appeal" that uses dramatic pacing and over the top martial arts action to "maintain tension levels and keep a crippler crossface hold on the audience's attention spans".[3] Ridwan Khan from Animefringe.com commented on the manga to have a "chubby" art style but as the series continued it gets more refined with the characters leaner and more muscular. He also noted he preferred the manga versions of the series to their animated counterparts that makes the story slower and pointless.[76] Anime News Network praised the story and humor of the manga to be very good due conveying of all the characters's personalities. They also remarked Viz's translation to be one of the best ones of all the English editions of the series praising the lack of censor.[77] Rationalmagic.com remarked the first manga volume as "a superior humor title". They praised Goku's innocence and Bulma's insistence as one of the funniest parts of the series.[78] Writer Jason Thompson commented that the series popularity comes from a formula that Toriyama used in various story arcs from which he describes as "lots of martial arts, lots of training sequences, a few jokes." Yet, he noted that such formula became the model for other manga from the same genre such as *Naruto*.[79]

Anime

The anime adaptations have also received positive reviews. *Dragon Ball Z* was listed as the 78th best animated show in IGN's Top 100 Animated Series,[80] and was also listed as the 50th greatest cartoon in *Wizard* magazine's "Top 100 Greatest Cartoons" list.[81] T.H.E.M. Anime Reviews considered the series characters to be different from stereotypical stock characters and noted that they undergo much more development.[82] Despite praising *Dragon Ball Z* for its cast of characters, they criticized it for having long and repetitive fights.[83] Anime News Network considered Trunks's storyline to be one of the better story arcs of the series, with the characters having more motivation than in previous stories.[84] Some critics and fans have given FUNimation's English dub of *Dragon Ball Z* mixed reviews over the years. IGN critisised some of the voices to be "quite annoying" and also noted that Frieza's English voice "made him sound like a lady. This combined with Frieza's appearance left a lot of fans confused about Frieza's gender for a while." IGN commented *Dragon Ball GT* "is downright repellent" mentioning that the material and characters had lost their novelty and fun. They also criticized the character designs of Trunks and Vegeta as being goofy.[85] Anime News Network also gave negative comments about *Dragon Ball GT*, mentioning that the fights from the series were "a very simple childish exercise" and that many other anime were superior. The plot of the series has also been criticized for giving a formula that was already used in its predecessors.[86] On several occasions the *Dragon Ball* anime series has topped Japan's DVD sales.[87] [88] The first episode of *Dragon Ball Kai* earned a viewer ratings percentage of 11.3, ahead of *One Piece* and behind *Crayon Shin-Chan*.[89] Although following episodes had lower ratings, *Dragon Ball Kai* was among the top 10 anime in viewer ratings every week in Japan for most of its run.[90] [91]

References

[1] http://www.uk-anime.net/newsitem/Manga_Entertainment_to_release_Dragon_Ball_Z_in_the_UK.html

[2] Murakami, Takashi (May 15, 2005). "Earth in My Window". *Little Boy: The Art of Japan's Exploding Subculture*. Linda Hoaglund (translator). Yale University Press, Japan Society. pp. 105–106. ISBN 0300102852.

[3] "Anime Radar: News". *Animerica* (San Francisco, California: Viz Media) **9** (2): 36. March 2001. ISSN 1067-0831. OCLC 27130932.

[4] Wiedemann, Julius (September 25, 2004). "Akira Toriyama". In Amano Masanao (ed.). *Manga Design*. Taschen. p. 372. ISBN 3822825913.

[5] Clements, Jonathan; Helen McCarthy (September 1, 2001). *The Anime Encyclopedia: A Guide to Japanese Animation Since 1917* (1st ed.). Berkeley, California: Stone Bridge Press. pp. 101–102. ISBN 1-880656-64-7. OCLC 47255331.

[6] Toriyama, Akira (1995). *DRAGON BALL COMPLETE ILLUSTRATION*. Shueisha. pp. 206–207. ISBN 4-08-782754-2.

[7] "Interview with the Majin! Revisited". *Shonen Jump* **5** (11): 388. November 2007. ISSN 1545-7818.

[8] Toriyama, Akira (1995). *DRAGON BALL WORLD GUIDE*. Shueisha. pp. 164–169. ISBN 4-08-782754-2.

[9] Published in *Daizenshuu* Vol. 4 World Guide, p74–75.

[10] Toriyama, Akira (1995). *DRAGON BALL STORY GUIDE*. Shueisha. pp. 261–265. ISBN 4-08-782752-6.

[11] "Shenron's Newspaper", *Daizenshuu* vol.2 (limited edition)

[12] "Comipress News article on "The Rise and Fall of Weekly Shōnen Jump"" (http://comipress.com/article/2007/05/06/1923). *comipress.com*. May 6, 2007. . Retrieved June 2, 2008.

[13] "DRAGON BALL 1 " (http://books.shueisha.co.jp/CGI/search/syousai_put.cgi?isbn_cd=4-08-851831-4&mode=1) (in Japanese). Shueisha. . Retrieved June 2, 2008.

[14] "DRAGON BALL 42 " (http://books.shueisha.co.jp/CGI/search/syousai_put.cgi?isbn_cd=4-08-851090-9&mode=1) (in Japanese). Shueisha. . Retrieved June 2, 2008.

[15] " " (http://books.shueisha.co.jp/CGI/search/syousai_put.cgi?isbn_cd=4-08-851090-9&mode=1) (in Japanese). Shueisha. . Retrieved July 22, 2008.

[16] "Dragon Ball Manga: A History" (http://daizex.db-destiny.net/manganext_2007_panel_final.pdf). Daizenshuu EX. October 7, 2007. . Retrieved April 1, 2011.

[17] "Podcast Episode #0097 — Live From MangaNEXT" (http://www.daizex.com/old_updates/updates-2007_10.shtml). Daizenshuu EX. October 7, 2007. . Retrieved April 1, 2011.

[18] "Viz Media — Products: Dragon Ball Vol. 16" (http://www.viz.com/products/products.php?product_id=2236). Viz Media. . Retrieved June 2, 2008.

[19] "Dragon Ball Z, Vol. 26" (http://www.viz.com/products/products.php?product_id=5898). Viz Media. . Retrieved June 2, 2008.

[20] "Dragon Ball Z, Vol. 1 (VIZBIG Edition)" (http://www.viz.com/products/products.php?product_id=7414). Viz Media. . Retrieved June 3, 2008.

[21] "Dragon Ball, Vol. 1 (VIZBIG Edition)" (http://www.viz.com/products/products.php?product_id=7409). Viz Media. . Retrieved June 8, 2008.

[22] "*Dragon Ball: Yo! The Return of Son Goku and Friends!! Manga Adaptation*". *V Jump* **17** (2): 55. February 2009.

[23] "Dragon Ball SD Manga Spinoff to Be Printed in December" (http://www.animenewsnetwork.com/news/2010-11-18/dragon-ball-sd-manga-spinoff-to-be-printed-in-december). Anime News Network. November 18, 2010. . Retrieved June 2, 2008.

[24] "!" (http://www.shonenjump.com/j/news/detail101108_2.html). Shonenjump.com. . Retrieved 2011-08-10.

[25] "Daizenshuu EX - General - Podcast" (http://daizex.com/general/podcast/#0241). Daizex.com. . Retrieved 2011-08-10.

[26] Kanzentai (2011-06-20). "Kanzentai • Updated July 27th, 2011" (http://www.kanzentai.com/index.php?subaction=showcomments&id=1308587091&archive=&start_from=&ucat=&). Kanzentai.com. . Retrieved 2011-08-10.

[27] "Rough Air Date for Dragon Ball" (http://www.animenewsnetwork.com/news/2001-03-09/rough-air-date-for-dragonball). Anime News Network. March 9, 2001. . Retrieved July 19, 2008.

[28] "Dragon Ball on CN debut date confirmed" (http://www.animenewsnetwork.com/news/2001-05-02/dragonball-on-cn-debut-date-confirmed). Anime News Network. May 2, 2001. . Retrieved July 19, 2008.

[29] "DragonBall Re-dub" (http://www.animenewsnetwork.com/news/2001-08-21/dragonball-re-dub). Anime News Network. August 21, 2001. . Retrieved July 19, 2008.

[30] "Dragon Ball Returns to US TV" (http://www.animenewsnetwork.com/news/2006-11-12/dragon-ball-returns-to-us-tv). Anime News Network. November 12, 2006. . Retrieved July 19, 2008.

[31] "FUNimation 2005 Plans" (http://www.animenewsnetwork.com/news/2004-12-31/funimation-2005-plans). Anime News Network. December 31, 2004. . Retrieved September 14, 2008.

[32] "DBZ uncut on Cartoon Network" (http://www.animenewsnetwork.com/news/2005-06-09/dbz-uncut-on-cartoon-network). Anime News Network. June 9, 2005. . Retrieved September 14, 2008.

[33] "Funimation Entertainment Announces First U.S. Release of Dragon Box" (http://www.dragonballz.com/index.cfm?page=news&id=586) (Press release). Funimation. July 20, 2009. . Retrieved August 28, 2009.

[34] "Dragon Ball Z: Dragon Box One" (http://www.amazon.com/dp/B002KPINF2/). Amazon.com. . Retrieved 2011-08-10.

[35] "Funimation to Stream Dragon Ball Z in North America" (http://www.animenewsnetwork.com/news/2011-01-03/funimation-to-stream-dragon-ball-z-in-north-america). Anime News Network. January 3, 2011. . Retrieved January 3, 2011.

[36] Official FUNimation Anime Online Community - Rojas's Blog (http://www.funimation.com/rojas/blog/funimation-entertainment-announces-dragon-ball-z-series-on-blu-ray/351891)

[37] "FUNimation Entertainment Announces Dragon Ball Z Series on Blu-ray - FLOWER MOUND, Texas" (http://www.prnewswire.com/news-releases/funimation-entertainment-announces-dragon-ball-z-series-on-blu-ray-125982173.html). Texas: Prnewswire.com. July 21, 2011. . Retrieved 2011-08-10.

[38] "FUNimation Ent. Announces Dragon Ball Z Series on Blu-ray" (http://www.animemaki.net/2011/07/funimation-ent-announces-dragon-ball-z-series-on-blu-ray/). Anime Maki. 2011-07-21. . Retrieved 2011-08-10.

[39] "DBZ FAQ Update" (http://www.animenewsnetwork.com/news/2001-04-10/dbz-faq-update). April 10, 2001. . Retrieved June 5, 2008.

[40] "Dragon Ball GT: The Complete Series: Eric Vale, Sean Schemmel, Christopher Neel, Jeremy Carlile: Movies & TV" (http://www.amazon.com/dp/B003SPDS7G/). Amazon.com. . Retrieved 2011-08-10.

[41] "Dragon Ball Z to Rerun on Japanese TV in HD in April" (http://www.animenewsnetwork.com/news/2009-02-06/dragon-ball-z-to-rerun-on-japanese-tv-in-hd-in-april). Anime News Network. February 6, 2009. . Retrieved February 21, 2009.

[42] "Japan's Remastered DBZ to Be Called Dragon Ball Kai" (http://www.animenewsnetwork.com/news/2009-02-19/japan-remastered-dbz-to-be-called-dragon-ball-kai). Anime News Network. February 19, 2009. . Retrieved February 21, 2009.

[43] " Z 20! H D !!!! ... DRAGON BALL KAI" (in Japanese). *V Jump* (Japan: Shueisha): 10. February 9, 2009.

[44] "News: Toei Replaces Possibly Infringing Dragon Ball Kai Music" (http://www.animenewsnetwork.com/news/2011-03-10/toei-replaces-possibly-infringing-dragon-ball-kai-music). Anime News Network. March 10, 2011. . Retrieved March 14, 2011.

[45] Navarre Reveals Funimation's Dragon Ball Kai License (Updated) (http://www.animenewsnetwork.com/news/2010-02-02/navarre-reveals-funimation-dragon-ball-kai-license) -- Anime News Network (Received February 2, 2010)

[46] Nickelodeon Announces Fall Plans (http://www.ohiomm.com/blogs/heldenfiles/2010/03/nickelodeon-announces-fall-plans) - Ohio.com Retrieved March 11, 2010

[47] "Dragon Ball Z Kai to Air on TheCW4Kids on Saturdays" (http://www.animenewsnetwork.com/news/2010-04-26/dragon-ball-z-kai-to-air-on-thecw4kids-on-saturdays). Anime News Network. April 26, 2010. . Retrieved April 26, 2010.

[48] "Dragon Ball Z Kai: Part One Blu-ray" (http://www.amazon.com/dp/B0036ZA8G2/). Amazon.com. . Retrieved 2011-08-10.

[49] dragonballzkai.com

[50] "List of OVA produced by Toei in the 1990s" (http://web.archive.org/web/20080420082913/http://www.toei-anim.co.jp/tv/ov1990.html). Archived from the original (http://www.toei-anim.co.jp/tv/ov1990.html) on 2008-04-20. . Retrieved September 27, 2008.

[51] Beveridge, Chris (November 21, 2011). "'Dragonball Episode Of Bardock' Manga Gets Anime Adaptation" (http://www.fandompost.com/2011/11/21/dragonball-episode-bardock-manga-anime-adaptation/). Fandom Postr. . Retrieved November 22, 2011.

[52] "Dragon Ball: Daimaou Fukkatsu" (http://www.gamespot.com/nes/rpg/dragonballdaimaoufukkatsu/index.html?tag=result;title;6). GameSpot. . Retrieved September 26, 2008.

[53] "Dragon Ball Z Super Butouden" (http://www.gamespot.com/snes/action/dragonballz/index.html?tag=result;title;3). GameSpot. . Retrieved September 26, 2008.

[54] "Dragon Ball GT: Final Bout" (http://www.gamespot.com/ps/action/dragonballfinalbout/index.html?tag=result;title;0). GameSpot. . Retrieved January 10, 2009.

[55] "DBZ: Budokai to ship early" (http://www.animenewsnetwork.com/news/2002-11-13/dbz-budokai-to-ship-early). Anime News Network. November 13, 2002. . Retrieved September 26, 2008.

[56] Mueller, Greg (August 15, 2005). "Dragon Ball Z: Budokai Tenkaichi Hands-On" (http://www.gamespot.com/ps2/action/dragonballzbudokaitenkaichi/news.html?sid=6130926&mode=previews&tag=result;title;2). GameSpot. . Retrieved September 26, 2008.

[57] "Atari Announces the Next Generation Chapter for Dragon Ball Z(R) Video Game Series" (http://phx.corporate-ir.net/phoenix.zhtml?c=66845&p=irol-newsArticle&t=Regular&id=1097059&) (Press release). Atari. January 16. . Retrieved September 24, 2008.

[58] "DRAGONBALL ONLINE - -" (http://www.ntl-kk.com/dboreport.html) (in Japanese). Ntl-kk.com. . Retrieved 2011-08-10.

[59] "Dragonball Music Collection" (http://www.cdjapan.co.jp/detailview.html?KEY=COCC-72015). CDJapan.com. . Retrieved September 26, 2008.

[60] "DRAGONBALL Z Hit Kyoku Shu" (http://www.cdjapan.co.jp/detailview.html?KEY=COCX-33908). CDJapan.com. . Retrieved September 26, 2008.

[61] "DRAGONBALL Z Hit Kyoku Shu 18½" (http://www.cdjapan.co.jp/detailview.html?KEY=COCX-33927). CDJapan.com. . Retrieved September 26, 2008.

[62] "Dragon Ball Z Best Song Collection "Legend of Dragonworld"" (http://www.cdjapan.co.jp/detailview.html?KEY=COCX-33567). CDJapan.com. . Retrieved September 26, 2008.

[63] "Fox draws deal for *DragonBall* live-action pics". *The Hollywood Reporter* **372** (28). March 12, 2002.

[64] Tatiana Siegel (November 13, 2007). "*Dragonball* comes to bigscreen" (http://www.variety.com/article/VR1117975946.html?categoryid=13&cs=1). *Variety*. . Retrieved November 14, 2007.

[65] Claude Brodesser, Claude (June 24, 2004). "Ramsey rolls *Dragonball Z*" (http://www.variety.com/article/VR1117907037.html?categoryid=1238&cs=1). *Variety*. . Retrieved September 7, 2007.

[66] "*DragonBall* Movie" (http://www.foxjapan.com/movies/lineup/main.html#DB). Fox Japan. . Retrieved June 18, 2008.

[67] Castillo, Michelle (April 7, 2009). "Movies: 'Dragonball' star: 'No one wants to make a movie that people will hate'" (http://latimesblogs.latimes.com/herocomplex/2009/04/dragonball-star-no-one-wants-to-make-a-movie-that-people-will-hate.html). *L.A. Times*. . Retrieved January 23, 2010.

[68] "Dragonball Evolution" (http://www.boxofficemojo.com/movies/?page=main&id=dragonball.htm). Box Office Mojo. . Retrieved June 19, 2009.

[69] "GT (Vol.1) (·) ()" (in Japanese). Amazon.com. ASIN (amazon.jp): 4088740890.

[70] "GT (Vol.2) (·) ()" (in Japanese). Amazon.com. ASIN (amazon.jp): 4088740904.

[71] "Dragon Ball Movie to Film This Year for 2008 Release" (http://www.animenewsnetwork.com/news/2007-11-14/dragonball-movie-to-film-this-year-for-2008-release). . Retrieved November 22, 2011.

[72] "1000! "" ! " (http://rn-cdn.oricon.co.jp/news/ranking/45750/#rk) (in Japanese). . Retrieved October 28, 2007.

[73] Kishimoto, Masashi (2007). *Uzumaki: the Art of Naruto*. Viz Media. pp. 138–139. ISBN 1-4215-1407-9.

[74] *One Piece Color Walk* - Vol.1 - Interview with Eiichiro Oda and Akira Toriyama ((JP) ISBN 978-4-08-859217-6)

[75] "Japans Favorite TV Anime" (http://www.animenewsnetwork.com/news/2006-10-13/japan's-favorite-tv-anime). Anime News Network. October 13, 2006. . Retrieved July 9, 2008.

[76] Khan, Ridwan (July 2003). "Dragon Ball Vol.1 review" (http://www.animefringe.com/magazine/2003/07/reviews/08/). Animefringe.com. . Retrieved September 27, 2008.

[77] Divers, Allen (November 18, 2001). "Dragon Ball (manga) Graphic Novel vol 5" (http://www.animenewsnetwork.com/review/dragon-ball-gn-5). Anime News Network. . Retrieved September 27, 2008.

[78] "Dragon Ball Volume 1 review" (http://www.rationalmagic.com/Comics/DragonBall1.html). Rationalmagic.com. . Retrieved October 3, 2008.

[79] Thompson, Jason (April 8, 2009). "What is Dragon Ball?" (http://io9.com/5201656/what-is-dragon-ball). Io9. . Retrieved December 9, 2009.

[80] "78. Dragon Ball Z" (http://tv.ign.com/top-100-animated-tv-series/78.html). IGN. January 23, 2009. . Retrieved September 5, 2009.

[81] "Wizard Magazine's Top 100 cartoons list" (http://www.listal.com/list/wizard-magazines-top-100-cartoons). Listal. . Retrieved April 30, 2011.

[82] Jones, Tim. "Dragon Ball anime review" (http://www.themanime.org/viewreview.php?id=565). themanime.org. . Retrieved October 3, 2008.

[83] Ross, Carlos. "Dragon Ball Z anime review" (http://www.themanime.org/viewreview.php?id=245). themanime.org. . Retrieved October 3, 2008.

[84] Shepard, Chris. "The History of Trunks review" (http://www.animenewsnetwork.com/review/dbz-history-of-trunks-dvd). Anime News Network. . Retrieved October 3, 2008.

[85] Harris, Jeffrey (November 12, 2007). "Dragon Ball GT — The Lost Episodes DVD Box Set Review" (http://dvd.ign.com/articles/834/834547p1.html). IGN. . Retrieved October 3, 2008.

[86] Bertschy, Zac (June 6, 2004). "Dragon Ball GT DVD 8: Salvation" (http://www.animenewsnetwork.com/review/dragon-ball-gt/dvd-8). Anime News Network. . Retrieved October 4, 2008.

[87] "Japanese Animation DVD Ranking, September 10–16" (http://www.animenewsnetwork.com/news/2008-09-20/japanese-animation-dvd-ranking-september-10-16). Anime News Network. September 20, 2008. . Retrieved October 11, 2008.

[88] "Japanese Animation DVD Ranking, August 6–12" (http://www.animenewsnetwork.com/news/2008-08-14/japanese-animation-dvd-ranking-august-6-12). Anime News Network. August 14, 2008. . Retrieved October 11, 2008.

[89] "Japanese Anime TV Ranking, March 30-April 5" (http://www.animenewsnetwork.com/news/2009-04-13/japanese-anime-tv-ranking-march-30-april-5). Anime News Network. April 13, 2009. . Retrieved April 29, 2009.

[90] "Japanese Anime TV Ranking, April 6–12" (http://www.animenewsnetwork.com/news/2009-04-16/japanese-anime-tv-ranking-april-6-12). Anime News Network. April 16, 2009. . Retrieved April 29, 2009.

[91] "Japanese Anime TV Ranking, April 13–19" (http://www.animenewsnetwork.com/news/2009-04-22/japanese-anime-tv-ranking-april-13-19). Anime News Network. April 22, 2009. . Retrieved April 29, 2009.

External links

- Official Toei Animation's *Dragon Ball* website (http://www.toei-anim.co.jp/tv/dragon/) (**Japanese**)
- Official Toei Animation's *Dragon Ball Z* website (http://www.toei-anim.co.jp/tv/dragonz/) (**Japanese**)
- Official Toei Animation's *Dragon Ball GT* website (http://www.toei-anim.co.jp/tv/dragongt/) (**Japanese**)
- Official Toei Animation's *Dragon Ball Kai* website (http://www.toei-anim.co.jp/tv/dragon_kai/) (**Japanese**)
- Official FUNimation's *Dragon Ball Z* website (http://www.dragonballz.com/)
- Official FUNimation's *Dragon Ball Z Kai* website (http://www.dragonballzkai.com/)
- Official Madman's *Dragon Ball Z* website (http://www.dragonballz.com.au/)
- *Dragon Ball* (http://www.animenewsnetwork.com/encyclopedia/manga.php?id=297) (manga) at Anime News Network's Encyclopedia
- *Dragon Ball* (http://www.animenewsnetwork.com/encyclopedia/anime.php?id=243) (anime) at Anime News Network's Encyclopedia
- *Dragon Ball Z* (http://www.animenewsnetwork.com/encyclopedia/anime.php?id=244) (anime) at Anime News Network's Encyclopedia
- *Dragon Ball GT* (http://www.animenewsnetwork.com/encyclopedia/anime.php?id=616) (anime) at Anime News Network's Encyclopedia
- *Dragon Ball Kai* (http://www.animenewsnetwork.com/encyclopedia/anime.php?id=10877) (anime) at Anime News Network's Encyclopedia

Manga

The kanji for "manga" from *Seasonal Passersby* (*Shiki no Yukikai*), 1798, by Santō Kyōden and Kitao Shigemasa.

Manga (kanji: ; hiragana: ; katakana: ; listen; English /ˈmɑːŋɡə/ or /ˈmæŋɡə/) is the Japanese word for "comics" and consists of comics and print cartoons (sometimes also called *komikku*). In the West, the term "manga" has been appropriated to refer specifically to comics created in Japan, or by Japanese authors, in the Japanese language and conforming to the style developed in Japan in the late 19th century.[1] In their modern form, manga date from shortly after World War II, but they have a long, complex pre-history in earlier Japanese art.[2]

In Japan, people of all ages read manga. The medium includes works in a broad range of genres: action-adventure, romance, sports and games, historical drama, comedy, science fiction and fantasy, mystery, horror, sexuality, and business/commerce, among others.[3] Since the 1950s, manga has steadily become a major part of the Japanese publishing industry,[4] representing a ¥406 billion market in Japan in 2007 (approximately $3.6 billion). Manga have also gained a significant worldwide audience.[5] In 2008, the U.S. and Canadian manga market was valued at $175 million. Manga stories are typically printed in black-and-white,[6] although some full-color manga exist (e.g. *Colorful*). In Japan, manga are usually serialized in large manga magazines, often containing many stories, each presented in a single episode to be continued in the next issue. If the series is successful, collected chapters may be republished in paperback books called *tankōbon*.[7] A manga artist (*mangaka* in Japanese) typically works with a few assistants in a small studio and is associated with a creative editor from a commercial publishing company.[8] If a manga series is popular enough, it may be animated after or even during its run,[9] although sometimes manga are drawn centering on previously existing live-action or animated films[10] (e.g. *Star Wars*).

"Manga" as a term used outside Japan refers specifically to comics originally published in Japan.[11] However, manga-influenced comics, among original works, exist in other parts of the world, particularly in Taiwan ("manhua"), South Korea ("manhwa"),[12] and China, notably Hong Kong ("manhua").[13] In France, "la nouvelle manga" has developed as a form of *bande dessinée* (literally *drawn strip*) drawn in styles influenced by Japanese manga. In the United States, people refer to what they perceive as manga "styled" comics as Amerimanga, world manga, or original English-language manga (OEL manga). Still, the original term "manga" is primarily used in English-speaking countries solely to describe comics of Japanese origin.

Etymology

The Chinese characters used to write the word *manga* in Japanese can be translated as "whimsical drawings". The word first came into common usage in the late 18th century with the publication of such works as Santō Kyōden's picturebook *Shiji no yukikai* (1798), and in the early 19th century with such works as Aikawa Minwa's *Manga hyakujo* (1814) and the celebrated *Hokusai Manga* books (1814–1834) containing assorted drawings from the sketchbooks of the famous ukiyo-e artist Hokusai.[14] Rakuten Kitazawa (1876–1955) first used the word "manga" in the modern sense.[15]

History and characteristics

Historians and writers on manga history have described two broad and complementary processes shaping modern manga. Their views differ in the relative importance they attribute to the role of cultural and historical events following World War II versus the role of pre-War, Meiji, and pre-Meiji Japanese culture and art.

One view emphasizes events occurring during and after the U.S. Occupation of Japan (1945–1952), and stresses that manga strongly reflect U.S. cultural influences, including U.S. comics (brought to Japan by the GIs) and images and themes from U.S. television, film, and cartoons (especially Disney).[16] Alternately, other writers such as Frederik L. Schodt, Kinko Ito, and Adam L. Kern stress continuity of Japanese cultural and aesthetic traditions as central to the history of manga.[17]

Modern manga originated in the Occupation (1945–1952) and post-Occupation years (1952–early 1960s), while a previously militaristic and ultra-nationalist Japan rebuilt its political and economic infrastructure. An explosion of artistic creativity occurred in this period,[18] involving manga artists such as Osamu Tezuka (*Astro Boy*) and Machiko Hasegawa (*Sazae-san*).

A *kami-shibai* story teller from *Sazae-san* by Machiko Hasegawa. Sazae appears with her hair in a bun.

Astro Boy quickly became (and remains) immensely popular in Japan and elsewhere,[19] and the anime adaptation of *Sazae-san* continues to run as of 2011, regularly drawing more viewers than any other anime on Japanese television. Tezuka and Hasegawa both made stylistic innovations. In Tezuka's "cinematographic" technique, the panels are like a motion picture that reveals details of action bordering on slow motion as well as rapid zooms from distance to close-up shots. This kind of visual dynamism was widely adopted by later manga artists.[20] Hasegawa's focus on daily life and on women's experience also came to characterize later *shōjo manga*.[21] Between 1950 and 1969, an increasingly large readership for manga emerged in Japan with the solidification of its two main marketing genres, *shōnen manga* aimed at boys and *shōjo* manga aimed at girls.[22]

In 1969 a group of female manga artists (later called the *Year 24 Group*, also known as *Magnificent 24s*) made their *shōjo* manga debut ("year 24" comes from the Japanese name for the year 1949, the birth-year of many of these artists).[23] The group included Hagio Moto, Riyoko Ikeda, Yumiko Oshima, Keiko Takemiya, and Ryoko Yamagishi, and they marked the first major entry of female artists into manga.[7] Thereafter, primarily female manga artists would draw *shōjo* for a readership of girls and young women.[24] In the following decades (1975–present), *shōjo* manga continued to develop stylistically while simultaneously evolving different but overlapping subgenres.[25] Major subgenres include romance, superheroines, and "Ladies Comics" (in Japanese, *redisu* , *redikomi* , and *josei*).[26]

Modern *shōjo* manga romance features love as a major theme set into emotionally intense narratives of self-realization.[27] With the superheroines, *shōjo* manga saw releases such as Pink Hanamori's *Mermaid Melody Pichi Pichi Pitch* Reiko Yoshida's *Tokyo Mew Mew*, And, Naoko Takeuchi's *Pretty Soldier Sailor Moon*, which became internationally popular in both manga and anime formats.[28] Groups (or *sentais*) of girls working together have also been popular within this genre. Like Lucia, Hanon, and Rina singing together, and Sailor Moon, Sailor Mercury, Sailor Mars, Sailor Jupiter, and Sailor Venus working together.[29]

Manga for male readers sub-divides according to the age of its intended readership: boys up to 18 years old (*shōnen* manga) and young men 18- to 30-years old (*seinen* manga);[30] as well as by content, including action-adventure often involving male heroes, slapstick humor, themes of honor, and sometimes explicit sexuality.[31] The Japanese use different kanji for two closely allied meanings of "seinen"—for "youth, young man" and for "adult, majority"—the second referring to sexually overt manga aimed at grown men and also called *seijin* ("adult") manga.[32] *Shōnen*, *seinen*, and *seijin* manga share many features in common.

Boys and young men became some of the earliest readers of manga after World War II. From the 1950s on, *shōnen* manga focused on topics thought to interest the archetypal boy, including subjects like robots, space-travel, and heroic action-adventure.[33] Popular themes include science fiction, technology, sports, and supernatural settings. Manga with solitary costumed superheroes like Superman, Batman, and Spider-Man generally did not become as popular.[34]

The role of girls and women in manga produced for male readers has evolved considerably over time to include those featuring single pretty girls (*bishōjo*)[35] such as Belldandy from *Oh My Goddess!*, stories where such girls and women surround the hero, as in *Negima* and *Hanaukyo Maid Team*, or groups of heavily armed female warriors (*sentō bishōjo*)[36]

With the relaxation of censorship in Japan in the 1990s, a wide variety of explicit sexual themes appeared in manga intended for male readers, and correspondingly occur in English translations.[37] However, in 2010 the Tokyo Metropolitan Government passed a bill to restrict harmful content.[38]

The *gekiga* style of drawing—emotionally dark, often starkly realistic, sometimes very violent—focuses on the day-in, day-out grim realities of life, often drawn in gritty and unpretty fashions.[38] *Gekiga* such as Sampei Shirato's 1959–1962 *Chronicles of a Ninja's Military Accomplishments* (*Ninja Bugeichō*) arose in the late 1950s and 1960s partly from left-wing student and working-class political activism[39] and partly from the aesthetic dissatisfaction of young manga artists like Yoshihiro Tatsumi with existing manga.[40]

Publications

In Japan, manga constituted an annual 406 billion yen (approximately $3.6 billion USD) publication-industry by 2007.[41] Recently, the manga industry has expanded worldwide, where distribution companies license and reprint manga into their native languages.

After a series has run for a while, publishers often collect the stories together and print them in dedicated book-sized volumes, called *tankōbon*. These are the equivalent of U.S. trade paperbacks or graphic novels. These volumes use higher-quality paper, and are useful to those who want to "catch up" with a series so they can follow it in the magazines or if they find the cost of the weeklies or monthlies to be prohibitive. Recently, "deluxe" versions have also been printed as readers have gotten older and the need for something special grew. Old manga have also been reprinted using somewhat lesser quality paper and sold for 100 yen (about $1 U.S. dollar) each to compete with the used book market.

Marketeers primarily classify manga by the age and gender of the target readership.[42] In particular, books and magazines sold to boys (*shōnen*) and girls (*shōjo*) have distinctive cover art and are placed on different shelves in most bookstores. Due to cross-readership, consumer response is not limited by demographics. For example, male readers subscribing to a series intended for girls and so on.

Japan also has manga cafés, or *manga kissa* (*kissa* is an abbreviation of *kissaten*). At a *manga kissa*, people drink coffee and read manga, and sometimes stay there overnight.

There has been an increase in the amount of publications of original webmanga. It is internationally drawn by enthusiasts of all levels of experience, and is intended for online viewing. It can be ordered in graphic novel form if available in print.

The Kyoto International Manga Museum maintains a very large website listing manga published in Japanese.[43]

Magazines

Eshinbun Nipponchi; credited as the first manga magazine ever made.

Manga magazines usually have many series running concurrently with approximately 20–40 pages allocated to each series per issue. Other magazines such as the anime fandom magazine *Newtype* featured single chapters within their monthly periodicals. Other magazines like *Nakayoshi* feature many stories written by many different artists; these magazines, or "anthology magazines", as they are also known (colloquially "phone books"), are usually printed on low-quality newsprint and can be anywhere from 200 to more than 850 pages thick. Manga magazines also contain one-shot comics and various four-panel *yonkoma* (equivalent to comic strips). Manga series can run for many years if they are successful. Manga artists sometimes start out with a few "one-shot" manga projects just to try to get their name out. If these are successful and receive good reviews, they are continued. Magazines often have a short life.[44]

History

Kanagaki Robun and Kawanabe Kyosai created the first manga magazine in 1874: *Eshinbun Nipponchi*. The magazine was heavily influenced by *Japan Punch*, founded in 1862 by Charles Wirgman, a British cartoonist. *Eshinbun Nipponchi* had a very simple style of drawings and did not become popular with many people. *Eshinbun Nipponchi* ended after three issues. The magazine *Kisho Shimbun* in 1875 was inspired by *Eshinbun Nipponchi*, which was followed by *Marumaru Chinbun* in 1877, and then *Garakuta Chinpo* in 1879.[45] *Shōnen Sekai* was the first *shōnen* magazine created in 1895 by Iwaya Sazanami, a famous writer of Japanese children's literature back then. *Shōnen Sekai* had a strong focus on the First Sino-Japanese War.[46]

In 1905 the manga-magazine publishing boom started with the Russo-Japanese War,[47] *Tokyo Pakku* was created and became a huge hit.[48] After *Tokyo Pakku* in 1905, a female version of *Shōnen Sekai* was created and named *Shōjo Sekai*, considered the first *shōjo* magazine.[49] *Shōnen Pakku* was made and is considered the first children's manga magazine. The children's demographic was in an early stage of development in the Meiji period. *Shōnen Pakku* was influenced from foreign children's magazines such as *Puck* which an employee of Jitsugyō no Nihon (publisher of the magazine) saw and decided to emulate. In 1924, *Kodomo Pakku* was launched as another children's manga magazine after *Shōnen Pakku*.[48] During the boom, *Poten* (derived from the French "potin") was published in 1908. All the pages were in full color with influences from *Tokyo Pakku* and *Osaka Pakku*. It is unknown if there were any more issues besides the first one.[47] *Kodomo Pakku* was launched May 1924 by Tokyosha and featured high-quality art by many members of the manga artistry like Takei Takeo, Takehisa Yumeji and Aso Yutaka. Some of the manga featured speech balloons, where other manga from the previous eras did not use speech balloons and were silent.[48]

Published from May 1935 to January 1941, *Manga no Kuni* coincided with the period of the Second Sino-Japanese War. *Manga no Kuni* featured information on becoming a mangaka and on other comics industries around the world. *Manga no Kuni* handed its title to *Sashie Manga Kenkyū* in August 1940.[50]

Dōjinshi

Dōjinshi, produced by small publishers outside of the mainstream commercial market, resemble in their publishing small-press independently published comic books in the United States. Comiket, the largest comic book convention in the world with around 500,000 visitors gathering over three days, is devoted to *dōjinshi*. While they most often contain original stories, many are parodies of or include characters from popular manga and anime series. Some *dōjinshi* continue with a series' story or write an entirely new one using its characters, much like fan fiction. In 2007, *dōjinshi* sold for 27.73 billion yen (245 million USD).[41]

International markets

As of 2007 the influence of manga on international comics had grown considerably over the past two decades.[51] "Influence" is used here to refer to effects on the comics markets outside of Japan and to aesthetic effects on comics artists internationally.

Traditionally, manga stories flow from top to bottom and from right to left. Some publishers of translated manga keep to this original format. Other publishers mirror the pages horizontally before printing the translation, changing the reading direction to a more "Western" left to right, so as not to confuse foreign readers or traditional comics-consumers. This practice is known as "flipping".[52] For the most part, criticism suggests that flipping goes against the original intentions of the creator (for example, if a person wears a shirt that reads "MAY" on it, and gets flipped, then the word is altered to "YAM"). Flipping may also cause oddities with familiar asymmetrical objects or layouts, such as a car being depicted with the gas pedal on the left and the brake on the right, or a shirt with the buttons on the wrong side.

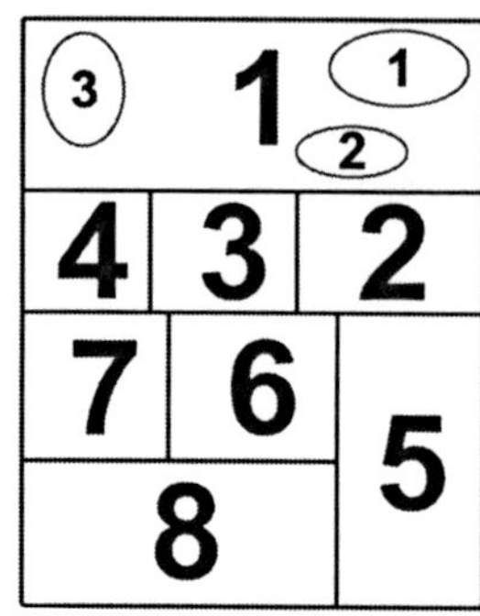

The reading direction in a traditional manga

United States

Manga made their way only gradually into U.S. markets, first in association with anime and then independently.[53] Some U.S. fans became aware of manga in the 1970s and early 1980s.[54] However, anime was initially more accessible than manga to U.S. fans,[55] many of whom were college-age young people who found it easier to obtain, subtitle, and exhibit video tapes of anime than translate, reproduce, and distribute *tankōbon*-style manga books.[56] One of the first manga translated into English and marketed in the U.S. was Keiji Nakazawa's *Barefoot Gen*, an autobiographical story of the atomic bombing of Hiroshima issued by Leonard Rifas and Educomics (1980–1982).[57] More manga were translated between the mid-1980s and 1990s, including *Golgo 13* in 1986, *Lone Wolf and Cub* from First Comics in 1987, and *Kamui*, *Area 88*, and *Mai the Psychic Girl*, also in 1987 and all from Viz Media-Eclipse Comics.[58] Others soon followed, including *Akira* from Marvel Comics' Epic Comics imprint and *Appleseed* from Eclipse Comics in 1988, and later *Iczer-1* (Antarctic Press, 1994) and Ippongi Bang's *F-111 Bandit* (Antarctic Press, 1995).

In the 1980s to the mid-1990s, Japanese animation, like *Akira*, *Dragon Ball*, *Neon Genesis Evangelion*, and *Pokémon*, made a bigger impact on the fan experience and in the market than manga.[59] Matters changed when translator-entrepreneur Toren Smith founded Studio Proteus in 1986. Smith and Studio Proteus acted as an agent and translator of many Japanese manga, including Masamune Shirow's *Appleseed* and Kōsuke Fujishima's *Oh My Goddess!*, for Dark Horse and Eros Comix, eliminating the need for these publishers to seek their own contacts in Japan.[60] Simultaneously, the Japanese publisher Shogakukan opened a U.S. market initiative with their U.S. subsidiary Viz, enabling Viz to draw directly on Shogakukan's catalogue and translation skills.[52]

A young boy reading *Black Cat* in a Barnes & Noble bookstore

The U.S. manga market took an upturn with mid-1990s anime and manga versions of Masamune Shirow's *Ghost in the Shell* (translated by Frederik L. Schodt and Toren Smith) becoming very popular among fans.[61] Another success of the mid-1990s was *Sailor Moon*.[62] By 1995–1998, the *Sailor Moon* manga had been exported to over 23 countries, including China, Brazil, Mexico, Australia, North America and most of Europe.[63] In 1997, Mixx Entertainment began publishing *Sailor Moon*, along with CLAMP's *Magic Knight Rayearth*, Hitoshi Iwaaki's *Parasyte* and Tsutomu Takahashi's *Ice Blade* in the monthly manga magazine *MixxZine*. Two years later, *MixxZine* was renamed to *Tokyopop* before discontinuing in 2000. Mixx Entertainment, later renamed Tokyopop, also published manga in trade paperbacks and, like Viz, began aggressive marketing of manga to both young male and young female demographics.[64]

In the following years, manga became increasingly popular, and new publishers entered the field while the established publishers greatly expanded their catalogues.[65] As of 2008, the U.S. and Canadian manga market generated $175 million in annual sales.[66] Simultaneously, mainstream U.S. media began to discuss manga, with articles in *The New York Times*, *Time* magazine, *The Wall Street Journal*, and *Wired* magazine.[67]

Europe

Manga has influenced European cartooning in a way somewhat different from the U.S. experience. Broadcast anime in Italy and France opened the European market to manga during the 1970s.[68] French art has borrowed from Japan since the 19th century (Japonisme),[69] and has its own highly developed tradition of bande dessinée cartooning.[70] In France, beginning in the mid-1990s,[71] manga has proven very popular to a wide readership, accounting for about one-third of comics sales in France since 2004.[72] According to the Japan External Trade Organization, sales of manga reached $212.6 million within France and Germany alone in 2006.[68] European publishers marketing manga translated into French include Glénat, Asuka, Casterman, Kana, and Pika Édition, among others.

European publishers also translate manga into German, Italian, Dutch, and other languages. Manga publishers based in the United Kingdom include Gollancz and Titan Books. Manga publishers from the United States have a strong marketing presence in the United Kingdom: for example, the Tanoshimi line from Random House.

Localized manga

A number of artists in the United States have drawn comics and cartoons influenced by manga. As an early example, Vernon Grant drew manga-influenced comics while living in Japan in the late 1960s and early 1970s.[73] Others include Frank Miller's mid-1980s *Ronin*, Adam Warren and Toren Smith's 1988 *The Dirty Pair*,[74] Ben Dunn's 1987 *Ninja High School*, Stan Sakai's 1984 *Usagi Yojimbo*, and *Manga Shi 2000* from Crusade Comics (1997).

By the 21st century several U.S. manga publishers had begun to produce work by U.S. artists under the broad marketing label of manga.[75] In 2002, I.C. Entertainment, formerly Studio Ironcat and now out of business, launched a series of manga by U.S. artists called *Amerimanga*.[76] In 2004 eigoMANGA launched *Rumble Pak* and *Sakura Pakk* anthology series. Seven Seas Entertainment followed suit with *World Manga*.[77] Simultaneously, TokyoPop introduced original English-language manga (OEL manga) later renamed *Global Manga*.[78] TokyoPop is currently the largest U.S. publisher of original English language manga.[79]

Francophone artists have also developed their own versions of manga, like Frédéric Boilet's *la nouvelle manga*. Boilet has worked in France and in Japan, sometimes collaborating with Japanese artists.[80]

Awards

The Japanese manga industry grants a large number of awards, mostly sponsored by publishers, with the winning prize usually including publication of the winning stories in magazines released by the sponsoring publisher. Examples of these awards include:

- the Akatsuka Award for humorous manga
- the Dengeki Comic Grand Prix for one-shot manga
- the Kodansha Manga Award (multiple genre awards)
- the Seiun Award for best science fiction comic of the year
- the Shogakukan Manga Award (multiple genres)
- the Tezuka Award for best new serial manga
- the Tezuka Osamu Cultural Prize (multiple genres)

The Japanese Ministry of Foreign Affairs has awarded the International Manga Award annually since May 2007.[81]

See also

- Emakimono / Etoki (horizontal, illustrated narrative form)
- Japanese popular culture
- Lianhuanhua (small Chinese picture book)
- List of films based on manga
- List of licensed manga in English
- List of manga distributors
- Manga iconography
- Oekaki (act of drawing)
- Q-version (cartoonification)
- Fan service

Footnotes

[1] Lent 2001, pp. 3–4, Tchiei 1998, Gravett 2004, p. 8
[2] Kern 2006, Ito 2005, Schodt 1986
[3] Gravett 2004, p. 8
[4] Kinsella 2000, Schodt 1996
[5] Wong 2006, Patten 2004
[6] Katzenstein & Shiraishi 1997
[7] Gravett 2004, p. 8, Schodt 1986
[8] Kinsella 2000
[9] Kittelson 1998
[10] Johnston-O'Neill 2007
[11] Merriam-Webster 2009
[12] Webb 2006
[13] Wong 2002
[14] Bouquillard & Marquet 2007
[15] Shimizu 1985, pp. 53–54, 102–103
[16] Kinsella 2000, Schodt 1986
[17] Schodt 1986, Ito 2004, Kern 2006, Kern 2007
[18] Schodt 1986, Schodt 1996, Schodt 2007, Gravett 2004
[19] Kodansha 1999, pp. 692–715, Schodt 2007
[20] Schodt 1986
[21] Gravett 2004, p. 8, Lee 2000, Sanchez 1997–2003
[22] Schodt 1986, Toku 2006
[23] Gravett 2004, pp. 78–80, Lent 2001, pp. 9–10
[24] Schodt 1986, Toku 2006, Thorn 2001

[25] Ōgi 2004
[26] Gravett 2004, p. 8, Schodt 1996
[27] Drazen 2003
[28] Allison 2000, pp. 259–278, Schodt 1996, p. 92
[29] Poitras 2001
[30] Thompson 2007, pp. xxiii–xxiv
[31] Brenner 2007, pp. 31–34
[32] Schodt 1996, p. 95, Perper & Cornog 2002
[33] Schodt 1986, pp. 68–87, Gravett 2004, pp. 52–73
[34] Schodt 1986, pp. 68–87
[35] Perper & Cornog 2002, pp. 60–63
[36] Gardner 2003
[37] Perper & Cornog 2002
[38] Schodt 1986, pp. 68–73, Gravett 2006
[39] Schodt 1986, pp. 68–73, Gravett 2004, pp. 38–42, Isao 2001
[40] Isao 2001, pp. 147–149, Nunez 2006
[41] Cube 2007
[42] Schodt 1996
[43] Manga Museum 2009
[44] Schodt 1996, pp. 101
[45] Eshinbun Nipponchi
[46] Griffiths 2007
[47] Poten
[48] Shonen Pakku
[49] Lone 2007, p. 75
[50] Manga no Kuni
[51] Pink 2007, Wong 2007
[52] Farago 2007
[53] Patten 2004
[54] In 1987, "...Japanese comics were more legendary than accessible to American readers", Patten 2004, p. 259
[55] Napier 2000, pp. 239–256, Clements & McCarthy 2006, pp. 475–476
[56] Patten 2004, Schodt 1996, pp. 305–340, Leonard 2004
[57] Schodt 1996, p. 309, Rifas 2004, Rifas adds that the original EduComics titles were *Gen of Hiroshima* and *I SAW IT* [*sic*].
[58] Patten 2004, pp. 37, 259–260, Thompson 2007, p. xv
[59] Leonard 2004, Patten 2004, pp. 52–73, Farago 2007
[60] Schodt 1996, pp. 318–321, Dark Horse Comics 2004
[61] Kwok Wah Lau, Jenny (2003). "4". *Multiple modernities: cinemas and popular media in transcultural East Asia*. Philadelphia: Temple University Press. pp. 78.
[62] Patten 2004, pp. 50, 110, 124, 128, 135, Arnold 2000
[63] Schodt 1996, p. 95
[64] Arnold 2000, Farago 2007, Bacon 2005
[65] Schodt 1996, pp. 308–319
[66] Reid 2009
[67] Glazer 2005, Masters 2006, Bosker 2007, Pink 2007
[68] Fishbein 2007
[69] Berger 1992
[70] Vollmar 2007
[71] Mahousu 2005
[72] Mahousu 2005, ANN 2004, Riciputi 2007
[73] Stewart 1984
[74] Crandol 2002
[75] Tai 2007
[76] ANN 2002
[77] ANN May 10, 2006
[78] ANN May 5, 2006
[79] ICv2 2007, Reid 2006
[80] Boilet 2001, Boilet & Takahama 2004
[81] ANN 2007, Ministry of Foreign Affairs of Japan 2007

References

- Allison, Anne (2000). "Sailor Moon: Japanese superheroes for global girls". In Craig, Timothy J.. *Japan Pop! Inside the World of Japanese Popular Culture*. Armonk, New York: M.E. Sharpe. ISBN 978-0765605610.
- Arnold, Adam (2000). "Full Circle: The Unofficial History of MixxZine" (http://www.animefringe.com/magazine/00.06/feature/1/index.php3). Retrieved December 19, 2007.
- Bacon, Michelle (April 14, 2005). "Tangerine Dreams: Guide to Shoujo Manga and Anime" (http://tangerine.astraldream.net/tokyopop.html). Retrieved April 1, 2008.
- Berger, Klaus (1992). *Japonisme in Western Painting from Whistler to Matisse*. Cambridge: Cambridge University Press. ISBN 0521373212.
- Boilet, Frédéric (2001). *Yukiko's Spinach*. Castalla-Alicante, Spain: Ponent Mon. ISBN 8493309346.
- Boilet, Frédéric; Takahama, Kan (2004). *Mariko Parade*. Castalla-Alicante, Spain: Ponent Mon. ISBN 84-933409-1-X.
- Bosker, Bianca (August 31, 2007). "Manga Mania" (http://online.wsj.com/article/SB118851157811713921.html). *The Wall Street Journal*. Retrieved April 1, 2008.
- Bouquillard, Jocelyn; Marquet, Christophe (June 1, 2007). *Hokusai: First Manga Master*. New York: Abrams. ISBN 0810993414.
- Brenner, Robin E. (2007). *Understanding Manga and Anime*. Westport, Connecticut: Libraries Unlimited/Greenwood. ISBN 978-1591583325.
- Clements, Jonathan; McCarthy, Helen (2006). *The Anime Encyclopedia: A Guide to Japanese Animation Since 1917, Revised and Expanded Edition*. Berkeley, California: Stone Bridge Press. ISBN 1933330104.
- Crandol, Mike (January 14, 2002). "The Dirty Pair: Run from the Future" (http://www.animenewsnetwork.com/review/the-dirty-pair/run-from-the-future). *Anime News Network*. Retrieved March 4, 2008.
- Cube (December 18, 2007). "20071866—" (http://www.inside-games.jp/article/2007/12/18/25855.html) (in Japanese). Inside for All Games. Retrieved December 18, 2007.
- "Dark Horse buys Studio Proteus" (http://www.animenewsnetwork.com/press-release/2004-02-06/dark-horse-buys-studio-proteus) (Press release). Dark Horse Comics. February 6, 2004.
- Drazen, Patrick (2003). *Anime Explosion! The What? Why? & Wow! of Japanese Animation*. Berkeley, California: Stone Bridge. ISBN 978-1880656723.
- Farago, Andrew (September 30, 2007). "Interview: Jason Thompson" (http://www.tcj.com/index.php?option=com_content&task=view&id=697&Itemid=70). The Comics Journal. Retrieved March 4, 2008.
- Fishbein, Jennifer (December 26, 2007). "Europe's Manga Mania" (http://www.businessweek.com/globalbiz/content/dec2007/gb20071226_346610.htm). *BusinessWeek*. Retrieved December 29, 2007.
- Gardner, William O. (November 2003). "Attack of the Phallic Girls" (http://www.depauw.edu/sfs/review_essays/gardner88.htm). *Science Fiction Studies* (88). Retrieved April 5, 2008.
- Glazer, Sarah (September 18, 2005). "Manga for Girls" (http://www.nytimes.com/2005/09/18/books/review/18glazer.html). *The New York Times*. Retrieved March 4, 2008.
- Gravett, Paul (2004). *Manga: Sixty Years of Japanese Comics*. New York: Harper Design. ISBN 1856693910.
- Gravett, Paul (October 15, 2006). "Gekiga: The Flipside of Manga" (http://www.paulgravett.com/index.php/articles/article/gekiga/). Retrieved March 4, 2008.
- Griffiths, Owen (September 22, 2007). "Militarizing Japan: Patriotism, Profit, and Children's Print Media, 1894–1925" (http://www.japanfocus.org/-Owen-Griffiths/2528). *Japan Focus*. Retrieved December 16, 2008.
- Isao, Shimizu (2001). "Red Comic Books: The Origins of Modern Japanese Manga". In Lent, John A.. *Illustrating Asia: Comics, Humor Magazines, and Picture Books*. Honolulu, Hawaii: University of Hawai'i Press. ISBN 978-0824824716.
- Ito, Kinko (2004). "Growing up Japanese reading manga". *International Journal of Comic Art* (6): 392–401.
- Ito, Kinko (2005). "A history of manga in the context of Japanese culture and society" (http://www.ingentaconnect.com/content/bpl/jpcu/2005/00000038/00000003/art00002). *The Journal of Popular Culture*

38 (3): 456–475. doi:10.1111/j.0022-3840.2005.00123.x. Retrieved April 5, 2008.

- Johnston-O'Neill, Tom (August 3, 2007). "Finding the International in Comic Con International" (http://parobs.org/index.php?module=article&view=279&lay_quiet=1&8dced886a4bd24eb30fc46843fb4287a=23679635f7832235dae9949749a76f35). The San Diego Participant Observer. Retrieved April 5, 2008.
- Katzenstein, Peter J.; Shiraishi, Takashi (1997). *Network Power: Japan in Asia*. Ithaca, New York: Cornell University Press. ISBN 978-0801483738.
- Kern, Adam (2006). *Manga from the Floating World: Comicbook Culture and the Kibyōshi of Edo Japan*. Cambridge: Harvard University Press. ISBN 978-0674022669.
- Kern, Adam (2007). "Symposium: Kibyoshi: The World's First Comicbook?". *International Journal of Comic Art* (9): 1–486.
- Kinsella, Sharon (2000). *Adult Manga: Culture and Power in Contemporary Japanese Society*. Honolulu, Hawaii: University of Hawai'i Press. ISBN 978-0824823184.
- Kittelson, Mary Lynn (1998). *The Soul of Popular Culture: Looking at Contemporary Heroes, Myths, and Monsters*. Chicago: Open Court. ISBN 978-0812693638.
- Lee, William (2000). "From Sazae-san to Crayon Shin-Chan". In Craig, Timothy J.. *Japan Pop!: Inside the World of Japanese Popular Culture*. Armonk, New York: M.E. Sharpe. ISBN 978-0765605610.
- Lent, John A. (2001). *Illustrating Asia: Comics, Humor Magazines, and Picture Books*. Honolulu, Hawaii: University of Hawaii Press. ISBN 0824824717.
- Leonard, Sean (September 12, 2004). "Progress Against the Law: Fan Distribution, Copyright, and the Explosive Growth of Japanese Animation" (http://web.mit.edu/seantek/www/papers/progress-columns.pdf). Retrieved December 19, 2007.
- Lone, Stewart (2007). *Daily Lives of Civilians in Wartime Asia: From the Taiping Rebellion to the Vietnam War* (http://books.google.com/?id=qmSnTIAPnoUC&pg=PA75&lpg=PA75&dq=shonen+sekai). Westport, Connecticut: Greenwood Publishing Group. ISBN 0313336849.
- Mahousu (January 2005). "Les editeurs des mangas" (http://home.comcast.net/~mahousu/editeurs.html). self-published. Retrieved December 19, 2007.
- Masters, Coco (August 10, 2006). "America is Drawn to Manga". *Time Magazine*.
- "First International MANGA Award" (http://www.mofa.go.jp/announce/announce/2007/6/1174276_828.html) (Press release). Ministry of Foreign Affairs of Japan. June 29, 2007.
- Napier, Susan J. (2000). *Anime: From Akira to Princess Mononoke*. New York: Palgrave. ISBN 0312238630.
- Nunez, Irma (September 24, 2006). "Alternative Comics Heroes: Tracing the Genealogy of Gekiga" (http://search.japantimes.co.jp/cgi-bin/fb20060924a1.html). *The Japan Times*. Retrieved December 19, 2007.
- Ōgi, Fusami (2004). "Female subjectivity and *shōjo* (girls) manga (Japanese comics): *shōjo* in Ladies' Comics and Young Ladies' Comics". *The Journal of Popular Culture* **36** (4): 780–803. doi:10.1111/1540-5931.00045.
- Patten, Fred (2004). *Watching Anime, Reading Manga: 25 Years of Essays and Reviews*. Berkeley, California: Stone Bridge Press. ISBN 978-1880656921.
- Perper, Timothy; Cornog, Martha (2002). "Eroticism for the masses: Japanese manga comics and their assimilation into the U.S.". *Sexuality & Culture* **6** (1): 3–126. doi:10.1007/s12119-002-1000-4.
- Perper, Timothy; Cornog, Martha (2003). "Sex, love, and women in Japanese comics". In Francoeur, Robert T.; Noonan, Raymond J.. *The Comprehensive International Encyclopedia of Sexuality*. New York: Continuum. ISBN 978-0826414885.
- Pink, Daniel H. (October 22, 2007). "Japan, Ink: Inside the Manga Industrial Complex" (http://www.wired.com/techbiz/media/magazine/15-11/ff_manga). *Wired* **15** (11). Retrieved December 19, 2007.
- Poitras, Gilles (2001). *Anime Essentials: Every Thing a Fan Needs to Know*. Berkeley, California: Stone Bridge. ISBN 978-1880656532.

- Reid, Calvin (March 28, 2006). "HarperCollins, Tokyopop Ink Manga Deal" (http://www.publishersweekly.com/pw/by-topic/book-news/comics/article/17182-harpercollins-tokyopop-ink-manga-deal.html). *Publishers Weekly*. Retrieved March 4, 2008.
- Reid, Calvin (February 6, 2009). "2008 Graphic Novel Sales Up 5%; Manga Off 17%" (http://www.publishersweekly.com/pw/by-topic/1-legacy/24-comic-book-reviews/article/9362-2008-graphic-novel-sales-up-5--manga-off-17-.html). *Publishers Weekly*. Retrieved September 7, 2009.
- Riciputi, Marco (October 25, 2007). "Komikazen: European comics go independent" (http://www.cafebabel.com/eng/article/22681/komikazen-european-comics-go-independent.html). Cafebabel.com. Retrieved March 4, 2008.
- Rifas, Leonard (2004). "Globalizing Comic Books from Below: How Manga Came to America". *International Journal of Comic Art* **6** (2): 138–171.
- Sanchez, Frank (1997–2003). "Hist 102: History of Manga" (http://web.archive.org/web/20080205123443/http://www.animeinfo.org/animeu/hist102.html). AnimeInfo. Archived from the original (http://www.animeinfo.org/animeu/hist102.html) on February 5, 2008. Retrieved September 11, 2007.
- Schodt, Frederik L. (1986). *Manga! Manga! The World of Japanese Comics*. Tokyo: Kodansha. ISBN 978-0870117527.
- Schodt, Frederik L. (1996). *Dreamland Japan: Writings on Modern Manga*. Berkeley, California: Stone Bridge Press. ISBN 978-1880656235.
- Schodt, Frederik L. (2007). *The Astro Boy Essays: Osamu Tezuka, Mighty Atom, and the Manga/Anime Revolution*. Berkeley, California: Stone Bridge Press. ISBN 978-1933330549.
- Shimizu, Isao (June 1985) (in Japanese). *: (Nihon Manga no Jiten - Dictionary of Japanese Manga)*. Sun lexica. ISBN 4385155860.
- Stewart, Bhob (October 1984). "Screaming Metal". *The Comics Journal* (94).
- Tai, Elizabeth (September 23, 2007). "Manga outside Japan" (http://thestar.com.my/lifestyle/story.asp?file=/2007/9/23/lifebookshelf/18898783&sec=lifebookshelf). *Star Online*. Retrieved December 19, 2007.
- Tchiei, Go (1998). "Characteristics of Japanese Manga" (http://www.dnp.co.jp/museum/nmp/nmp_i/articles/manga/manga2.html). Retrieved April 5, 2008.
- Thompson, Jason (2007). *Manga: The Complete Guide*. New York: Ballantine Books. ISBN 978-0345485908.
- Thorn, Matt (July–September 2001). "Shôjo Manga—Something for the Girls" (http://matt-thorn.com/shoujo_manga/japan_quarterly/). *The Japan Quarterly* **48** (3). Retrieved April 5, 2008.
- Toku, Masami (Spring 2006). "Shojo Manga: Girl Power!" (http://www.csuchico.edu/pub/cs/spring_06/feature_03.html). *Chico Statements* (California State University, Chico). ISBN 1886226105. Retrieved April 5, 2008.
- Vollmar, Rob (March 1, 2007). "Frederic Boilet and the Nouvelle Manga revolution" (http://goliath.ecnext.com/coms2/gi_0199-6318937/Frederic-Boilet-and-the-Nouvelle.html). *World Literature Today*. Retrieved September 14, 2007.
- Webb, Martin (May 28, 2006). "Manga by any other name is..." (http://search.japantimes.co.jp/cgi-bin/fl20060528x1.html). *The Japan Times*. Retrieved April 5, 2008.
- Wong, Wendy Siuyi (2002). *Hong Kong Comics: A History of Manhua*. New York: Princeton Architectural Press. ISBN 978-1568982694.
- Wong, Wendy Siuyi (2006). "Globalizing manga: From Japan to Hong Kong and beyond". *Mechademia: an Annual Forum for Anime, Manga, and the Fan Arts*: pp. 23–45.
- Wong, Wendy (September 2007). "The Presence of Manga in Europe and North America" (http://www.rthk.org.hk/mediadigest/20070913_76_121564.html). *Media Digest*. Retrieved December 19, 2007.
- "About Manga Museum: Current situation of manga culture" (http://www.kyotomm.com/HP/international/english/about_genjo.html). Kyoto Manga Museum. Retrieved September 6, 2009.

- "Correction: World Manga" (http://www.animenewsnetwork.com/news/2006-05-10/correction-world-manga). *Anime News Network*. May 10, 2006. Retrieved December 19, 2007.
- "I.C. promotes AmeriManga" (http://www.animenewsnetwork.com/news/2002-11-27/i.c-promotes-amerimanga). *Anime News Network*. November 11, 2002. Retrieved March 4, 2008.
- "Interview with Tokyopop's Mike Kiley" (http://www.icv2.com/articles/home/11249.html). *ICv2*. September 7, 2007. Retrieved December 19, 2007.
- *Japan: Profile of a Nation, Revised Edition*. Tokyo: Kodansha International. 1999. ISBN 4770023847.
- "Japan's Foreign Minister Creates Foreign Manga Award" (http://www.animenewsnetwork.com/news/2007-05-22/international-manga-award). *Anime News Network*. May 22, 2007. Retrieved October 5, 2009.
- "manga" (http://www.merriam-webster.com/dictionary/manga). Merriam-Webster Online Dictionary. Retrieved September 6, 2009.
- "Manga-mania in France" (http://www.animenewsnetwork.com/news/2004-02-04/manga-mania-in-france). *Anime News Network*. February 4, 2004. Retrieved December 19, 2007.
- "'Manga no Kuni': A manga magazine from the Second Sino-Japanese War period" (http://mmsearch.kyotomm.jp/infolib/search/CsvSearch.cgi?DEF_XSL=eng&GRP_ID=G0000002&DB_ID=G0000002GALLERY&IS_DB=G0000002GALLERY&IS_TYPE=csv&IS_STYLE=eng&SUM_KIND=CsvSummary&SUM_NUMBER=10&IS_SCH=CSV&META_KIND=NOFRAME&IS_KIND=CsvDetail&IS_NUMBER=1&SUM_TYPE=normal&IS_START=4&IS_KEY_A1="GALLERY"&IS_TAG_A1=Cul11&IS_ADDSCH_CNT=1&VIEW_FLG=0). Kyoto International Manga Museum. Retrieved December 21, 2008.
- "'Poten': a manga magazine from Kyoto" (http://mmsearch.kyotomm.jp/infolib/search/CsvSearch.cgi?DEF_XSL=eng&GRP_ID=G0000002&DB_ID=G0000002GALLERY&IS_DB=G0000002GALLERY&IS_TYPE=csv&IS_STYLE=eng&SUM_KIND=CsvSummary&SUM_NUMBER=10&IS_SCH=CSV&META_KIND=NOFRAME&IS_KIND=CsvDetail&IS_NUMBER=1&SUM_TYPE=normal&IS_START=3&IS_KEY_A1="GALLERY"&IS_TAG_A1=Cul11&IS_ADDSCH_CNT=1&VIEW_FLG=0). Kyoto International Manga Museum. Retrieved December 21, 2008.
- "'Shonen Pakku'; Japan's first children's manga magazine" (http://mmsearch.kyotomm.jp/infolib/search/CsvSearch.cgi?DEF_XSL=eng&GRP_ID=G0000002&DB_ID=G0000002GALLERY&IS_DB=G0000002GALLERY&IS_TYPE=csv&IS_STYLE=eng&SUM_KIND=CsvSummary&SUM_NUMBER=10&IS_SCH=CSV&META_KIND=NOFRAME&IS_KIND=CsvDetail&IS_NUMBER=1&SUM_TYPE=normal&IS_START=2&IS_KEY_A1="GALLERY"&IS_TAG_A1=Cul11&IS_ADDSCH_CNT=1&VIEW_FLG=0). Kyoto International Manga Museum. Retrieved December 21, 2008.
- "The first Japanese manga magazine: Eshinbun Nipponchi" (http://mmsearch.kyotomm.jp/infolib/search/CsvSearch.cgi?DEF_XSL=eng&GRP_ID=G0000002&DB_ID=G0000002GALLERY&IS_DB=G0000002GALLERY&IS_TYPE=csv&IS_STYLE=eng&SUM_KIND=CsvSummary&SUM_NUMBER=10&IS_SCH=CSV&META_KIND=NOFRAME&IS_KIND=CsvDetail&IS_NUMBER=1&SUM_TYPE=normal&IS_START=1&IS_KEY_A1="GALLERY"&IS_TAG_A1=Cul11&IS_ADDSCH_CNT=1&VIEW_FLG=0). Kyoto International Manga Museum. Retrieved December 21, 2008.
- "Tokyopop To Move Away from OEL and World Manga Labels" (http://www.animenewsnetwork.com/news/2006-05-05/tokyopop-to-move-away-from-oel-and-world-manga-labels). *Anime News Network*. May 5, 2006. Retrieved December 19, 2007.

Further reading

- "Japanese Manga Market Drops Below 500 Billion Yen" (http://comipress.com/news/2007/03/10/1622). ComiPress. March 10, 2007.
- "Un poil de culture - Une introduction à l'animation japonaise" (http://www.metalchroniques.fr/guppy/articles.php?lng=fr&pg=437) (in French). July 11, 2007.

External links

- Manga (http://www.dmoz.org/Arts/Comics/Manga/) at the Open Directory Project

Anime

Anime (, Japanese pronunciation: [anime] (listen); English: /ˈænɨmeɪ/ (listen) or English pronunciation: /ˈɑːnɨmeɪ/) is the Japanese abbreviated pronunciation of "animation". The definition sometimes changes depending on the context.[1] In English-speaking countries, the term most commonly refers to Japanese animated cartoons.[2]

While the earliest known Japanese animation dates to 1917, and many original Japanese cartoons were produced in the ensuing decades, the characteristic anime style developed in the 1960s—notably with the work of Osamu Tezuka—and became known outside Japan in the 1980s.

Anime, like manga, has a large audience in Japan and recognition throughout the world. Distributors can release anime via television broadcasts, directly to video, or theatrically, as well as online.

Both hand-drawn and computer-animated anime exist. It is used in television series, films, video, video games, commercials, and internet-based releases, and represents most, if not all, genres of fiction. As the market for anime increased in Japan, it also gained popularity in East and Southeast Asia. Anime is currently popular in many different regions around the world.

History

A cell from the earliest known Japanese animated short from 1917.

Anime began at the start of the 20th century, when Japanese filmmakers experimented with the animation techniques also pioneered in France, Germany, the United States, and Russia.[3] The oldest known anime in existence first screened in 1917 – a two-minute clip of a samurai trying to test a new sword on his target, only to suffer defeat.[4] [5] Early pioneers included Shimokawa Oten, Jun'ichi Kouchi, and Seitarō Kitayama.[6]

By the 1930s animation became an alternative format of storytelling to the live-action industry in Japan. But it suffered competition from foreign producers and many animators, such as Noburō Ōfuji and Yasuji Murata still worked in cheaper cutout not cel animation, although with masterful results.[7] Other creators, such as Kenzō Masaoka and Mitsuyo Seo, nonetheless made great strides in animation technique, especially with increasing help from a government using animation in education and propaganda.[8] The first talkie anime was *Chikara to Onna no Yo no Naka*, produced by Masaoka in 1933.[9] [10] The first feature length animated film was *Momotaro's Divine Sea*

Screenshot from *Momotaro's Divine Sea Warriors* (1944), the first feature-length anime film

Warriors directed by Seo in 1945 with sponsorship by the Imperial Japanese Navy.[11]

The success of The Walt Disney Company's 1937 feature film *Snow White and the Seven Dwarfs* influenced Japanese animators.[12] In the 1960s, manga artist and animator Osamu Tezuka adapted and simplified many Disney animation-techniques to reduce costs and to limit the number of frames in productions. He intended this as a temporary measure to allow him to produce material on a tight schedule with inexperienced animation-staff.

The 1970s saw a surge of growth in the popularity of manga – many of them later animated. The work of Osamu Tezuka drew particular attention: he has been called a "legend"[13] and the "god of manga".[14][15] His work – and that of other pioneers in the field – inspired characteristics and genres that remain fundamental elements of anime today. The giant robot genre (known as "Mecha" outside Japan), for instance, took shape under Tezuka, developed into the Super Robot genre under Go Nagai and others, and was revolutionized at the end of the decade by Yoshiyuki Tomino who developed the Real Robot genre. Robot anime like the *Gundam* and *The Super Dimension Fortress Macross* series became instant classics in the 1980s, and the robot genre of anime is still one of the most common in Japan and worldwide today. In the 1980s, anime became more accepted in the mainstream in Japan (although less than manga), and experienced a boom in production. Following a few successful adaptations of anime in overseas markets in the 1980s, anime gained increased acceptance in those markets in the 1990s and even more at the turn of the 21st century.

Terminology

Japanese write the English term "animation" in *katakana* a s (*animēshon*, pronounced Japanese pronunciation: [animeːɕoɴ]), and the term (*anime*, pronounced Japanese pronunciation: [anime] (listen) in Japanese) emerged in the 1970s as an abbreviation.[16] Others claim that the word derives from the French phrase *dessin animé*.[3] Japanese-speakers use both the original and abbreviated forms interchangeably, but the shorter form occurs more commonly.

The pronunciation of *anime* in Japanese, [anime], differs significantly from the Standard English English pronunciation: /ˈænɪmeɪ/, which has different vowels and stress. (In Japanese each mora carries equal stress.) As with a few other Japanese words such as *saké*, *Pokémon*, and *Kobo Abé*, English-language texts sometimes spell *anime* as *animé* (as in French), with an acute accent over the final *e*, to cue the reader to pronounce the letter, not to leave it silent as English orthography might suggest.

Word usage

In Japan, the term *anime* does not specify an animation's nation of origin or style; instead, it serves as a blanket term to refer to all forms of animation from around the world.[17] [18] English-language dictionaries define *anime* as "a Japanese style of motion-picture animation" or as "a style of animation developed in Japan".[19] [20]

Non-Japanese works that borrow stylization from anime are commonly referred to as "anime-influenced animation" but it is not unusual for a viewer who does not know the country of origin of such material to refer to it as simply "anime". Some works result from co-productions with non-Japanese companies, such as most of the traditionally animated Rankin/Bass works, the Cartoon Network and Production I.G series *IGPX* or *Ōban Star-Racers*; different viewers may or may not consider these anime.

In English, *anime*, when used as a common noun, normally functions as a mass noun (for example: "Do you watch anime?", "How much anime have you collected?").[21] However, in casual usage the word also appears as a count noun. *Anime* can also be used as a suppletive adjective or classifier noun ("The anime *Guyver* is different from the movie *Guyver*").

Synonyms

English-speakers occasionally refer to anime as "Japanimation", but this term has fallen into disuse. "Japanimation" saw the most usage during the 1970s and 1980s, but the term "anime" supplanted it in the mid-1990s as the material became more widely known in English-speaking countries.[22] In general, the term now only appears in nostalgic contexts.[22] Since "anime" does not identify the country of origin in Japanese usage, "Japanimation" is used to distinguish Japanese work from that of the rest of the world.[22]

In Japan, "manga" can refer to both animation and comics. Among English speakers, "manga" has the stricter meaning of "Japanese comics", in parallel to the usage of "anime" in and outside of Japan. The term "ani-manga" is used to describe comics produced from animation cels.[23]

Visual characteristics

Many commentators refer to anime as an art form.[24] As a visual medium, it can emphasize visual styles. The styles can vary from artist to artist or from studio to studio. Some titles make extensive use of common stylization: *FLCL*, for example, has a reputation for wild, exaggerated stylization. Other titles use different methods: *Only Yesterday* or *Jin-Roh* take much more realistic approaches, featuring few stylistic exaggerations; *Pokémon* uses drawings which specifically do not distinguish the nationality of characters.[25]

Anime artists use many distinct visual styles.

While different titles and different artists have their own artistic styles, many stylistic elements have become so common that describe them as definitive of anime in general. However, this does not mean that all modern anime share one strict, common art-style. Many anime have a very different art style from what would commonly be called "anime style", yet fans still use the word "anime" to refer to these titles. Generally, the most common form of anime drawings include "exaggerated physical features such as large eyes, big hair and elongated limbs... and dramatically shaped speech bubbles, speed lines and onomatopoeic, exclamatory typography."[26]

The influences of Japanese calligraphy and Japanese painting also characterize linear qualities of the anime style. The round ink brush traditionally used for writing kanji and for painting, produces a stroke of widely varying thickness.

Anime also tends to borrow many elements from manga, including text in the background and panel layouts. For example, an opening may employ manga panels to tell the story, or to dramatize a point for humorous effect. See for example the anime *Kare Kano*.

Character design

Proportions

Body proportions emulated in anime come from proportions of the human body. The height of the head is considered by the artist as the base unit of proportion. Head heights can vary as long as the remainder of the body remains proportional. Most anime characters are about seven to eight heads tall, and extreme heights are set around nine heads tall.[27]

Variations to proportion can be modified by the artist. Super-deformed characters feature a non-proportionally small body compared to the head. Sometimes specific body parts, like legs, are shortened or elongated for added emphasis. Most super deformed characters are two to four heads tall. Some anime works like *Crayon Shin-chan* completely disregard these proportions, such that they resemble Western cartoons. For exaggeration, certain body features are increased in proportion.[27]

Eye styles

Many anime and manga characters feature large eyes. Osamu Tezuka, who is believed to have been the first to use this technique, was inspired by the exaggerated features of American cartoon characters such as Betty Boop, Mickey Mouse, and Disney's *Bambi*.[3] [28] Tezuka found that large eyes style allowed his characters to show emotions distinctly. When Tezuka began drawing *Ribbon no Kishi*, the first manga specifically targeted at young girls, Tezuka further exaggerated the size of the characters' eyes. Indeed, through *Ribbon no Kishi*, Tezuka set a stylistic template that later *shōjo* artists tended to follow.

Coloring is added to give eyes, particularly to the cornea, some depth. The depth is accomplished by applying variable color shading. Generally, a mixture of a light shade, the tone color, and a dark shade is used.[29] [30] Cultural anthropologist Matt Thorn argues that Japanese animators and audiences do not perceive such stylized eyes as inherently more or less foreign.[31]

However, not all anime have large eyes. For example, some of the work of Hayao Miyazaki and Toshiro Kawamoto are known for having realistically proportioned eyes, as well as realistic hair colors on their characters.[32]

Facial expressions

Anime characters may employ a variety of predetermined facial expressions to denote moods and thoughts.[33] These techniques are often different in form than their counterparts in western animation, and they include a fixed iconography that's used as shorthand for certain emotions and moods.[34]

There are a number of other stylistic elements that are common to conventional anime as well but more often used in comedies. Characters that are shocked or surprised will perform a "face fault", in which they display an extremely exaggerated expression. Angry characters may exhibit a "vein" or "stress mark" effect, where lines representing bulging veins will appear on their forehead. Angry women will sometimes summon a mallet from nowhere and strike another character with it, mainly for the sake of slapstick comedy. Male characters will develop a bloody nose around their female love interests (typically to indicate arousal, which is a play on an old wives' tale).[34] Embarrassed or stressed characters either produce a massive sweat-drop (which has become one of the most widely recognized motifs of conventional anime) or produce a visibly red blush or set of parallel (sometimes squiggly) lines beneath the eyes, especially as a manifestation of repressed romantic feelings. Characters who want to childishly taunt someone may pull an akanbe face (by pulling an eyelid down with a finger to expose the red underside). Characters may also have large "X" eyes to show a knockout, or in some cases, even illness. This is typically used for comedic purposes. Vacant, non-reflecting eyes can be used to indicate a state of semi-consciousness.

Animation technique

Like all animation, the production processes of storyboarding, voice acting, character design, cel production and so on still apply. With improvements in computer technology, computer animation increased the efficiency of the whole production process.

Anime is often considered a form of limited animation. That means that stylistically, even in bigger productions the conventions of limited animation are used to fool the eye into thinking there is more movement than there is.[3] Many of the techniques that are used comprise cost-cutting measures while working under a set budget.

Anime scenes place emphasis on achieving three-dimensional views. Backgrounds depict the scenes' atmosphere.[3] For example, anime often puts emphasis on changing seasons, as can be seen in numerous anime, such as *Tenchi Muyo!*. Sometimes actual settings have been duplicated into an anime. The backgrounds for the *Melancholy of Haruhi Suzumiya* are based on various locations within the suburb of Nishinomiya, Hyogo, Japan.[35]

Camera angles, camera movement, and lighting play an important role in scenes. Directors often have the discretion of determining viewing angles for scenes, particularly regarding backgrounds. In addition, camera angles show perspective.[36] Directors can also choose camera effects within cinematography, such as panning, zooming, facial closeup, and panoramic.[37]

The large majority of anime uses traditional animation, which better allows for division of labor, pose to pose approach and checking of drawings before they are shot – practices favored by the anime industry.[38] Other mediums are mostly limited to independently made short films,[39] examples of which are the silhouette and other cutout animation of Noburō Ōfuji,[38] [40] the stop motion puppet animation of Tadahito Mochinaga, Kihachirō Kawamoto[41] and Tomoyasu Murata[42] and the computer animation of Satoshi Tomioka[43] (most famously *Usavich*).[44]

Story themes

A wide variety of stories have been adapted into anime. They are sourced from Japanese history, classical literature, and even adult-oriented themes. While animation for children exists, most anime are intended for an older audience.[45]

Distribution

While anime had entered markets beyond Japan in the 1960s, it grew as a major cultural export during its market expansion during the 1980s and 1990s. The anime market for the United States alone is "worth approximately $4.35 billion, according to the Japan External Trade Organization".[46] Anime has also had commercial success in Asia, Europe and Latin America, where anime has become more mainstream than in the United States. For example, the *Saint Seiya* video game was released in Europe due to the popularity of the show even years after the series has been off-air.

Anime distribution companies handled the licensing and distribution of anime outside Japan. Licensed anime is modified by distributors through dubbing into the language of the country and adding language subtitles to the Japanese language track. Using a similar global distribution pattern as Hollywood, the world is divided into five regions.

Some editing of cultural references may occur to better follow the references of the non-Japanese culture.[47] Certain companies may remove any objectionable content, complying with domestic law. This editing process was far more prevalent in the past (e.g. *Voltron*), but its use has declined because of the demand for anime in its original form. This "light touch" approach to localization has favored viewers formerly unfamiliar with anime. *Robotech* and *Star Blazers* were the earliest attempts to present anime (albeit still modified) to North American television audiences without harsh censoring for violence and mature themes.

With the advent of DVD, it became possible to include multiple language tracks into a simple product. This was not the case with VHS cassette, in which separate VHS media were used and with each VHS cassette priced the same as a single DVD. The "light touch" approach also applies to DVD releases as they often include both the dubbed audio and the original Japanese audio with subtitles, typically unedited. Anime edited for television is usually released on DVD "uncut", with all scenes intact.

The Internet has played a significant role in the exposure of anime beyond Japan. Prior to the 1990s, anime had limited exposure beyond Japan's borders. Coincidentally, as the popularity of the Internet grew, so did interest in anime. Much of the fandom of anime grew through the Internet. The combination of internet communities and increasing amounts of anime material, from video to images, helped spur the growth of fandom.[48] As the Internet gained more widespread use, Internet advertising revenues grew from 1.6 billion yen to over 180 billion yen between 1995 and 2005.[49]

Some fan groups add subtitles to anime on their own and distribute the episodes. These are known as fansubs. Before the popularity of the Internet, fansubbing used VHS as a means of distribution. Often, people will collect these fansubs and upload them to websites which they also put advertisements on so as to earn money, which violates copyright laws in many countries. The ethical implications of distributing or watching fansubs are topics of much controversy even when fansub groups do not profit from their activities. Once the series has been licensed outside of Japan, fansub groups often cease distribution of their work. In one case, Media Factory Incorporated requested that no fansubs of their material be made, which was respected by the fansub community.[50] In another instance, Bandai specifically thanked fansubbers for their role in helping to make *The Melancholy of Haruhi Suzumiya* popular in the English speaking world.[51]

Broadcasting

TV networks regularly broadcast anime programming. In Japan, major national TV networks, such as TV Tokyo broadcast anime regularly. Smaller regional stations broadcast anime under the UHF. In the United States, cable TV channels such as Cartoon Network, Disney, Syfy, and others dedicate some of their timeslots to anime. Some, such as the Anime Network and the FUNimation Channel, specifically show anime. Sony-based Animax and Disney's Jetix channel broadcast anime within many countries in the world. AnimeCentral solely broadcasts anime in the UK.

Influence on world culture

Anime has become commercially profitable in western countries, as early commercially successful western adaptations of anime, such as *Astro Boy*, have revealed.[52] The phenomenal success of Nintendo's multi-billion dollar *Pokémon* franchise[53] was helped greatly by the spin-off anime series that, first broadcast in the late 1990s, is still running worldwide to this day. In doing so, anime has made significant impacts upon Western culture. Since the 19th century, many Westerners have expressed a particular interest towards Japan. Anime dramatically exposed more Westerners to the culture of Japan. Aside from anime, other facets of Japanese culture increased in popularity.[54] Worldwide, the number of people studying Japanese increased. In 1984, the Japanese Language Proficiency Test was devised to meet increasing demand.[55]

Even domestic animation industries had made attempts at emulating anime. Anime-influenced animation refers to non-Japanese works of animation that emulate the visual style of anime.[56] Most of these works are created by studios in the United States, Europe, and non-Japanese Asia; and they generally incorporate stylizations, methods, and gags described in anime physics, as in the case of *Avatar: The Last Airbender*. Often, production crews either are fans of anime or are required to view anime.[57] Some creators cite anime as a source of inspiration with their own series.[58] [59] Furthermore, a French production team for *Ōban Star-Racers* moved to Tokyo to collaborate with a Japanese production team from Hal Film Maker.[60] Critics and the general anime fanbase do not consider them as anime.[61]

Some American animated television-series have singled out anime styling with satirical intent, for example *South Park* (with "Chinpokomon" and with "Good Times with Weapons"). *South Park* has a notable drawing style, itself parodied in "Brittle Bullet", the fifth episode of the anime *FLCL*, released several months after "Chinpokomon" aired. This intent on satirizing anime is the springboard for the basic premise of *Kappa Mikey*, a Nicktoons Network original cartoon. Even clichés normally found in anime are parodied in some series, such as *Perfect Hair Forever*.

Anime conventions began to appear in the early 1990s, during the Anime boom, starting with Anime Expo, Animethon, Otakon, and JACON. Currently anime conventions are held annually in various cities across the Americas, Asia, and Europe.[62] Many attendees participate in cosplay, where they dress up as anime characters. Also, guests from Japan ranging from artists, directors, and music groups are invited. In addition to anime conventions, anime clubs have become prevalent in colleges, high schools, and community centers as a way to publicly exhibit anime as well as broadening Japanese cultural understanding.[63]

Viewers may also pick up on Japanese terms either within or related to anime, though at times those words may take on different connotations. For instance, the Japanese term *otaku* is used as a term for anime fans beyond Japan, more particularly the obsessive ones. The negative connotations associated with the word in Japan have lessened in foreign context, where it instead connotes the pride of the fans.

See also

- List of anime conventions
- List of anime theatrically released in the United States
- List of animated feature films
- List of animated television series
- Original video animation
- Q-version (anime extras)
- Seiyū - Voice acting in Japan
- Wikt:Appendix:Japanese film credit terms
- Hentai (sexually explicit material)

References

[1] Brown, Steven T. *Cinema Anime*. Palgrave Macmillan, 2006, p. 7

[2] "anime - Definition from the Merriam-Webster Online Dictionary" (http://www.merriam-webster.com/dictionary/anime). Merriam-webster.com. 2010-08-13. . Retrieved 2010-11-01.

[3] Schodt, Frederik L. (Reprint edition (August 18, 1997)). *Manga! Manga!: The World of Japanese Comics*. Tokyo, Japan: Kodansha International. ISBN 0-87011-752-1.

[4] "Japan's oldest animation films" (http://imprinttalk.com/?p=1557). *ImprintTALK*. 2008-03-31. .

[5] "Historic 91-year-old anime discovered in Osaka" (http://web.archive.org/web/20080402012234/http://www.hdrjapan.com/japan/japan-news/historic-91yearold-anime-discovered-in-osaka/). HDR Japan. 2008-03-30. Archived from the original (http://www.hdrjapan.com/japan/japan-news/historic-91yearold-anime-discovered-in-osaka/) on 2008-04-02. . Retrieved 2008-05-12.

[6] Yamaguchi, Katsunori; Yasushi Watanabe (1977). *Nihon animēshon eigashi*. Yūbunsha. pp. 8–11.

[7] Sharp, Jasper (September 23, 2004). "Pioneers of Japanese Animation (Part 1)" (http://www.midnighteye.com/features/pioneers-of-anime.shtml). Midnight Eye. . Retrieved 11 December 2009.

[8] Yamaguchi, Katsunori; Yasushi Watanabe (1977). *Nihon animēshon eigashi*. Yūbunsha. pp. 26–37.

[9] Baricordi, Andrea; de Giovanni, Massimiliano; Pietroni, Andrea; Rossi, Barbara; Tunesi, Sabrina (December 2000). *Anime: A Guide to Japanese Animation (1958-1988)*. Montreal, Quebec, Canada: Protoculture Inc.. p. 12. ISBN 2-9805759-0-9.

[10] *Japan: An Illustrated Encyclopedia*. Tokyo, Japan: Kodansha. 1993. ISBN 9784062064897.

[11] Official booklet, *The Roots of Japanese Anime*, DVD, Zakka Films, 2009.

[12] "A Brief History of Anime" (http://www.corneredangel.com/amwess/papers/history.html). *Michael O'Connell, Otakon 1999 Program Book*. 1999. . Retrieved 2007-09-11.

[13] Ohara, Atsushi (2006-05-11). "5 missing manga pieces by Osamu Tezuka found in U.S." (http://web.archive.org/web/20060520053910/http://www.asahi.com/english/Herald-asahi/TKY200605110157.html). Asahi Shimbun. Archived from the original (http://www.asahi.com/english/Herald-asahi/TKY200605110157.html) on 2006-05-20. . Retrieved 2006-08-29.

[14] "Dr. Osamu Tezuka" (http://www.abcb.com/ency/t/tezuka_osamu.htm). *The Anime Encyclopedia*. The Anime Café. 2000-03-14. . Retrieved 2006-08-29.

[15] Gravett, Paul (2003). "Osamu Tezuka: The God of Manga" (http://web.archive.org/web/20071231013227/http://www.paulgravett.com/articles/006_tezuka/006_tezuka.htm). Archived from the original (http://www.paulgravett.com/articles/006_tezuka/006_tezuka.htm) on 2007-12-31. . Retrieved 2006-08-29.

[16] "Etymology Dictionary Reference: Anime accessdate=2007-09-13" (http://www.etymonline.com/index.php?term=anime). *Etymonline*. .

[17] "What is Anime?" (http://www.bellaonline.com/articles/art4260.asp). *Lesley Aeschliman*. Bellaonline. . Retrieved 2007-10-28.

[18] "Tezuka: The Marvel of Manga - Education Kit" (http://web.archive.org/web/20070830033821/http://artgallery.nsw.gov.au/__data/page/9842/Tezuka_Kit_1.pdf) (PDF). Art Gallery New South Wales. 2007. Archived from the original (http://www.artgallery.nsw.gov.au/__data/page/9842/Tezuka_Kit_1.pdf) on 2007-08-30. . Retrieved 2007-10-28.

[19] "Anime Dictionary Definition" (http://dictionary.reference.com/browse/anime). *Dictionary.com*. . Retrieved 2006-10-09.

[20] "Merriam-Webster:anime" (http://www.merriam-webster.com/dictionary/anime). *Merriam-Webster*. . Retrieved 2010-11-18.

[21] American Heritage Dictionary, 4th ed.; Dictionary.com Unabridged (v 1.1).

[22] Patten, Fred (2004). *Watching Anime, Reading Manga: 25 Years of Essays and Reviews*. Stone Bridge Press. ISBN 1880656922.

[23] "Inu Yasha Ani-MangaGraphic Novels" (http://www.animecornerstore.com/inuyaanno.html). Animecornerstore.com. 1999-11-01. . Retrieved 2010-11-01.

[24] "Ask John: Do Japanese Viewers Treat Anime Shows as Fads?" (http://www.animenation.net/blog/2006/04/07/ask-john-do-japanese-viewers-treat-anime-shows-as-fads/). *Ask John*. AnimeNation. 2006-04-07. . Retrieved 2008-01-23.

[25] Tobin, Joseph Jay (2004). *Pikachu's Global Adventure: The Rise and Fall of Pokémon*. Duke University Press. p. 88. ISBN 0-822-33287-6.

[26] "Japan Times" (http://search.japantimes.co.jp/print/fl20060528x1.html). . Retrieved 2008-02-06.

[27] "Body Proportion" (http://animeworld.com/howtodraw/bodies1.html). *Akemi's Anime World*. . Retrieved 2007-08-16.

[28] Schodt, Frederik L. (1996). *Dreamland Japan: Writings on Modern Manga*. Berkeley, California: Stone Bridge Press. ISBN 1-8806562-3-X.

[29] "Basic Anime Eye Tutorial" (http://www.biorust.com/tutorials/detail/141/en/). *Centi, Biorust.com*. . Retrieved 2007-08-22.

[30] Carlus (2007-06-06). "How to color anime eye" (http://www.youtube.com/watch?v=VyJ9yfYl_Fc). YouTube. . Retrieved 2007-08-22.

[31] "Do Manga Characters Look "White"?" (http://web.archive.org/web/20060517194357sh_re_/www.matt-thorn.com/mangagaku/faceoftheother.html). . Retrieved 2005-12-11.

[32] Poitras, Gilles (1998). *Anime Companion*. Berkeley, California: Stone Bridge Press. ISBN 1-880656-32-9.

[33] "Manga Tutorials: Emotional Expressions" (http://www.mangatutorials.com/tut/expressions.php). *Rio*. . Retrieved 2008-08-22.

[34] University of Michigan Animae Project (Current). "Emotional Iconography in Animae" (http://www.umich.edu/~anime/info_emotions.html). . Retrieved 2009-08-08.

[35] "Reference pictures to actual places" (http://www.rinku.zaq.ne.jp/p_v/haruhi.html). . Retrieved 2007-01-25.

[36] "Anime production process - feature film" (http://www.huitula.com/productionIG2_page2.htm). *PRODUCTION I.G*. 2000. . Retrieved 2007-08-27.

[37] "Cinematography: Looping and Animetion Techniques" (http://web.archive.org/web/20080613154025/http://www.understandinganime.com/cinematography.php). *Understanding Anime*. 1999. Archived from the original (http://www.understandinganime.com/cinematography.php) on 2008-06-13. . Retrieved 2007-08-29.

[38] Jouvanceau, Pierre; Clare Kitson (translator) (2004). *The Silhouette Film* (http://www.heeza.fr/description.php?lang=2&path=64&sort=Article&page=0&id=296). Genoa: Le Mani. p. 103. ISBN 88-8012-299-1. . Retrieved 2009-08-08.

[39] Sharp, Jasper (2003). "Beyond Anime: A Brief Guide to Experimental Japanese Animation" (http://www.midnighteye.com/features/beyond_anime.shtml). *Midnight Eye*. . Retrieved 2008-07-21.

[40] "Tribute to Noburō Ōfuji" (http://www.cinematheque.qc.ca/animation_japonaise.pdf) (PDF). *To the Source of Anime: Japanese Animation*. Cinémathèque québécoise. 2008. . Retrieved 2008-07-21.

[41] Sharp, Jasper (2004). "Interview with Kihachirō Kawamoto" (http://www.midnighteye.com/interviews/kihachiro_kawamoto.shtml). *Midnight Eye*. . Retrieved 2008-07-21.

[42] Munroe Hotes, Catherine (2008). "Tomoyasu Murata and Company" (http://www.midnighteye.com/features/tomoyasu-murata-and-company.shtml). *Midnight Eye*. . Retrieved 2008-07-21.

[43] Walters, Helen (2004). *Animation Unlimited: Innovative Short Films Since 1940* (http://lib.leeds.ac.uk/record=b2662684). London: Laurence King. ISBN 18-5669-346-5. . Retrieved 2009-08-08.

[44] "Works" (http://www.mtvjapan.com/usavich/about.html). *KANABAN-Web*. Kanaban Graphics. 2008. . Retrieved 2008-07-21.

[45] "An Anime Explosion" (http://www.utexas.edu/features/archive/2004/anime.html). University of Texas-Austin. 2008-10-09. . Retrieved 2011-10-04.

[46] "Manga Mania" (http://online.wsj.com/article/SB118851157811713921.html?mod=googlenews_wsj). *Bianca Bosker (Wall Street Journal)*. 2007-08-31. . Retrieved 2007-08-31.

[47] "Pokemon Case Study" (http://w3.salemstate.edu/~poehlkers/Emerson/Pokemon.html). W3.salemstate.edu. . Retrieved 2010-11-01.

[48] "100 Questions About Anime & Manga Overseas" (http://comipress.com/article/2006/07/20/489). *Comipress*. 2006-07-20. . Retrieved 2007-08-23.

[49] "Free Anime: Providers Bear Losses to Build Business" (http://en.j-cast.com/2005/12/21000171.html). *J-Cast Business News*. 2005-12-21. . Retrieved 2007-08-27.

[50] "Anxious times in the cartoon underground" (http://news.cnet.com/Anxious-times-in-the-cartoon-underground/2100-1026_3-5557177.html). *CNet*. 2005-02-01. . Retrieved 2007-09-06.

[51] "Adventures of the ASOS Brigade Episode 00: Made by Fans for Fans" (http://asosbrigade.com/). . Retrieved 2006-12-23.

[52] "Progress Against the Law: Fan Distribution, Copyright, and the Explosive Growth of Japanese Animation" (http://groups.csail.mit.edu/mac/classes/6.805/student-papers/fall03-papers/Progress_Against_the_Law.html). . Retrieved 2006-05-01.

[53] "Pokemon Franchise Approaches 150 Million Games Sold" (http://www2.prnewswire.com/cgi-bin/stories.pl?ACCT=104&STORY=/www/story/10-04-2005/0004159206&EDATE=). PR Newswire. Nintendo. 4 October 2005. .

[54] Faiola, Anthony (2003-12-27). "Japan's Empire of Cool" (http://www.washingtonpost.com/ac2/wp-dyn/A33261-2003Dec26?language=printer). *The Washington Post* (Washington Post Company): p. A1. . Retrieved 2007-08-17.

[55] "Introduction" (http://www.jlpt.jp/e/about/index.html). The Japan Foundation. . Retrieved 2009-05-01.

[56] "What is anime?" (http://www.animenewsnetwork.com/editorial/2002-07-26). *ANN*. 2002-07-26. . Retrieved 2007-08-18.

[57] "SciFi Channel Anime Review" (http://web.archive.org/web/20080117145951/http://www.scifi.com/sfw/anime/sfw12366.html). *SciFi*. Archived from the original (http://www.scifi.com/sfw/anime/sfw12366.html) on 2008-01-17. . Retrieved 2006-10-16.

[58] "Aaron McGruder - The Boondocks Interview" (http://www.ugo.com/ugo/html/article/?id=17924). *Troy Rogers*. UnderGroundOnline. . Retrieved 2007-10-14.

[59] "Ten Minutes with "Megas XLR"" (http://www.g4tv.com/screensavers/features/49962/Ten_Minutes_with_Megas_XLR.html). 2004-10-13. .

[60] "STW company background summary" (http://www.savtheworld.com/eng/company.php). .

[61] "How should the word *Anime* be defined?" (http://www.animenation.net/blog/2006/05/15/ask-john-how-should-the-word-anime-be-defined/). *AnimeNation*. 2006-05-15. . Retrieved 2008-09-26.

[62] "Convention Schedule" (http://www.animecons.com/events/). *AnimeCons*. . Retrieved 2007-09-06.

[63] "Anime achieves growing popularity among Stanford students" (http://web.archive.org/web/20071021222201/http://daily.stanford.edu/article/2002/5/21/animeAchievesGrowingPopularityAmongStanfordStudents). Archived from the original (http://daily.stanford.edu/article/2002/5/21/animeAchievesGrowingPopularityAmongStanfordStudents) on 2007-10-21. .

External links

- Anime (http://www.dmoz.org/Arts/Animation/Anime/) at the Open Directory Project

Spike_(company)

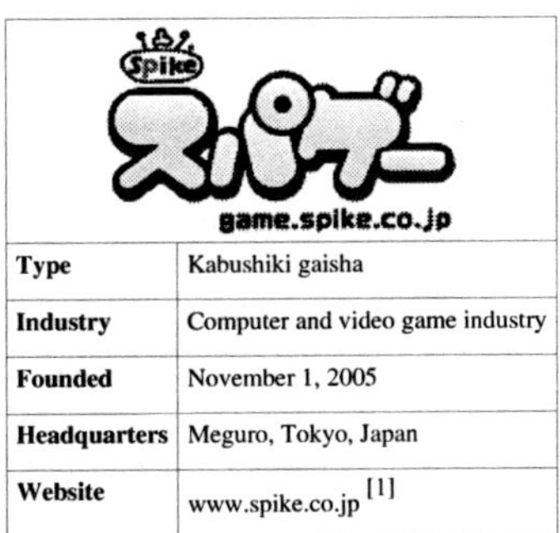

Type	Kabushiki gaisha
Industry	Computer and video game industry
Founded	November 1, 2005
Headquarters	Meguro, Tokyo, Japan
Website	www.spike.co.jp [1]

Spike (*Supaiku*) is a Japanese video game developer and publisher. Most of the staff were part of Human Entertainment. Human's *Fire Pro Wrestling* series is owned by Spike.

Games developed

- *Crimson Tears*
- *Dangan Ronpa: Kibou no Gakuen to Zetsubou no Koukousei*
- *Dragon Ball: Raging Blast*
- *Dragon Ball: Raging Blast 2*
- *Dragon Ball Z: Budokai Tenkaichi*
- *Dragon Ball Z: Budokai Tenkaichi 2*
- *Dragon Ball Z: Budokai Tenkaichi 3*
- *Dragon Ball Z: Ultimate Tenkaichi*
- *Elvandia Story*
- *Escape from Bug Island* (*Necro-Nesia* in Japan)
- *Fire Pro Wrestling*
- *Michigan*
- *King of Colosseum*
- *King of Colosseum II*
- *LifeSigns: Surgical Unit*
- *Twilight Syndrome: Kinjiratera Toshi Densetsu*
- *Shinjyuku no Ōkami*
- *428: Fūsa Sareta Shibuya de* (PlayStation 3 and PlayStation Portable ports)

Games published in Japan

- *Astro Tripper*
- *BioShock*
- *Call of Duty 3*
- *Conflict: Denied Ops* (*Double Clutch* in Japan)
- *Dead Island*
- *Dragon Age: Origins*
- *Fushigi no Dungeon: Fūrai no Shiren 3: Karakuri Yashiki no Nemuri Hime* (PSP version)
- *Formation Soccer 2002*
- *Fushigi no Dungeon: Fūrai no Shiren 4: Kami no Me to Akuma no Heso*
- *Gachitora!: Abarenbou Kyoushi in High School*
- *Greed Corp*
- *Haze*
- *Homefront*
- *Kane & Lynch: Dead Men*
- *Kenka Bancho*
- *Kenka Bancho 2: Full Throttle*
- *Kenka Bancho 3: Zenkoku Seiha* (*Kenka Bancho: Badass Rumble* in North America)
- *Kenka Bancho 4: Ichinen Sensō*
- *Kenka Bancho 5: Otoko no Rule*
- *MadWorld*
- *Metro 2033*
- *Midnight Club: Los Angeles*
- *Red Faction: Guerrilla*
- *Sacred 2: Fallen Angel*
- *Samurai Western*
- *The Darkness*
- *Tomb Raider: Legend*
- *Tomb Raider: Anniversary*
- *Tomb Raider: Underworld*
- *True Crime: New York City*
- *Urban Chaos: Riot Response*
- *Wanted: Weapons of Fate*
- *Way of the Samurai*
- *Way of the Samurai 2*
- *Way of the Samurai 3*
- *Way of the Samurai 4*
- *999: Nine Hours, Nine Persons, Nine Doors*

External links

- IGN's company profile [2]

Namco_Bandai_Games

Type	Subsidiary of Namco Bandai Holdings
Industry	Video game industry
Founded	2006
Headquarters	Shinagawa, Tokyo, Japan
Area served	Worldwide
Products	List of Namco Bandai games
Parent	Namco Bandai Holdings
Website	Bandai Namco Games in English site [1]

Namco Bandai Games Inc. (**NBGI**) (*Kabushiki Gaisha Bandai Namuko Gēmusu*) is an arcade, mobile and home video game developer and publisher based in Japan which is the product of a merger between the video game development divisions of Bandai and Namco. Namco Bandai Games is a wholly owned subsidiary of Namco Bandai Holdings (NBHD) and specializes in production and sales of home video games. It is the head company of Bandai Namco Group's Game Contents Strategic Business Unit (SBU). In North America, the U.S.-based Namco Bandai Games America represents the company as a regional publisher, while in PAL territories, publishing is handled by Namco Bandai Games Europe with distribution through Namco Bandai Partners. The company has its headquarters in Shinagawa, Tokyo.[1]

History

NBGI Headquarters
(formerly Shinagawa Panasonic Building)

In early 2005, Namco Ltd. and Bandai Co. Ltd. announced plans to combine their operations. The merger took effect in September of that same year and both companies came under the umbrella of the newly created Namco Bandai Holdings. On March 31, 2006, the game division of Bandai was merged to Namco and renamed Namco Bandai Games. In the United States, the Namco Hometek subsidiary was renamed to Namco Bandai Games America.

On April 1, 2008, Banpresto's video game operations were absorbed by Namco Bandai Games as well.[2] On August 6, 2008, it was announced that Bandai Networks Co. Ltd, Namco Bandai's mobile phone business, would be dissolved and absorbed into Namco Bandai Games on April 1, 2009.[3]

Surge label

At the end of 2008, just prior to the release of *Afro Samurai*, Namco Bandai Games announced the *Surge* publishing label.[4] The new label was an attempt at creating a separate identity for the company, aimed at a Western audience desiring more mature themes in game content.[5] Following *Afro*, early press materials indicated that *Dead to Rights: Retribution* and *Splatterhouse* would also be released under the *Surge* label, but neither game ultimately did; both were released with a retro *Namco* label instead to reflect the series' legacy. As of 2010, *Afro* is still the only game to have been released as a *Surge* title. However, more games will be on the way when Surge announced that they will publish 2 Days to Vegas by Steel Monkeys.

PlayStation Home

Namco Bandai Games has released a Home Space for the Asian, Japanese, and North American versions of PlayStation Home. There are two rooms. The first room, the "Game Arena" (Asia & Japan), or "Namco Arcade Center" (North America), has all the *Namco Museum.comm/Essentials* arcade machines, a "Vending Machine" that gives users a Namco Museum Soda Can for their personal space, a "Money Exchange Machine" that give users a Namco Museum Coin for their personal space, "*The Idolmaster SP* Wandering Star, Perfect Sun, & Missing Moon Vending Machines" (lets users play roulette to get cans that features the various idols and a Haruka costume), two video screens, a "Love Seat" and seating for the avatars. The second room, the "Community Arena" (Asia & Japan), or the "Namco Theatre" (North America), is a mini-museum featuring displays for *Soul Calibur IV*, *Ridge Racer 7*, and *Tekken 6*, as well as video screens behind each display, "*The Idolmaster SP* 961 Production Vending Machine", and seats for the avatars. This space was released in Japans version of Home on December 11, 2008, in the Asian version on March 26, 2009, and in the North American version on July 16, 2009.

See also

- List of Namco Bandai games

References

[1] " Corporate Overview (http://www.bandainamcogames.co.jp/english/overview/index.php)." Namco Bandai Games. Retrieved on March 16, 2010.

[2] IGN.com: Sayonara, Banpresto (http://games.ign.com/articles/833/833693p1.html)

[3] KatanaXtreme.com - Namco Bandai Games Inc. to Take Over Bandai Networks Co., Ltd. (http://www.katanaxtreme.com/anime_news/details.aspx?id=li1vddzCy06DZUz3hzp97g)

[4] McWhertor, Michael. "Namco Bandai Needs Westerners, Plans "Surge"" (http://kotaku.com/5102115/namco-bandai-needs-westerners-plans-surge). Kotaku. .

[5] Sheffield, Brandon. "Q&A: Namco Bandai's Iwai Talks Tricky Topics In Western Markets" (http://www.gamasutra.com/php-bin/news_index.php?story=22589). Gamasutra. .

External links

- Namco Bandai Games, Inc. (http://www.bandainamcogames.co.jp/english/)
- Namco Bandai Games America, Inc. (http://www.namcobandaigames.com/)
- Namco Bandai Games Europe (http://www.namcobandaigames.eu/)
- Namco Bandai Games, Inc. (http://www.mobygames.com/company/namco-bandai-games-inc) profile on MobyGames

PlayStation_3

WARNING: Article could not be rendered - ouputting plain text.

Potential causes of the problem are: (a) a bug in the pdf-writer software (b) problematic Mediawiki markup (c) table is too wide

""PS3 Slim gains ability to bitstream Dolby TrueHD, DTS-HD MA"". Joystiq. 2009-08-21. . Retrieved 2010-03-22.PlayStation 3 logoOriginal model, DualShock 3 controller and Slim modelCounter-clockwise from the top: Original logo, new logo, original "fat" model, DualShock#DualShock 3DualShock 3 controller, current "slim" model.DeveloperSony Computer EntertainmentSCEIManufacturerSony EMCS, Foxconn, AsusASUSTeKShilov, Anton (2006-07-18). "Asustek Computer Ships PlayStation 3 Consoles". X-bit labs. . Retrieved 2007-05-05.Product familyPlayStationHistory of video gamesGenerationHistory of video game consoles (seventh generation)Seventh generationRetail availability November 11, 2006 Units sold 55.5 million (as of September 30, 2011) "PlayStation®3 Worldwide Hardware Unit Sales (Unit: million)". SCEI. .MediaBlu-ray Disc (PlayStation 3 game disc) DVDCompact DiscPlayStation (console)PlayStation game disc PlayStation 2 game disc (#Model comparison1st & 2nd generations only)Super Audio CD (#Model comparison1st & 2nd generations only)Digital distributionOperating systemXrossMediaBarPlayStation 3 system softwareSystem software version 4.00(November 29, 2011) "PS3 firmware 4.00 now available". Shacknews. 2011-11-30. . Retrieved 2011-12-04.Central processing unitCPU 3.2 GHz Cell (microprocessor)Cell Broadband Engine with 1 Cell (microprocessor)#Power_Processor_Element (PPE)PPE & 7 Cell (microprocessor)#Synergistic Processing Elements (SPE)SPEsStorage capacity 2.5-inch SATA hard diskhard drive(20 GB, 40 GB, 60 GB, 80 GB, 120 GB, 160 GB, 250 GB, or 320 GB included) (upgradeable)Memory 256 MB system and 256 MB videoDisplayGraphics 550 MHz NVIDIA/Sony Computer EntertainmentSCEI RSX 'Reality Synthesizer'SoundController inputSixaxis, DualShock#DualShock 3DualShock 3, Logitech Driving Force GT, PlayStation 3 accessories#Logitech Cordless Precision ControllerLogitech Cordless Precision controller, standard PlayStation 3 accessories#USB ControllersUSB controllers, GT Force, Rhythm game controllers, PlayStation Move, GunConGunCon 3, PlayStation Portable, Keyboard and MouseConnectivityOnline servicesPlayStation NetworkList of best-selling video gamesBest-selling gameCall of Duty: Black Ops (no sales figures)Welsey Yin-Poole (2011-03-03). "Black Ops is the best-selling PS3 game ever". Eurogamer. . Retrieved 2011-07-22.Backward compatibilityBackwardcompatibilityPlayStation (console)PlayStation (all models)PlayStation 2 (20 GB, 60 GB and some (CECHExx) 80 GB models)PredecessorPlayStation 2The PlayStation 3 (3 Pureisutēshon Surī, officially abbreviated as PS3. "Entertainment on PS3 has a new look". Sony Computer EntertainmentSony Computer Entertainment Europe. 2009-08-18. . Retrieved 2009-09-01.) is the third home video game console produced by Sony Computer Entertainment and the successor to the PlayStation 2 as part of the PlayStationPlayStation series. The PlayStation 3 competes with Microsoft's Xbox 360 and Nintendo's Wii as part of the History of video game consoles (seventh generation)seventh generation of video game consoles. It was first released on November 11, 2006, in Japan, "PS3 sells out at launch". BBC. 2006-11-11. . Retrieved 2008-01-14. with international markets following shortly thereafter. "Sony's PS3 makes U.S. debut". USA Today. 2006-11-17. . Retrieved 2008-01-14. "SCEE official press release". Sony Computer Entertainment Europe. 2007-01-24. Archived from the original on January 19, 2008. . Retrieved 2008-01-14. "Sony Confirms PS3 Euro Launch Details". Next Gen. 2007-01-24. . Retrieved 2008-01-14.Major features of the console include its unified online gaming service, the PlayStation Network, "PlayStation Direct.net". PlayStation Direct.net. . Retrieved 2010-08-06. its multimedia capabilities, "PlayStation.com - PLAYSTATION3 Features: Multimedia". Sony Computer Entertainment America. . Retrieved 2010-08-06. connectivity with the PlayStation Portable,De Leon, Al (2007-12-26). "PlayStation Blog — PS3 Tips —

Remote Play and PlayStation Store". blog.us.playstation.com. PlayStation.Blog. . Retrieved 2008-01-14. and its use of the Blu-ray Disc as its primary storage medium. "PlayStation 3 Features - Blu-ray". SCEA. . Retrieved 2010-03-04.HistorySony officially unveiled the PlayStation 3 (then marketed as PLAYSTATION 3. "PLAYSTATION is in capitals". Kotaku. 2006-08-10. . Retrieved 2008-01-18.) to the public on May 16, 2005 at the E3 2005E3 2005 conference,Sulic, Ivan (2005-05-16). "IGN: E3 2005: PS3 Official". IGN. . Retrieved 2008-01-14. along with a 'boomerang' shaped prototype design of the SixaxisSixaxis controller. "E3 2005: PlayStation 3 - The ("boomerang") Controller". IGN Entertainment, Inc. 2005-05-16. . Retrieved 2010-02-07. A functional version of the system was not present there, "TGS 2005 booth round-up". Gamasutra. 2005-09-15. . Retrieved 2008-01-14. nor at the Tokyo Game Show in September 2005, "Sony E3 2005 press conference video". IGN. 2005-05-17. . Retrieved 2008-01-14. although demonstrations (such as Metal Gear Solid 4: Guns of the Patriots) were held at both events on software development kits and comparable personal computer hardware. Video footage based on the predicted PlayStation 3 specifications was also shown (notably a Final Fantasy VII tech demo).Allen, Jason (2005-05-15). "E3 2005: Eyes-on the Final Fantasy VII Tech Demo". IGN. . Retrieved 2006-07-24. The initial prototype shown in May 2005 featured two HDMI ports, three Ethernet ports and six USB ports; "PlayStation 3 announced for 2006". GameSpot. 2005-05-16. . Retrieved 2008-01-14. however, when the system was shown again a year later at E3 2006, these were reduced to one HDMI port, one Ethernet port and four USB ports, presumably to cut costs. "IGN: E3 2006: The final word on PlayStation". IGN. 2006-05-24. . Retrieved 2008-01-14. "IGN's Official PlayStation 3 FAQ". IGN PlayStation 3. 2007-03-26. pp. 1 of 4. . Retrieved 2007-04-23. Two hardware configurations were also announced for the console: a 20 GB model and a 60 GB model, priced at US$499 (€499) and US$599 (€599), respectively. The 60 GB model was to be the only configuration to feature an HDMI port, Wi-Fi internet, flash card readers and a chrome trim with the logo in silver. Both models were announced for a simultaneous worldwide release: November 11 for Japan and November 17 for North America and Europe. "Sony PlayStation 3 launch details". Arstechnica. 2006-05-09. . Retrieved 2008-01-18.On September 6, 2006, Sony announced that the PAL region PlayStation 3 launch would be delayed until March 2007, due to a shortage of materials used in the Blu-ray drive. "PlayStation 3 Euro launch delayed". BBC. 2006-09-06. . Retrieved 2008-01-14. At the Tokyo Game Show on September 22, 2006, Sony announced that it would include an HDMI port on the 20 GB system, but a chrome trim, flash card readers, silver logo and Wi-Fi would not be included. "TGS 2006 - TGS 06: No US PS3 price drop; HDMI in 20 GB model". GameSpot. 2006-09-23. . Retrieved 2008-01-14. Also, the launch price of the Japanese 20 GB model was reduced by over 20%,Anoop Gantayat (2006-09-22). "TGS 2006: Price Drop For Japanese PS3". IGN PlayStation 3. . Retrieved 2007-01-15. and the 60 GB model was announced for an open pricing scheme in Japan. During the event, Sony showed 27 playable List of PlayStation 3 gamesPS3 games running on final hardware. "Sony announces 27 playable titles for the September Tokyo Game Show.". IGN. 2006-08-16. . Retrieved 2006-08-16.LaunchSilver PlayStation 3 consoles on showcase in 2006. The PlayStation 3 was first released in Japan on November 11, 2006 at 07:00. According to Media Create, 81,639 PS3 systems were sold within 24 hours of its introduction in Japan. "Japanese PS3 Sales Data Also Depressing". Media Create Co. via Kotaku. 2006-11-14. . Retrieved 2008-10-24. Soon after its release in Japan, the PS3 was released in North America on November 17, 2006. Reports of violence surrounding the release of the PS3 include a customer shot, "1 Shot in Conn. PlayStation Waiting Line". CBS. 2006-11-17. . Retrieved 2010-03-27. campers robbed at gunpoint, "Camping PlayStation gamers robbed at gunpoint". KMTR. . Retrieved 2006-11-17. customers shot in a drive-by shooting with BB guns, "Four People Shot While Waiting for PlayStation Console, Including 27 NEWSFIRST Reporter". WKYT News. 2006-11-16. . Retrieved 2006-11-17. and 60 campers fighting over 10 systems. "PlayStation3 Crowd Gets More Than It Bargained for in Tysons". WTOP News. 2006-11-17. . Retrieved 2006-11-17.The console was originally planned for a global release through November, but the European and rest-of-the-world's release was delayed "until March" at the start of September.Ellie Gibson (2006-09-06). "PS3 delayed in Europe News | PS3". Eurogamer. . Retrieved 2010-08-25. With it being a somewhat last-minute delay, some companies had taken deposit-based pre-orders, to which Sony informed customers that they were eligible for full refunds or could continue the pre-order. "Official PS3 News Thread - Page 61 - NTSC-uk.co.uk Forums". Ntsc-uk.co.uk. . Retrieved 2010-08-25.

On January 24, 2007, Sony announced that the PlayStation 3 would go on sale on March 23, 2007 in Europe, Australia, the Middle East, Africa and New Zealand. The system sold about 600,000 units in its first two days. "Sony's PS3 has record launch in Europe". Financial Times. March 28, 2007. . Retrieved 2007-09-01. On March 7, 2007, the 60 GB PlayStation 3 launched in Singapore with a price of Singapore DollarS$799. "Singapore's PS3 launch party". CNET. 2007-03-08. Archived from the original on February 19, 2008. . Retrieved 2008-01-14. The console was launched in South Korea on June 16, 2007 in a single version equipped with an 80 GB hard drive and IPTV. "PlayStation 3 (80 GB) Korean". SCEI. . Retrieved 2008-01-18.PS3 Slim and console rebranding Following speculation that a 'slim' model was in the pipeline Sony officially announced the PS3 CECH-2000 model on August 18, 2009 at the Sony Gamescom press conference. "Entertainment on PS3 has a new look". PlayStation UK. 2009-08-18. . Retrieved 2009-08-18. "Sony Announces Slim PS3 - It Lives This September!". Kotaku. 2009-08-18. . Retrieved 2009-08-18. Among its features are a slimmer form factor and quieter noise when powered on. It was released in major territories by September 2009. As part of the release for the slim model, the logo was changed from the "Spider-Man in filmSpider-Man font" and capitalized PLAYSTATION 3 to a more traditional PlayStation (console)PlayStation and PlayStation 2 like 'PlayStation 3' logo with "PS3" imprinted on the console. Along with the console and logo redesign, the boot screen of all consoles changed from "Sony Computer Entertainment" to "PS3 PlayStation 3", with a new chime and the game start splashscreen being dropped. The cover art and packaging of games has also been changed to reflect the redesign.Console configurationsSystem unit The PlayStation 3 is convex on its left side, with the PlayStation logo upright, when vertical (the top side is convex when horizontal) and has a glossy black finish. PlayStation designer Teiyu Goto stated that the Spider-Man (film)Spider-Man-font-inspired logo "was one of the first elements SCEI president Ken Kutaragi decided on and the logo may have been the motivating force behind the shape of PS3". "Kutaragi "insisted" on Spider-Man font for PS3". computerandvideogames.com. 2007-03-29. . Retrieved 2008-01-18.The PlayStation 3 features a slot-loading 2x speed Blu-ray Disc drive for games, Blu-ray movies, DVDs, CDs and other optical media. "PlayStation 3 Technical Specifications". SCEA. Archived from the original on January 13, 2008. . Retrieved 2008-01-14. It was originally available with hard drives of 20 and 60 GB (20 GB model was not available in PAL regions) "Reeves: No PAL 80GB PS3 Plans - Edge Magazine". Next-gen.biz. 2007-06-11. . Retrieved 2011-06-28. "PS3 60 GB review". CNET. 2006-11-13. . Retrieved 2008-01-14. but various sizes up to 320 GB "Sony Computer Entertainment Japan to introduce PlayStation3 (PS3) with 160GB and 320GB hard disk drive, available from July 29, 2010 at a Recommended Retail Price of 29,980 yen and 34,980 yen respectively". Sony. 2010-07-06. . Retrieved 2010-07-06. "White PS3 Slim out in Japan this month". Eurogamer. 2010-07-06. . Retrieved 2010-07-06. have been made available since then (see: #Model comparisonmodel comparison). All PS3 models have user-upgradeable 2.5" SATA hard drives.Rubenstein, Jeff (2007-12-28). "PlayStation Blog — Upgrading the PS3's hard drive". blog.us.playstation.com. PlayStation.Blog. . Retrieved 2008-01-14.The PlayStation 3 uses the Sony, Toshiba, IBM-designed Cell microprocessor as its CPU, which is made up of one 3.2 GHz PowerPC-based "Power Processing Element" (PPE) and eight Synergistic Processing Elements (SPEs). "Cell Introduction" (PDF). IBM. . Retrieved 2008-01-14. The eighth SPE is disabled to improve chip yields. "Sony PlayStation 3 Cell Processor". North Carolina State University. . Retrieved 2008-01-14.Martin Linklater. "Optimizing Cell Code". Game Developer Magazine, April 2007: pp. 15–18. "To increase fabrication yelds, Sony ships PlayStation 3 Cell processors with only seven working SPEs. And from those seven, one SPE will be used by the operating system for various tasks, This leaves six SPEs for game programmer to use." Only six of the seven SPEs are accessible to developers as the seventh SPE is reserved by the console's operating system. Computer graphicsGraphics processing is handled by the NVIDIA RSX 'Reality Synthesizer', which can output resolutions from 480i/576i Standard definitionSD up to 1080p high definition televisionHD. The PlayStation 3 has 256 MB of XDR DRAM main memory and 256 MB of GDDR3 video memory for the RSX. "PS3 Specs". IGN. 2005-05-16. . Retrieved 2008-01-14.The system has BluetoothBluetooth 2.0 (with support for up to 7 bluetooth devices quick reference manual, for 20 GB US PlayStation 3 page 14), gigabit Ethernet, Universal Serial BusUSB 2.0 and HDMI 1.4Initially used HDMI 1.3a specification, but was upgraded to 1.4 with the introduction of stereoscopic 3D gaming and Blu-ray playback (via firmware update). The PlayStation 3 does not currently support any other HDMI 1.4 capabilities. built in on all currently shipping models. Wi-Fi networking is also built-in on all but the 20 GB models, while a flash card reader (compatible with Memory Stick, Secure Digital cardSD/MultiMediaCardMMC and CompactFlash/Microdrive media) is built-in on 60 GB and CECHExx 80 GB models.The PS3's hardware has also been used to PlayStation 3 clusterbuild supercomputers for high-performance computing. "Building Supercomputer Using PlayStation 3". Console Watcher.com. 2006-08-28. . Retrieved 2006-08-28. Fixstars Solutions sell a version of Yellow Dog Linux for the PlayStation 3 (originally sold by Terra Soft Solutions). "Fixstars: Products". Fixstars Solutions | Yellow Dog Linux Get YDL. . Retrieved 2010-02-04. RapidMind produced a stream programming package for the PS3, "RapidMind and Terra Soft partner to unleash PlayStation 3 for Linux". RapidMind. Archived from the original on 2007-08-22. . but were acquired by Intel in 2009. Also, on January 3, 2007, Dr. Frank Mueller, ProfessorAssociate Professor of Computer science at North Carolina State UniversityNCSU, clustered 8 PS3s. Mueller commented that the 256 MB of system RAM is a limitation for this particular application and is considering attempting to retrofit more RAM. Software includes: Fedora (operating system)Fedora Core 5 Linux ppc64, MPICH2, OpenMP v 2.5, GNU Compiler Collection and CellSDK 1.1. "Engineer Creates First Academic PlayStation 3 Computing Cluster". PhysOrg.com. . "NC State Engineer Creates First Academic PlayStation 3 Computing Cluster". College of Engineering, North Carolina State University. . "Sony PS3 Cluster (IBM Cell BE)". Frank Mueller, Associate Professor, College of Engineering, North Carolina State University. . As a more cost-effective alternative to conventional supercomputers, the U.S. military has purchased clusters of PS3 units for research purposes. "Sony still subsidizing US military supercomputer efforts". Ars Technica. . Retrieved 2009-12-18. Retail PS3 Slim units cannot be used for supercomputing, because the PS3 Slim lacks the ability to boot into a third-party OS. On March 22, 2007, SCE and Stanford University released the Folding@home project for the PlayStation 3. "Folding@home". Sony Computer Entertainment. . Retrieved 2007-03-18. This program allows PS3 owners to lend the computing power of their consoles to help study the physical process of protein folding. In December 2008, a group of hackers used a cluster of 200 PlayStation 3's to hack the security protocol Transport Layer SecuritySSL. "PlayStation 3 used to hack SSL". Engadget. . Retrieved 2009-10-31.Original model60 GB model There are several original PlayStation 3 hardware models, which are commonly referred to by the size of their included hard disk drive: 20, 40, 60, 80 or 160 GB. "PlayStation 3 - Things to Know — PlayStation 3 model comparison". SCEA. Archived from the original on 2008-02-13. . Retrieved 2008-01-14. Although referred to by their HDD size, the capabilities of the consoles vary by region and release date. The only difference in the appearance of the first five models was the color of the trim, number of USB ports, the presence or absence of a door (which covers the flash card readers on equipped models) and some minor changes to the air vents. All retail packages include one or two Sixaxis controllers and/or a PlayStation 3 accessories#DualShock 3DualShock 3 controller (beginning June 12, 2008 "Metal Gear Solid 4 PS3 Bundle Announced | Kotaku Australia". Kotaku.com.au. . Retrieved 2008-10-27. "Metal Gear Solid 4: The PS3 Bundle: The Cardboard Box". Kotaku.com. . Retrieved 2008-10-27.), one miniUSB to USB cable (for connecting the controller and PlayStation Portable to the system), one composite video/stereo audio output cable, one Ethernet cable (20, 60 and CECHExx 80 GB only) and one power cable. "PS3 80GB bundle contents". SCEA. . Retrieved 2008-10-24. "Compare PLAYSTATION3 Systems". SCEA. . Retrieved 2009-08-14. All models support software emulation of the original PlayStation,Boyes, Emma (2007-10-05). "UK PS3 drops price, 40GB finally official". GameSpot. CNET. . Retrieved 2007-10-06.Plunkett, Luke. "Customer Service: Sony Could Run BC on 40GB PS3s. They Just Don't Want To". . Retrieved 2007-10-09. but support for PlayStation 2 backward compatibility has continually diminished with later models and the last model to advertise integrated backward compatibility was the 80GB Metal Gear Solid 4 Bundle. "Can I play PlayStation game console and PlayStation2 computer entertainment system format software titles on the PLAYSTATION3 computer entertainment system?". Sony. . Retrieved 2008-01-14. Compatibility issues with games for both systems are detailed in a public database hosted by the manufacturer. "Backwards Game Compatibility | Support - PlayStation.com". SCEA. . Retrieved 2011-01-01. All models, excluding the 20 GB model, include IEEE 802.11802.11 b/g Wi-Fi. In addition to all of the features of the 20 GB model, the 60 GB model has internal IEEE 802.11 b/g Wi-Fi, multiple flash card readers (SD/MultiMedia Card,

CompactFlash Type I/Type II, Microdrive, "PlayStation 3 Safety & Support manual" (PDF). SCEI. . Retrieved 2008-01-15. Memory Stick/PRO/Duo) and a chrome colored trim. "About PlayStation3 - Technical Specifications". PlayStation.com. Archived from the original on 2008-02-13. . Retrieved 2008-02-25. In terms of hardware, the 80 GB model released in South Korea is identical to the 60 GB model released in the PAL regions, except for the difference in hard drive size. "South Korean PS3". gamesindustry.biz. 2007-05-21. . Retrieved 2008-01-14.Like the South Korean and European models, the North American 80 GB (2007) model also excludes the PlayStation 2 "Emotion Engine" CPU chip. However, it retains the "Graphics Synthesizer" GPU. "PS2 Software Emulation Unlikely for 40GB PS3 - Edge Magazine". Next-gen.biz. 2007-10-08. . Retrieved 2011-06-28. Due to PlayStation 3 games#Removal of hardware supportthe elimination of the "Emotion Engine", the level of compatibility was reduced. The 40 GB, 80 GB (2008) and 160 GB models have two USB ports instead of the four USB ports on other models and do not include multiple flash card readers, Super Audio CDSACD support, "PS3 SACD FAQ". ps3sacd.com. . Retrieved 2009-08-22. or any backward compatibility with PlayStation 2 games. "Sony PlayStation 3 review". CNET. 2007-11-16. . Retrieved 2008-01-14. This was due to the removal of "Graphics Synthesizer" GPU, which stripped the units of all PlayStation 2 based hardware. "New PLAYSTATION3 Model to Take Holiday Season by Storm". Sony Computer Entertainment. October 5, 2007. . Retrieved 2007-10-05.Tretton, Jack (speaker) (2008-07-15). E3 2008: Sony Press Conference (Part 3). Sony Computer Entertainment America (via IGN). Event occurs at 17:38. . Retrieved 2008-07-16.No official Wi-Fi or flash memory card readers were ever released by Sony for the 20 GB system, although Sony had plans to do so.Ellie Gibson (2006-05-16). "20 GB PlayStation 3 will be upgradeable, says Sony". gamesindustry.biz. . Retrieved 2006-05-16. As of September 2009, Sony have placed no further emphasis on these proposed add-ons. Nevertheless, as the model features four USB 2.0 ports, wireless networking and flash memory card support can already be obtained through the use of widely available external USB adapters and third-party PS3-specific media hubs.It was rumored that the Cell processors in the third-generation PS3s (40 GB, 2008 80 GB and 160 GB) would move from a 90 nm process to the newer 65 nm process,Conrad Quilty Harper (2007-10-30). "40GB PS3 features 65nm chips, lower power consumption (30% less)". Engadget. . Retrieved 2007-10-31. which SCEI CEO Kaz Hirai later confirmed,Anoop Gantayat (2007-11-08). "New PS3 Uses 65nm Process". ign. . Retrieved 2007-11-10. and later to 45 nm. This change lowers the Energy conservationpower consumption of the console and makes it less expensive to produce.Jon Stokes (2008-02-07). "IBM shrinks Cell to 45nm". Ars Technica. . Retrieved 2009-01-07.Slim model120 GB Slim model The redesigned, slimmer version of the PlayStation 3 (commonly referred to as the "PS3 Slim" and officially branded "PS3") is currently the only model in production. It features an upgradeable 120 GB, 160 GB, 250 GB or 320 GB hard drive and is 33% smaller, 36% lighter and consumes 34% (CECH-20xx) or 45% (CECH-21xx) less power than the previous model,Paul Miller (2009-08-18). "Sony unveils slimmer PS3: $300, lands in September (updated!)". Engadget. . Retrieved 2009-08-18. "PlayStation 3 Slim review". Engadget.com. . Retrieved 2010-08-25. or one third of the original PS3 model. The Cell microprocessor has moved to a 45 nm manufacturing process, which lets it run cooler and quieter than previous models, and the cooling system has been redesigned. "Sony answers our questions about the new PlayStation 3". Ars Technica. August 18, 2009. . Retrieved August 19, 2009. The RSX 'Reality Synthesizer'RSX moved to a 40 nm process "Sony PS3 upgraded with cooler 40-nm RSX graphics chip, profits await (updated)". Engadget. . Retrieved 2010-08-25. in the latest revision. The PS3 slim also includes support for High-Definition Multimedia Interface#CECCEC (more commonly referred to by its manufacturer brandings of BraviaSync, VIERA Link, EasyLink etc.) which allows control of the console over HDMI by using the TV's remote control. The PS3 Slim no longer has the "main power" switch like the previous PS3 models, similar to redesigned PlayStation 2#slimlineslimline PlayStation 2. Support for emulation to play PS2 titles is not present in the Slim version, however shortly after the release of the PS3 slim, Sony announced a new series of PS2 remasters called Classics HD as in PS2 and PSP titles remastered in HD for the PS3 with trophies added and sometimes PlayStation Move compatibility. As of October 2011, PS2 classics are available for purchase in the Playstation Store.http://www.next-gen.biz/news/ps2-classics-hit-psn-week Edge Magazine Retrieved 12-6-2011The PS3 slim was officially released on September 1, 2009 in North America and Europe and on September 3, 2009 in Japan,

Australia and New Zealand. "Entertainment on PS3 has a new look". PlayStation AU. 2009-08-18. . Retrieved 2009-08-24.David Carnoy (2009-08-18). "Sony officially announces $299 PS3 Slim". cnet. . Retrieved 2009-08-18. However, some retailers such as Amazon.com, Best Buy and GameStop started to sell the PS3 slim on August 25, 2009.Stephen Totilo (2009-08-21). "Sony, GameStop Indicate PS3 Slim Available In U.S. Before September". Kotaku. . Retrieved 2009-08-22.Ross Miller (2009-08-25). "PS3 Slim popping up all over the US (update: it's officially on sale)". Engadget. . Retrieved 2009-08-26. The PS3 Slim sold in excess of a million units in its first 3 weeks on sale. "New PS3 has sold over one million". Eurogamer. . Retrieved 2009-11-07. A 250 GB Final Fantasy XIII-themed PS3 Slim, which was white in color with pink designs, was officially announced on September 24, 2009 at the Tokyo Game Show as part of a bundle in Japan for Final Fantasy XIII, it was initially revealed in U.S. Federal Communications Commission (FCC) filings as the PS3 CECH-2000B.CJPC (2009-08-18). "PS3 Slim CECH-2000B 250GB Model Uncovered at FCC Site". ps3news. . Retrieved 2009-08-22. "First Look At The Final Fantasy XIII PlayStation 3". Kotaku. 2009-09-24. . Retrieved 2009-09-24. Sony Computer Entertainment Australia also announced later that day that it would be bringing the 250 GB PS3 slim to Australia which would be bundled with other games and will not feature the Final Fantasy XIII theme. Although no North American bundles have been announced for the 250 GB PS3 slim, it is sold as a stand-alone console in North America. "250GB PS3 Slim bundles coming to Australia". qj.net. 2009-09-24. . Retrieved 2009-09-25.In July 2010, Sony announced two new sizes of Slim PS3, 160 GB and 320 GB, with the 120 GB model being discontinued in Japan. These were launched on July 29, 2010 in Japan, with the 160 GB version available in "Classic White" as well as the standard "Charcoal Black". The black 160 GB version was also made available as a bundle with the Japan-only DVR accessory PlayStation 3 accessories#tornetorne. It was later announced that the new sizes were to be launched in other regions, with the 160 GB model available from August 2010 in North AmericaFrucci, Adam (2010-08-17). "New, Higher-Capacity PS3 Models Coming This Fall". Gizmodo, Gawker Media. . Retrieved 2010-08-18. and October 2010 in Europe. "Sony PlayStation Move Bundle Arrives in September". mashable.com. 2010-08-18. . Retrieved 2010-08-19. The 320 GB model is to be available in North America only as part of a bundle with PlayStation Move, a PlayStation Eye and a copy of Sports Champions, and in Europe with PlayStation Move, a PlayStation Eye and a demo disc. The bundles were released on September 19, 2010 and September 15, 2010 respectively, to coincide with the launch of PlayStation Move.Model comparison "Limited "Dragon" PlayStation 3 Console Is For Japanese Eyes Only". Kotaku. 2009-01-06. . Retrieved 2010-04-29. "PS3 250GB bundles priced, detailed". Eurogamer. 2009-09-09. . Retrieved 2010-05-20.Ashcraft, Brian (2010-08-18). "Gran Turismo 5 Gets Blue PS3 In Japan". Kotaku, Gawker Media. . Retrieved 2010-08-18.Bender, Tim. "Gobble Up Great PS3 Deals on Black Friday". blog.us.playstation.com. PlayStation.Blog. . Model Features Available Colors First Available In production Available bundles 20 GBCECHBxxNTSC 4 USB 2.0 ports Hardware-based PS2 emulatoremulationSuper Audio CDSACD playbackLinux support[a]Sixaxis controller N/A Piano Black, Black trim JapanJPNorth AmericaNA November 2006 No "Sony Scraps Sale of Priciest PlayStation 3s in Japan". 2008-01-10. . Retrieved 2008-01-10.N/A60 GBCECHAxxNTSCmemory cardFlash memory card readers 802.11b/g Wi-Fi Piano Black, Chrome trim No 60 GBCECHCxxPAL802.11b/g Wi-FiFlash memory card readers 4 USB 2.0 ports Partially software-based PS2 Video game console emulatoremulation "Sony's Euro PS3 will play fewer older games". IDG Network. . Retrieved 2009-11-07.Super Audio CDSACD playback Linux support[a]Sixaxis controller (MGS4 bundles sold with DualShock 3 controller)United KingdomUKRepublic of IrelandROI March 16, 2007EuropeEUAustralasiaAUS March 23, 2007MotorStorm, Resistance: Fall of Man and an extra Sixaxis gamepad (UK) "BBC Confirms European PS3 Bundle Deal". Kotaku. 2007-07-12. . Retrieved 2010-05-20.80 GBCECHExxNTSCNorth AmericaNA August 2007 Metal Gear Solid 4: Guns of the PatriotsMGS4MotorStorm40 GBCECHGxx CECHHxx CECHJxx PAL, NTSC802.11b/g Wi-Fi 2 USB 2.0 ports Linux support[a]Sixaxis controller Piano Black Ceramic White[b] Satin Silver[c]Tanaka, John (2008-02-04). "New PS3 colour in Japan". IGN. . Retrieved 2008-02-05.(All with Satin Silver trim) Gun-Metal Grey, Gun-Metal Grey trim[d]GamingBits (2008-05-19). "Metal Gear Solid 4 Limited Edition PLAYSTATION3 Bundle". Gamingbits.com. . Retrieved 2010-08-25.EuropeEUAustralasiaAUS October 2007 North AmericaNAJapanJP November 2007

NoRandolph Ramsay and Luke Anderson (2008-07-16). "E3 08: 80GB PS3 coming to Europe, Australia on August 27". GameSpot. . Retrieved 2008-07-17.MGS4 (Gun-metal grey console)[d]Grand Theft Auto IVGTA IV (Europe)Jem Alexander (2008-04-23). "GTA IV bundle confirmed for Europe". Joystiq. . Retrieved 2010-04-28.80 GBCECHKxx CECHLxx CECHMxxPAL, NTSCDualShock 3 controller Piano Black Ceramic White Satin Silver[c](All with Satin Silver trim)North AmericaNAEuropeEUAustralasiaAUS August 2008 JapanJP October 2008 No 160 GBCECHPxx CECHQxxPAL, NTSC "company press release". SCEE. 2008-08-20. . Retrieved 2008-08-20. "Live from Leipzig: North American Hardware Announcements". blog.us.playstation.com. PlayStation.Blog. 2008-08-20. . Retrieved 2008-08-20. Piano Black Cloud Black[f]North AmericaNA November 2008 EuropeEU October 2008 Uncharted: Drake's FortuneUncharted (NTSC region)Final Fantasy VII: Advent Children Blu-ray (Japan; "cloud black" console with custom design)[f]North, Dale. "Final Fantasy VII Advent Children PS3 bundle is real". Destructoid. .120 GB slimCECH-20xxA CECH-21xxA PAL, NTSC802.11b/g Wi-Fi 2 USB 2.0 ports BRAVIA Sync XMB control (HDMI-CECCEC) Slimmer form factor Dolby TrueHD and DTS-HD Master Audio bitstreamingDualShock 3 controller Charcoal Black North AmericaNAEuropeEU September 1, 2009 AustralasiaAUSJapanJP September 3, 2009 Sony Computer Entertainment Europe - Virtual Press OfficeLittleBigPlanet and Ratchet & Clank Future: A Crack in Time (NTSC region)Michael Jackson's This Is ItThis Is It Blu-rayEllie Gibson (2010-01-06). "Michael Jackson PS3 bundle on the way PlayStation 3 News - Page 1". Eurogamer.net. . Retrieved 2010-08-25.250 GB slimCECH-20xxBCECH-21xxBPAL, "TGS 09: Motion Controller Details, Other New Info". blog.eu.playstation.com. PlayStation.Blog. September 24, 2009. . Retrieved October 2, 2009. NTSC Charcoal Black White and Pink[g]JapanJP December 17, 2009 North AmericaNA November 3, 2009 "NEW 250GB PlayStation 3 System Available November 3 – PlayStation.Blog". blog.us.playstation.com. PlayStation.Blog. 2009-10-15. . Retrieved 2010-08-25.EuropeEU October 2009AustralasiaAUS October 15, 2009160 GB slimCECH-25xxAPAL, NTSC Charcoal Black Classic White "White PS3 Slim with 160GB HDD hits Japan July 29th, new 320GB HDD option". Engadget. 2010-07-06. . Retrieved 2010-07-06. Titanium BlueJapanJP July 29, 2010North AmericaNA August 2010EuropeEU October 2010Yes 320 GB slimCECH-25xxBCECH-30xxB "Greener, lighter, 320GB PS3 confirmed". Eurogamer. .PAL, NTSC Charcoal Black Classic WhiteGilbert, Ben. "White 320GB PS3 coming to Japan October 21". Joystiq. .Splash Blue and Scarlet RedSvetlik, Joe. "PS3 Slim coming in red and blue, joins white". CNET UK. .JapanJP July 29, 2010North AmericaNA September 19, 2010EuropeEU September 15, 2010PlayStation Move, PlayStation Eye and Sports Champions bundle (North-America)PlayStation Move, PlayStation Eye and demo disc bundle (Europe)Gran Turismo 5 (Europe)Ashcraft, Brian. "Europe Getting Gran Turismo 5 Bundle [Update"]. Kotaku, Gawker Media. .PlayStation 3 accessories#tornetorne (Japan) "PlayStation®3 HDD 320GB | ® [PlayStation® 3 HDD Recorder 320GB Pack Charcoal Black | PlayStation® Official Site]" (in Japanese). SCEI. .Key:"1st Generation" "2nd Generation" "3rd Generation" "4th Generation" All Piano Black and Ceramic White models have a glossy finish "PlayStation 3 40GB model official specifications". SCEI. Archived from the original on 2008-01-23. . Retrieved 2008-01-15.All models include: Blu-ray DiscBlu-ray/DVD/CD drive, High-Definition Multimedia InterfaceHDMI 1.3a, "PLAYSTATION3 HDD 20 GB EQUIPPED WITH HDMI AS STANDARD" (PDF) (Press release). Sony Computer Entertainment. 2006-09-22. . Retrieved 2007-05-24. Bluetooth 2.0, Gigabit Ethernet (10BASE-T, 100BASE-TX, 1000BASE-T) and PlayStation (console)PlayStation backward compatibility through software emulation.Model numbers differ by region. See PlayStation 3 hardware#Model numbersPlayStation 3 hardware – model numbers for details.a Linux support removed in firmware version 3.21. See #Removal of "OtherOS" supportRemoval of "OtherOS" support for detailsb Ceramic white model available in Asia and Japan only.c Satin silver model available in Asia and Japan only.d Gun-Metal Gray model is only available as part of the Metal Gear Solid 4: Guns of the PatriotsMGS4 bundle.e Yakuza 3 bundle features a Ceramic White model with custom grey dragon designs on its case. This version had a limited run of 10,000 units. "Yakuza 3 PS3 Special Edition unveiled". Quick Jump Network (QJ.net). 2009-01-06. . Retrieved 2010-04-29.f "Cloud Black" (dark grey) console is only available as part of a Japanese limited edition Final Fantasy VII: Advent Children bundle and features a custom white design on the consoleg White and Pink model is only available as part of the Japanese Final

Fantasy XIII bundle and features a pink design of Final Fantasy XIII character "Lightning (Final Fantasy)Lightning" on its case.Controllers and accessoriesDualShock 3 controller in hand Numerous accessories for the console have been developed. These accessories include the wireless Sixaxis and DualShock 3 controllers, the Logitech Driving Force GT, the PlayStation 3 accessories#Logitech Cordless Precision ControllerLogitech Cordless Precision Controller, the PlayStation 3 accessories#Official PS3 Bluetooth Blu-ray remoteBD Remote, the PlayStation Eye camera, and the PlayTV DVB-T tuner/digital video recorder accessory. "PlayStation 3 Accessories". SCEA. Archived from the original on 2008-01-13. . Retrieved 2008-01-14. "PlayTV on PlayStation 3 explained". Sony Computer Entertainment Europe via Kotaku. August 22, 2007. . Retrieved 2008-10-14.At Sony's 2006 E3 press conference, the (then) standard wireless Sixaxis controller was announced. The controller was based on the same basic design as the PlayStation 2PlayStation 2's DualShock 2 controller but was wireless, lacked vibration capabilities, had a built-in accelerometer (that could detect motion in three directional and three rotational axes; six in total, hence the name Sixaxis) and had a few cosmetic tweaks. At its press conference at the 2007 Tokyo Game Show, Sony announced the DualShock 3 (trademarked DUALSHOCK 3), a PlayStation 3 controller with the same function and design as the Sixaxis, but with vibration capability included. "DualShock 3 Wireless Controller With Rumble Feature To Be Introduced For PlayStation 3". Sony Computer Entertainment. 2007-09-20. . Retrieved 2007-09-20. Hands-on accounts describe the controller as being noticeably heavier than the standard Sixaxis controller and capable of vibration forces comparable to the DualShock 2. "TGS '07: Spot On — The Dual Shock 3". GameSpot. 2007-09-19. . Retrieved 2007-09-20. It was released in Japan on November 11, 2007; "DualShock 3 will launch on November 11 in Japan". Kotaku. 2007-10-08. . Retrieved 2008-01-14. in North America on April 5, 2008; "Sony Announces New 80GB PS3 Bundle, DualShock 3, Red PSP". Daily Tech. 2008-02-26. . Retrieved 2008-03-17. in Australia on April 24, 2008; in New Zealand on May 9, 2008; in mainland Europe on July 2, 2008 "TGS 07: DualShock 3 announced". IGN. 2007-09-19. . Retrieved 2008-01-18. and in the United Kingdom and Ireland on July 4, 2008. During E3 09E3 2009, Sony unveiled plans to release a motion controller later to be named PlayStation Move at GDC 2010. It was released on September 15, 2010 in Europe; September 19, 2010 in North America and October 21, 2010 in Japan. "PlayStation Move motion controller to hit worldwide market starting this September". Sony Computer Entertainment. 2010-06-16. . Retrieved 2010-06-30.On October 13, 2010, Sony announced an official surround sound system for the PS3 through the official PlayStation YouTube channel. "The official surround sound system for your PS3". YouTube. . Retrieved 2010-11-11.ReliabilityThe PlayStation 3 illuminating the yellow light, indicating a non-specific failureA 2009 study by SquareTrade, a warranty provider, found a two-year failure rate of 10% for PlayStation 3s.Sands A, Tseng V. (2009). Game Console Failure Rates: Wii 9 times more reliable than Xbox 360, 4 times more than PS3. SquareTrade. According to Ars Technica, the number of PlayStation 3 consoles that have experienced failure is well within the normal failure rates in the consumer electronics industry.Kuchera, Ben (2008-02-14). "Xbox 360 failure rates worse than most consumer electronics". Arstechnica.com. . Retrieved 2010-08-25.Yellow light of death In September 2009, the BBC television programme Watchdog (TV series)Watchdog aired a report investigating the "yellow light of death" (YLOD) issue, a yellow light that indicates a non-specific hardware failure which renders the system unusable. According to the report the failure was likely to occur after 18-24 months, whilst the standard Sony warranty is one year in duration, after which PlayStation 3 owners can pay Sony a set fee for a refurbished console.Rob Unsworth (2009-09-17). "BBC "Watchdog" report on "Yellow Light of Death"". BBC. . Retrieved 2010-05-31.Sony claimed that, according to its statistics of returned consoles approximately only 0.5% of consoles were reported as showing the YLOD. In response to the program Sony issued a document criticizing the program's accuracy and conclusions; specifically that the faults were evidence of a manufacturing defect. The document also complained that the report had been innapropiate in tone, and might do damage to Sony's brand. "Sony tackles BBC over 'PS3 failure' report", www.gameindustry.biz, September 17, 2009, "Sony rebuts BBC PlayStation claim". news.bbc.co.uk (BBC News). September 18, 2009. .Leap year bug On March 1, 2010 (UTC), many of the original (non-Slim) PlayStation 3 models worldwide were experiencing errors related to their internal system clock. The error had a multitude of symptoms. Initially, the main problem seemed to be the inability to connect to the PlayStation Network. However, the root cause of the problem was unrelated to the PlayStation Network, since even users who had never been online also had problems playing installed offline games (which queried the system timer as part of startup) and using system themes. At the same time many users noted that the console's clock had gone back to December 31, 1999. The event was nicknamed the ApocalyPS3, a play on the word apocalypse.Previous post Next post. "ApocalyPS3: Buggy Clock Drives PlayStation 3 Haywire | GameLife". Wired.com. . Retrieved 2010-08-25.The error code displayed was typically 8001050F and affected users were unable to sign in, play games, use dynamic themes and view/sync trophies. Sony PS3 network glitch affects thousands The Daily Telegraph, March 1, 2010 The problem only resided within the 1st through to the 3rd generation original PS3 units while the newer "Slim" models were unaffected due to different internal hardware for the clock. Sony confirmed that there was an error and stated that they were narrowing down the issue and were continuing to work to restore service. "PlayStation Network Status Update – PlayStation.Blog.Europe". blog.eu.playstation.com. PlayStation.Blog. . Retrieved 2010-08-25. By March 2 (UTC), 2010, owners of original PS3 models could connect to PSN successfully and the clock no longer showed December 31, 1999. "PS3 issues apparently resolved, game on". Engadget. . Retrieved 2010-08-25. Sony stated that the affected models incorrectly identified 2010 as a leap year, due to a bug in the Binary-coded decimalBCD method of storing the date.Jeff Bakalar (2010-03-01). "PS3 console errors fixed, leap year bug to blame". CNET.com. . Retrieved 2010-03-02.Hiroko Tabuchi (2010-03-01). "Error Blocked PlayStation 3 From Using Its Network". New York Times. . Retrieved 2010-03-02. However, for some users, the hardware's operating system clock (mainly updated from the internet and not associated with the internal clock) needed to be updated manually or by re-syncing it via the internet. On June 29, 2010, Sony released PS3 system software update 3.40, which improved the functionality of the internal clock to properly account for leap years. "PlayStation 3 Features Software Updates". Sony. . Retrieved 2010-09-03.PlayStation Network outage On April 20, 2011, Sony shut down the PlayStation Network and Qriocity for a prolonged interval, revealing on April 23 that this was due to "an external intrusion on our system". Sony later revealed that the personally identifiable informationpersonal information of 77 million users might have been taken, including: names; addresses; countries; email addresses; birthdates; PSN/Qriocity logins, passwords and handles/PSN online IDs. "PSN/Qriocity Service Update — PlayStation.Blog.Europe". blog.eu.playstation.com. PlayStation.Blog. . They also stated that it was possible that users' profile data, including purchase history and billing address, and PlayStation Network/Qriocity password security answers may have been obtained. There was no evidence that any credit card data had been taken, but the possibility could not be ruled out, and Sony advised customers that their credit card data may have been obtained. "PlayStation Network and Qriocity Outage FAQ — PlayStation.Blog.Europe". blog.eu.playstation.com. PlayStation.Blog. . Additionally, the credit card numbers were encrypted and Sony never collected the three digit CVC or CSC number from the back of the credit cards which is required for authenticating some transactions.Chloe Albanesius (2011-04-28). "PlayStation Network Credit Card Data Encrypted, Sony Says". . In response to the incident, Sony announced a "Welcome Back" program, 30 days free membership of PlayStation Plus for all PSN members, two free downloadable PS3 games, and a free one-year enrollment in an identity theft protection program.Wesley Yin-Poole (2011-05-01). "PSN: Sony outlines "Welcome Back" gifts News - PlayStation 3 - Page 1". Eurogamer.net. . Retrieved 2011-05-09. "PlayStation Network nears restoration; users get ID theft protection - Game Hunters: In search of video games and interactive awesomeness - USATODAY.com". Content.usatoday.com. 2011-05-05. . Retrieved 2011-05-09.Operating systemSystem softwareSony has included the ability for the operating system, referred to as System Software, to be updated. "PlayStation.com - PLAYSTATION3 - Features - SystemUpdate". SCEA. . Retrieved 2010-02-04. The updates can be acquired in several ways: If the PlayStation 3 has an active InternetworkingInternet connection, updates may be downloaded directly from the PlayStation Network to the PS3 and subsequently installed. Systems with active Internet will automatically check online for software updates each time the console is started. Using an external PC, a user may download the update from the official PlayStation website, transfer it to portable Computer storagestorage media and install it on the System. Some game discs come with system software updates on the disc. This may be due to the game requiring the update in order to run. If so, the software may be installed from the disc.The original PlayStation 3 also included the ability to install other operating systems, "Install Other OS".

PlayStation 3 User's Guide. Sony Computer Entertainment. 2007. . Retrieved 2007-04-25. such as Linux on the PlayStation 3Linux.Boyes, Emma (2006-11-27). "Yellow Dog Linux launches for PS3". GameSpot. . Retrieved 2006-11-30. This was not included in the newer slim models and was removed from all older PlayStation 3 consoles with the release of firmware update 3.21 in April 2010. The functionality is now only available to users of original consoles who choose not to update their system software beyond version 3.15.Andrew Yoon (2010-03-29). "Linux support dropped from PS3 firmware 3.21". Joystiq. . Retrieved 2010-03-29.Graphical user interfaceThe standard PlayStation 3 version of the XrossMediaBar (pronounced Cross Media Bar, or abbreviated XMB) includes nine categories of options. These are: Users, Settings, Photo, Music, Video, Game, Network, PlayStation Network and Friends (similar to the PlayStation Portable media bar). A tenth TV category is displayed between Music and Video if PlayTV or torne is installed or if the console meets certain criteria to access select Internet televisioncatch-up television services. By default, the #What's NewWhat's New section of PlayStation Network is displayed when the system starts up. The PS3 includes the ability to store various master and secondary user profiles, manage and explore photos with or without a musical Slideshowslide show, play music and copy Red Book (audio CD standard)audio CD tracks to an attached data storage device, play movies and video files from the hard disk drive, an optical disc (Blu-ray Disc or DVD-Video) or an optional USB mass storage or Flash card, compatibility for a USB keyboard and mouse and a web browser supporting in/compatible file download function. "PlayStation 3 Features Connectivity". Sony. . Retrieved 2010-09-03. Additionally, Universal Plug and PlayUPnP media will appear in the respective audio/video/photo categories if a UPnP AV MediaServerscompatible media server or Digital Living Network AllianceDLNA server is detected on the local network. The Friends menu allows mail with emoticon and attached picture features and Videoconferencingvideo chat which requires an optional PlayStation Eye or EyeToy webcam. "About the PlayStation Network". Sony. Archived from the original on 2008-01-12. . Retrieved 2008-01-13. The Network menu allows online shopping through the PlayStation Store and connectivity to the PlayStation Portable via Remote Play.Digital rights management The PlayStation 3 console protects certain types of data and uses digital rights management to limit the data's use. Purchased games and content from the PlayStation Network store are governed by PlayStation's Network Digital Rights Management (NDRM). The NDRM allows users to access the data from up to 5 different PlayStation 3's that have been activated using a user's PlayStation Network ID. "PlayStation Knowledge Center". Sony. . Retrieved 2010-07-06. "Can multiple PlayStation®Network accounts access PlayStation®Store content that was purchased by a different account?" PlayStation 3 also limits the transfer of copy protected videos downloaded from its store to other machines and states that copy protected video "may not restore correctly" following certain actions after making a backup such as downloading a new copy protected movie. "PlayStation Knowledge Center". Sony. . Retrieved 2010-07-06. "How do I back up and restore videos that were downloaded from the PlayStation®Store?"Photo managementPhoto Gallery Photo Gallery main menu Photo Gallery is an optional application to view, create and group photos from the PS3, which is installed separately from the system software at 105 MB. It was introduced in system software version 2.60 and provides a range of tools for sorting through and displaying the system's pictures. The key feature of this application is that it can organize photos into groups according to various criteria. Notable categorizations are colors, ages, or facial expressions of the people in the photos. Slideshows can be viewed with the application, along with music and playlists. The software was updated with the release of system software version 3.40 allowing users to upload and browse photos on Facebook and Picasa. "PS3 System Software Update – 3.40". blog.eu.playstation.com. PlayStation.Blog. . Retrieved 2010-06-28.PlayMemories PlayMemories is an optional stereoscopic 3D (and also standard) photo viewing application, "What is the function of PlayMemories application?". . Retrieved September 22, 2010. which is installed from the PlayStation Store at 182 MB. The application is dedicated specifically to 3D photos and features the ability to zoom into 3D environments and change the angle and perspective of panoramas. "Sony launches PlayMemories app for 3D photos". . Retrieved September 22, 2010. It requires system software 3.40 or higher; 3D photos; a 3D HDTV, and an HDMI cable for the 3D images to be viewed properly. Video servicesVideo editor and uploader A new application was released as part of system software version 3.40 which allows users to edit videos on the PlayStation 3 and upload them to the Internet. The software features basic video editing tools including the ability to cut videos and add music and captions. Videos can then be rendered and uploaded to video sharing websites such as Facebook and YouTube.Video on demand In addition to the video service provided by the Sony Entertainment Network the PlayStation 3 console has access to a variety of third party video services, dependant on region: Since June 2009 VidZone has offered a free music video streaming service in Europe,Stevie Smith (12 June 2009), "VidZone gives PlayStation 3 a free music video jukebox", www.thetechherald.com, Australia and New Zealand.Mike Bantick (12 June 2009), "VidZone review - The online music video jukebox", www.itwire.com, In October 2009, Sony Computer Entertainment and Netflix announced that Netflix would also be available on the PlayStation 3 for in the United States; initially had to use a free Blu-ray disc to access the streaming video. A paid Netflix subscription was required for the service. "Coming Soon: Netflix Members Can Instantly Watch Movies and TV Episodes Streamed to TVs Via the PlayStation(R)3 Computer Entertainment System (Press Release)", www.reuters.com (Netflix Inc. via Reuters), 26 Otober 2009, The service became available in November 2009. "Available Now: Netflix Members can Instantly Watch Movies and TV Episodes Streamed to TVs Via the PlayStation(R)3 Computer Entertainment System", www.prnewswire.com (Netflix Inc. via PRNewswire), 10 November 2009, In October 2010 the requirement to use a disc to gain access was removed.Greg Peters (Netflix) (14 October 2010), "Netflix on PS3: Disc-free Next Week", blog.us.playstation.com (PlayStation.Blog), In November 2010 access to the video and social networking site Mubi (website)MUBI was enabled for European, New Zealand, and Australian users; the service integrates elements of social networking with rental or subscription video streaming, allowing users to watch and discuss films with other users.Sources: Efe Cakarel (18 May 2010). "PS3 + MUBI + YOU". blog.eu.playstation.com. PlayStation.Blog. .Efe Cakarel (1 November 2010), "MUBI Is Launching On PS3 This Week", blog.eu.playstation.com (PlayStation.Blog), "MUBI Goes Live This Week", uk.ps3.ign.com (IGN), 1 November 2010, Also in November 2010 the video rental service VUDU, and subscription service Hulu Plus launched on the PlayStation 3 in the United States.Sources: John P. Falcone (16 November 2010), "Vudu coming to PS3 with updated interface", news.cnet.com (CNET), Susan Panico (16 November 2010), "VUDU Comes to the PS3, Adding Another Choice for PSN Members to Experience Media Instantly", blog.us.playstation.com (PlayStation.Blog), Sources: Susan Panico (10 November 2010), "Get Hulu Plus on your PS3 Today!", blog.us.playstation.com (PlayStation.Blog), PlayStation Portable connectivity The PlayStation Portable can connect with the PlayStation 3 in many ways, including in-game connectivity. For example, Formula One Championship Edition, a racing game, was shown at E3 2006 using a PSP as a real-time rear-view mirror.Haynes, Jeff (2006-05-08). "E3 2006: Formula One 06 Resurfaces". IGN PlayStation 3. . Retrieved 2007-01-21. In addition, users are able to download original PlayStation format games from the PlayStation Store, transfer and play them on the PSP as well as the PS3 itself.Deleon, Nicholas (2006-12-07). "Downloadable PS1 Games Hit The PSP (Provided You Have a PS3)". Gizmodo. . Retrieved 2007-01-21.Kennedy, Sam (2006-10-20). "Phil Harrison Talks Downloadable PS1 Games on PS3". 1up.com. . Retrieved 2007-04-03. It is also possible to use the Remote Play feature to play these and some PlayStation Network games, remotely on the PSP over a network or internet connection.Sony has also demonstrated the PSP playing back video content from the PlayStation 3 hard disk across an Mobile ad-hoc networkad-hoc wireless network. This feature is referred to as Remote Play located under the browser icon on both the PlayStation 3 and the PlayStation Portable. Remote play has since expanded to allow remote access to the PS3 via PSP from any wireless access point in the world.Nix, Marc (2006-10-20). "PSP and PS3 Play Together". IGN PSP. . Retrieved 2007-01-21.OtherOS support The PlayStation 3 initially shipped with the ability to install an alternative operating system alongside the main system software; Linux and other Unix based operating systems were available. The hardware allowed access to six of the seven Synergistic Processing Elements of the Cell microprocessor, but not the RSX 'Reality Synthesizer' graphics chip.The 'OtherOS' functionality was not present in the updated PS Slim models, and the feature was subsequently removed from previous versions of the PS3 as part of the machine's firmware update version 3.21 which was released on April 1, 2010; "PSA: PS3 Firmware Update 3.21 available now". 2010-04-01. . Retrieved 2010-04-01. Sony cited security concerns as the rationale. The firmware update 3.21 was mandatory for access to the PlayStation Network. "PS3 Firmware (v3.21) Update". blog.us.playstation.com (PlayStation.Blog). 2010-03-28. . Retrieved 2010-03-28. ".. but due to security concerns,

Sony Computer Entertainment will remove the functionality through the 3.21 system software update" The removal caused some controversy; as the update removed officially advertised features from already sold products, and gave rise to several class action lawsuits aimed at making Sony return the feature or provide compensation. "Sony Steals Feature From Your PlayStation 3". 2010-03-30. . Retrieved 2010-03-30.Sources: Oli Welsh (2010-05-11). "Sony hit by two more Other OS lawsuits". Eurogamer. . Retrieved 2010-05-11. "Man Files class action lawsuit over Other OS support removal". Joystiq. 2010-04-30. . "PlayStation 3 Owner files class action lawsuit against Sony for dropping 'Other OS'". Kotaku. 2010-04-29. .PlayStation NetworkPlayStation Network logoPlayStation Network is the unified online multiplayer gaming and digital media delivery service provided by Sony Computer Entertainment for the PlayStation 3 and PlayStation Portable, announced during the 2006 PlayStation Business Briefing meeting in Tokyo. The service is always connected, "PlayStation 3 announced for 2006". GameSpot. 2005-05-16. . Retrieved 2007-04-04. free, "Official PlayStation Website — PlayStation Network". SCEA. Archived from the original on 2008-03-15. . Retrieved 2008-01-15. and includes multiplayer support. The network enables online gaming, the PlayStation Store, PlayStation Home and other services. PlayStation Network uses real currency and PlayStation Network Cards as seen with the PlayStation Store and PlayStation Home.PlayStation PlusPlayStation Plus (commonly abbreviated PS+ and occasionally referred to as PSN Plus) is a premium PlayStation Network subscription service that was officially unveiled at E3 2010 by Jack Tretton, President and CEO of SCEA. Rumors of such service had been in speculation since Kaz Hirai's announcement at TGS 2009 of a possible paid service for PSN but with the current PSN service still available. Launched alongside PS3 firmware 3.40 and PSP firmware 6.30 on June 29, 2010, the paid-for subscription service provides users with enhanced services on the PlayStation Network, on top of the current PSN service which is still available with all of its features. These enhancements include the ability to have demos, game and system software updates download automatically to the PlayStation 3. Subscribers also get early or exclusive access to some betas, game demos, premium downloadable content and other PlayStation Store items. North American users also get a free subscription to Qore (PlayStation Network)Qore. Users may choose to purchase either a one-year or a three-month subscription to PlayStation Plus.PlayStation StoreThe European PlayStation Store on the PlayStation 3. The PlayStation Store is an online virtual market available to users of Sony Computer EntertainmentSony's PlayStation 3 (PS3) and PlayStation Portable (PSP) Video game consolegame consoles via the PlayStation Network. The Store offers a range of downloadable content both for purchase and available free of charge. Available content includes full games, add-on content, Game demoplayable demos, Theme (computing)themes and game and movie Trailer (film)trailers. The service is accessible through an icon on the XrossMediaBarXMB on the PS3 and PSP. The PS3 store can also be accessed on the PSP via a Remote Play connection to the PS3. The PSP store is also available via the PC application, Media Go. As of September 24, 2009, there have been over 600 million downloads from the PlayStation Store worldwide. "TGS 2009: Keynote Address". IGN. .The PlayStation Store is updated with new content each Tuesday in North America, and each Wednesday in PAL regions. "PSN Thursday no more: say hello to PSN Tuesday and Wednesday". . In May 2010 this was changed from Thursdays to allow PSP games to be released digitally, closer to the time they are released on Universal Media DiscUMD. "PSN Store Day Move Not Due To 360". Thesixthaxis.com. 2010-04-23. .What's NewThe What's New screen What's New was announced at Gamescom 2009 and was released on September 1, 2009, with PlayStation 3 system software 3.0. "PlayStation Games & Media News: – Entertainment on PS3 has a new look". SCEE. . The feature was to replace the existing [Information Board], which displayed news from the PlayStation website associated with the user's region. The concept was developed further into a major PlayStation Network feature, which interacts with the [Status Indicator] to display a ticker of all content, excluding recently played content (currently in North America and Japan only). The system displays the What's New screen by default instead of the [Games] menu (or [Video] menu, if a movie was inserted) when starting up. What's New has four sections: "Our Pick", "Recently Played", latest information and new content available in PlayStation Store. There are four kinds of content the What's New screen displays and links to, on the sections. "Recently Played" displays the user's recently played games and online services only, whereas, the other sections can contain website links, links to play videos and access to selected sections of the PlayStation Store. The PlayStation Store icons in the [Game] and

[Video] section act similarly to the What's New screen, except that they only display and link to games and videos in the PlayStation Store, respectively.PlayStation HomePlayStation Home is a virtual 3D social networking service for the PlayStation Network. "GDC: Phil Harrison's Keynote Speech". Games Industry. May 16, 2007. . Home allows users to create a custom avatar (computing)avatar, which can be groomed realistically. "PlayStation Home unveiled". Engadget. 2007-03-07. . Retrieved 2008-01-15. "GDC 2007 Sony keynote speech". Engadget. 2007-03-08. . Retrieved 2008-01-15. Users can edit and decorate their personal apartments, avatars or club houses with free, premium or won content. Users can shop for new items or win prizes from PS3 games, or Home activities. Users interact and connect with friends and customise content in a virtual world. "PlayStation Home, the free virtual world of PlayStation 3". JoyStiq. 2007-03-07. . Retrieved 2008-01-15. Home also acts as a meeting place for users that want to play multiplayer games with others.A closed beta began in Europe from May 2007 and expanded to other territories soon after. "IGN: Europeans Go Home". IGN PlayStation 3. 2007-05-01. . Retrieved 2007-05-02. Home was delayed and expanded "PlayStation Home Beta Extended Today". blog.us.playstation.com. PlayStation.Blog. . several times before initially releasing. "Virtual Worlds News: Sony Home Delayed to Spring 2008". Virtual Worlds News. September 20, 2007. . Retrieved 2009-07-01. "PlayStation Home not open till fall of 08". Platform Nation. 22 April 2007. . The Open Development stageBeta test was started on December 11, 2008. Home is available directly from the PlayStation 3 XrossMediaBar#PlayStation 3 XMBXrossMediaBar. Membership is free and requires a PSN account.Home features places to meet and interact, dedicated game spaces, developer spaces, company spaces and events. The service undergoes a weekly maintenance and frequent updates. As of August 2011, Home has been downloaded by over 23 million users.Contreras, Paulmichael (2011-08-23). "PlayStation Home – Over 23 Million Strong, Evolving With A New Major Redesign". PlayStation LifeStyle. . Retrieved 2011-11-29.Life with PlayStationThe Life with PlayStation application showing weather forecasts and news headlines for New York City. Screenshot taken at approximately 8pm pacific standard timePST. Life with PlayStation, released on September 18, 2008 "Life with PlayStation". PlayStation.com. September 18, 2009. . Retrieved October 7, 2009. to succeed Folding@home#PlayStation 3Folding@home. Life with PlayStation uses virtual globe data to display news and information by city. Along with Folding@home functionality, the application also provides the user with access to three other information "channels", the first of which being the Live Channel which offers news headlines and weather. Information is provided by Google News, The Weather Channel, the University of Wisconsin–Madison Space Science and Engineering Center, among other sources. "Life With PlayStation out now". Kotaku. . The second channel is the World Heritage channel which offers historical information about historical sites. The third channel is the United Village channel. United Village is a project designed to share information about communities and cultures worldwide. "United Village". . Retrieved October 1, 2009. A recent update has allowed video and photo viewing in the application. The fourth channel is the USA exclusive PlayStation Network Game Trailers Channel for direct streaming of game trailers.Games The PlayStation 3 launched in North America with 14 titles, with another three being released before the end of 2006 in video gaming2006. "IGN's Official PlayStation 3 FAQ". IGN PlayStation 3. 2007-03-26. pp. 3 of 4. . Retrieved 2007-05-02. After the first week of sales it was confirmed that Resistance: Fall of Man from Insomniac Games was the top-selling launch titlelaunch game in North America. "Gamestop Corp. Q3 2006 Earnings Conference Call Summary". Gamestop Corporation via alacrastore.com. 2006-11-21. . Retrieved 2009-01-08. "Saling The World: In Search of Zelda - week of November 24, 2006". Gamasutra. 2006-11-24. . Retrieved 2009-01-08. The game was heavily praised by numerous Computer and video gamesvideo game websites, including GameSpot and IGN, both of whom awarded it their PlayStation 3 Game of the Year award for 2006.GameSpot Staff (2006-12-21). "Best Games and Worst Games of 2006 at GameSpot — Best PlayStation 3 Game". GameSpot. . Retrieved 2007-04-30.IGN Staff (2006-12-21). "IGN.com presents The Best of 2006 - PlayStation 3: Game of the Year". IGN.com. . Retrieved 2007-04-30. Some titles missed the launch window and were delayed until early 2007 in video gaming2007, such as The Elder Scrolls IV: Oblivion, F.E.A.R. and Sonic the Hedgehog (2006 game)Sonic the Hedgehog. During the Japanese launch, Ridge Racer 7 was the top-selling game, while Mobile Suit Gundam: Crossfire also fared well in sales,Gantayat, Anoop (2006-11-13). "Sony Number Two in Japan". IGN. . Retrieved 2009-01-08. both of which were offerings from Namco Bandai Games. The PlayStation 3 launched in Europe with 24 titles, including ones that were not offered in the North American and Japanese launches, such as Formula One Championship Edition, MotorStorm and Virtua Fighter 5. Resistance: Fall of Man and MotorStorm were the most successful titles of 2007, "Cathedral row over video game". BBC. 2007-06-09. . Retrieved 2008-01-18. "Motorstorm completes hat-trick". Eurogamer. 2007-04-05. . Retrieved 2008-01-18. and both games subsequently received sequels in the form of Resistance 2 and MotorStorm: Pacific Rift.Evolution Studios (2007). "MotorStorm 2 sequel confirmed for PS3, in development at Evolution Studios". www.videogamesblogger.com. . Retrieved 2007-09-06.Insomniac Games (2007). "Full Moon Show Podcast". www.insomniacgames.com. Archived from the original on 2007-09-27. . Retrieved 2007-11-12.At History of E3#2007 (July 11–13)E3 2007, Sony was able to show a number of their upcoming video games for the PlayStation 3, including Heavenly Sword, Lair (video game)Lair, Ratchet & Clank Future: Tools of Destruction, Warhawk (PlayStation 3 game)Warhawk and Uncharted: Drake's Fortune; all of which were released in the third and fourth quarters of 2007. They also showed off a number of titles that were set for release in 2008 in video gaming2008 and 2009 in video gaming2009; most notably Killzone 2, Infamous (video game)Infamous, Gran Turismo 5 Prologue, LittleBigPlanet and SOCOM: U.S. Navy SEALs Confrontation. "Best of E3 2007 Winners". Kotaku. July 31, 2007. . Retrieved 2007-08-12. A number of third-party exclusives were also shown, including the highly anticipated Metal Gear Solid 4: Guns of the Patriots,Hatfield, Daemon (2007-07-11). "Haze now Fully PS3 Exclusive, 360 & PC Dropped". Gamespot. . Retrieved 2007-08-11. alongside other high-profile third-party titles such as Grand Theft Auto 4, Call of Duty 4: Modern Warfare, Assassin's Creed, Devil May Cry 4 and Resident Evil 5. Two other important titles for the PlayStation 3, Final Fantasy XIII and Final Fantasy Versus XIII, were shown at Tokyo Game ShowTGS 2007 in order to appease the Japanese market. "E308:No Changes on Final Fantasy Versus XIII Being PS3 Exclusive". Kotaku. 2008-07-14. . Retrieved 2008-11-29.Gibson, Ellie (2007-07-11). "E3: Square Enix's John Yamamoto". GamesIndustry.biz. . Retrieved 2007-08-11.Sony have since launched their budget range of PlayStation 3 titles, known as the List of Sony Greatest Hits gamesGreatest Hits range in North America, "PS3 Greatest Hits Launch Today". blog.us.playstation.com. PlayStation.Blog. 2008-07-28. . Retrieved 2009-08-15. the Platinum range in Europe and Australia "PS3 Platinum range in UK from August". Computerandvideogames.com. 2008-07-17. Archived from the original on 2008-07-17. . Retrieved 2009-08-15. and The Best range in Japan. "PlayStation 3 the best". Jp.playstation.com. . Retrieved 2009-08-15. Among the titles available in the budget range include Resistance: Fall of Man, MotorStorm, Uncharted: Drakes Fortune, Rainbow Six: Vegas, Call Of Duty 3, Assassin's Creed and Ninja Gaiden Sigma. As of October 2009 Metal Gear Solid 4: Guns of the Patriots, Ratchet & Clank Future: Tools of Destruction, Devil May Cry 4, Army of Two, Battlefield: Bad Company and Midnight Club: Los Angeles have also joined the list. When they are put on the "List of Sony Greatest Hits gamesGreatest Hits" list the new unused copies retail for $30 USD and are re-shipped in a new red case.As of September 30, 2010, there have been 350.6 million games sold for the PlayStation 3. "PLAYSTATION3 Worldwide Software Unit Sales". Sony Computer Entertainment Inc.. . Retrieved 2010-05-23.Stereoscopic 3D In December 2008, the Chief technology officerCTO of Blitz Games announced that it would bring stereoscopystereoscopic 3D gaming and movie viewing to the Xbox 360 and PlayStation 3 with its own technology. "Blitz Games Introduces True Stereoscopic 3D For Xbox 360 And PS3". Cinemablend.com. 2008-12-01. . Retrieved 2009-08-15. This technology was first demonstrated publicly on the PS3 in January 2009 at the Consumer Electronics Show. Journalists were shown Wipeout HD and Gran Turismo 5 Prologue in 3D as a demonstration of how the technology might work if it is implemented in the future. "Sony teases with high-quality PS3 HD". Ars Technica. 2009-01-08. . System software update 3.30 has prepared the PS3 for stereoscopic 3D gaming, while 3.50 prepared it for 3D films. "PS3 3.30 System Software Update". blog.us.playstation.com. PlayStation.Blog. April 21, 2010. . Firmware update 3.30 officially allows PS3 titles to be played in 3D, requiring a compatible display for use. "Sony dates PS3 3D update". MCV. June 1, 2010. . While the game itself must be programmed to take advantage of the 3D technology, titles may be patched to add in the functionality retroactively. Titles with such patches include Wipeout HD, Pain (video game)Pain, and Super Stardust HD.by JC Fletcher on May 31st 2010 3:15PM (2010-05-31). "First PS3 3D game updates available in Japan on June 10". Joystiq. . Retrieved 2010-08-25.HomebrewPlayStation Jailbreak In August 2010, PlayStation Jailbreak, a

Universal Serial BusUSB device that allows execution of unsigned code (such as backup games and Homebrew (video games)homebrew) on the PlayStation 3, was released. "PS3 Jailbreak is Finally Here". Techie Buzz. 2010-08-20. . Retrieved 2010-10-06. The device included a piece of software called Backup Manager, which allows users to copy original games from the Blu-ray Disc to either the internal hard disk drive (HDD) or an external HDD (File Allocation TableFAT32 formatted).The original device was soon Reverse engineeringreverse engineered and many clones, such as PS3 Key and PS3 Break, have since been released. "New PS3 Break PS3 Jailbreak Copy – USB Break is coming soon". Press Release Point. 2010-09-17. . Retrieved 2010-10-06. An open source implementation known as PS Groove was also released, which allows the same functionality to be achieved using various external devices, such as calculators and portable media players, via a USB connection. "PS3 Jailbreak Code Hits Internet, No Stopping It Now". Kotaku. 2010-09-01. . Retrieved 2010-10-06. An open source clone of Backup Manager, known as Open Manager, which mimics and expands upon the original's features has also been released. "Open Manager 1.0 PS3" (in spanish). Todo Soluciones. 2010-10-01. . Retrieved 2010-10-06.Private key compromised At the 2010 Chaos Communication Congress (CCC) in Berlin, a group calling itself fail0verflow announced it had succeeded in bypassing a number of the PlayStation 3's security measures, allowing unsigned code to run without a dongle. They also announced that it was possible to recover the private key used by Sony to sign software, due to an improper implementation of the Elliptic Curve DSA (ECDSA), but chose not to publish this key because it was not necessary to run homebrew software on the device.Bendel, Mike (2010-12-29). "Hackers Describe PS3 Security As Epic Fail, Gain Unrestricted Access". Exophase.com. . Retrieved 2011-01-05. The release of this key would allow anyone to sign their code and therefore be able to run it on any PlayStation 3 console. This would also mean that no countermeasures could be taken by Sony without rendering old software useless, as there is no distinction between official and homebrew software.Fildes, Jonathan (2011-01-06). "iPhone hacker publishes secret Sony PlayStation 3 key". BBC. . Retrieved 2011-03-20. On January 3, 2011, George Hotzgeohot published the aforementioned private key, as well as a Hello world program for the PS3. "Hacker Claims To Have The PS3's Front Door Keys". Kotaku. 2011-01-03. . Retrieved 2011-01-05. "Geohot: Here is your PS3 Root Key! - Now with "HELLO WORLD" proof!". (psx-scene administrator). . (original GeoHot posting to psx-scene with key and hello world program) On January 12, 2011, Sony Computer Entertainment America filed lawsuits against both fail0verflow and geohot for violations of the Digital Millennium Copyright ActDMCA and Computer Fraud and Abuse ActCFAA. "Sony to Geohot: See you in court". Eurogamer. . "Motion for TRO". Scribd.com. 2011-01-12. . Retrieved 2011-03-20.Custom firmware (CFW) To allow for homebrew using the newly discovered encryption keys, several modified versions of system update 3.55 have been released by Geohot and others. The most common feature is the addition of an "App Loader" that allows for the installation of homebrew apps as signed DLC-like packages. Although Backup Managers could run at that time, they could not load games at first even though some success had been made by making backups look like DLC games and then signing them. An LV2 patch was later released to allow Backup Managers to load game backups and was later integrated into the Managers themselves so that it doesn't have to be run whenever the PS3 is restarted. This CFW jailbreak method is now the trend in jailbreaking the PS3 because it doesn't need any special hardware and can be permanently installed onto any PS3. PS3 System Software updates v3.56 and v3.60 added security measures to prevent creation of custom firmwares, building on previous updates which blocked the PS Jailbreak exploit. However, users may choose not to update and games requiring a firmware version above 3.55 can be patched to run on v3.55 or lower. Soon after v3.60 was released, updates to the Playstation Network were conducted to block any methods known that allowed PSN access on firmwares older than the latest official firmware (v3.66 currently), thereby blocking users who chose not to update. A custom firmware known as "Rebug",The site rebug.me is credited in an April 26 Eurogamer.es report. released on March 31, gave retail PS3s most of the options and functionality of debug/developer PS3 units. One week later, tutorials became available allowing users to download PSN content for free, using fake (rather than stolen) credit card numbers. "PlayStation Network taken down to halt piracy?". computerandvideogames.com. 2011-04-25. . and linked articlesA possible example of such a tutorial at ps3hax.net was deleted approximately April 21. One April 12 report described hackers using the jailbroken firmware to access the dev-PSN to get back on games like Call of

Duty, with widespread reports of cheating. "PlayStation 3 Hackers Using Developer Network to Access PSN". attackofthefanboy.com. 2011-04-12. . While some sources blamed Rebug for the subsequent intrusion to Sony's private developer network, Time's "Techland" described such theories as "highly--as in looking down at the clouds from the tip-top of Mount Everest highly--speculative".Matt Peckham (2011-04-26). "Did Custom 'Rebug' Firmware Kill Sony's PlayStation Network?". techland.time.com. .Sales and production costs Region List of best-selling game consolesUnits sold First available Canada "about 1.5 million" as of October 6, 2010Peter Nowak (October 6, 2010). "Sony aims for PS3 holiday sales boost". Canadian Broadcasting CorporationCBC. . Retrieved October 31, 2010. November 17, 2006 Europe(May include UK & other PAL regions) 16 million as of August 17, 2010Shaun McInnis (2010-08-17). "New Ratchet and Clank, Resistance headline Sony conference". GameSpot. . Retrieved 2010-08-17. March 23, 2007 Japan 6,341,950 as of April 1, 2011 "PSP Japan's best-selling console in 2010". Eurogamer. . November 11, 2006 United Kingdom 3 million as of January 26, 2010Richard Mitchell (2010-01-25). "UK PS3 sales surpass 3 million". Joystiq. . Retrieved 2010-02-04. March 23, 2007 United States 13.5 million as of November 11, 2010 "Gamasutra - News - Fils-Aime: Holidays More Important To Nintendo Than Rivals". Gamasutra. . November 17, 2006 Worldwide 55.5 million as of September 30, 2011 November 11, 2006 The PlayStation 3's initial production cost is estimated by iSuppli to have been United States dollarUS$805.85 for the 20 GB model and US$840.35 for the 60 GB model.Sources: "iSuppli: 60GB PS3 Costs $840 to Produce", www.next-gen.biz, 16 November 2006, "iSuppli Corporation : Applied Market Intelligence : PlayStation 3 Offers Supercomputer Performance at PC Pricing, iSuppli's Teardown Analysis Reveals", www.isuppli.com (iSuppli), 16 November 2006, archived from the original on 27 September 2007, However, they were priced at US$499 and US$599 respectively, meaning that units may have been sold at an estimated loss of $306 or $241 depending on model, if the cost estimates were correct,Arik Hesseldahl (16 November 2006), "Nothing Cheap About the PS3", www.businessweek.com (Bloomberg L.P.Bloomberg), and contributing to Sony's games division posting an operating loss of ¥232.3 billion (US$1.97 billion) in the fiscal year ending March 2007. "PS3 launch hits Sony profits". GamesIndustry.biz. May 16, 2007. . Retrieved 2007-08-31. In April 2007, soon after these results were published, Ken Kutaragi, President of Sony Computer Entertainment, announced plans to retire. Various news agencynews agencies, including The TimesSabbagh, Dan; Kennedy, Siobhan (2007-04-27). "PlayStation boss pays the price of weak sales". London: The TimesTimes Online. . Retrieved 2007-04-27. and The Wall Street JournalChristopher Grant (2007-04-26). "Sony's Ken Kutaragi announces plans to retire". Joystiq. . Retrieved 2007-07-03. reported that this was due to poor sales, while SCEI maintains that Kutaragi had been planning his retirement for six months prior to the announcement.In January 2008, Kaz Hirai, chief executive officerCEO of Sony Computer Entertainment, suggested that the console may start making a profit by early 2009, stating that, "the next fiscal year starts in April and if we can try to achieve that in the next fiscal year that would be a great thing" and that "[profitability] is not a definite commitment, but that is what I would like to try to shoot for". "Sony's Hirai eyes PS3 profit as costs decline". Reuters. 2008-01-07. . Retrieved 2008-04-22. However, market analysts Nikko Citigroup have predicted that the PlayStation 3 could be profitable by August 2008. "PS3 profitable by August, say Nikko Citigroup analysts". GamesIndustry.biz. 2008-04-21. . Retrieved 2008-04-22. In a July 2008 interview, Hirai stated that his objective is for the PlayStation 3 to sell 150 million units by its ninth year, surpassing the PlayStation 2's sales of 140 million in its nine years on the market.Chris Nuttall (2008-07-20). "Sony sets 150m sales target for PS3". Financial Times. . Retrieved 2008-08-02. In January 2009 Sony announced that their gaming division was profitable in Q3 2008.nofi (2009-01-29). "Sony's Game Division In Profit". TheSixthAxis. . Retrieved 2009-03-03.Since the system's launch, production costs have been reduced significantly as a result of phasing out the Emotion Engine chip and falling hardware costs. "Sony cuts PS3 production costs in half". psu.com. 2008-01-11. . Retrieved 2008-01-22. "PlayStation 3 manufacturing costs to go down, way down". ps3fanboy.com. 2007-02-08. . Retrieved 2007-07-15. The cost of manufacturing Cell microprocessors has fallen dramatically as a result of moving to the 65 nanometer65 nm production process, "IBM Shrinks Cell Broadband Engine to 65nm Process". dailytech.com. 2007-03-13. . Retrieved 2007-07-15. and Blu-ray Disc diodes have become cheaper to manufacture. "PlayStation 3 Production Costs Drop". gameinformer.com. 2007-05-07. Archived from the original on 2007-05-09.

. Retrieved 2007-07-15. As of January 2008, each unit cost around $400 to manufacture; "Sony's Blu-Ray Breakthrough". BusinessWeek. 2008-01-08. . Retrieved 2008-01-14. "PS3 production costs halved, says analyst". www.gamesindustry.biz. 2008-01-14. . Retrieved 2008-01-14. by August 2009, Sony had reduced costs by a total of 70%, meaning it only costs Sony around $240 per unit. "PS3 70 percent cheaper to make". www.gamersyndrome.com. 2009-08-01. . Retrieved 2009-08-09.Kat Bailey (2009-08-22). "PlayStation 3 Production Costs Slashed 70 Percent". 1up. . Retrieved 2009-07-31.Brian Ashcraft (2009-08-22). "Sony: PS3 Manufacturing Costs Down 70 Percent". Kotaku. . Retrieved 2009-07-31.ReceptionOriginal model Early PlayStation 3 reviews soon after launch were critical of its high price and lack of quality launch games, but commended the system's hardware capabilities and potential.Riyad Emeran (2007-03-23). "Sony PlayStation 3". Trustedreviews.com. . Retrieved 2007-07-10.Danny Allen (2007-03-01). "First review: PlayStation 3". PC Advisor. . Retrieved 2007-07-10. However, after a series of price revisions, Blu-ray's victory over HD DVD, "Toshiba Announces Discontinuation of HD DVD Businesses" (Press release). Toshiba. 2008-02-19. . Retrieved 2008-08-31. and the release of several well received titles, the system received better reviews. IGN judged the PlayStation 3 to have the best game line-up of 2008, based on their review scores in comparison to those of the Wii and Xbox 360.Shea, Cam. "IGN: Aussie Console Showdown 08: PS3 vs. Wii vs. 360". Ps3.ign.com. . Retrieved 2009-08-07.The PS3 was given the number-eight spot on PC World (magazine)PC World magazine's list of "The Top 21 Tech Screwups of 2006", where it was criticized for being "Late, Expensive and Incompatible".Tynan, Dan (2006-12-20). "The Top 21 Tech Screwups of 2006". PC World. . Retrieved 2007-02-03. GamesRadar ranked the PS3 as the top item in a feature on game-related Public relationsPR disasters, asking how Sony managed to "take one of the most anticipated game systems of all time and — within the space of a year — turn it into a hate object reviled by the entire internet", but added that despite its problems the system has "untapped potential".Reparaz, Mikel (2007-03-21). "The Top 7... PR disasters". GamesRadar. . Retrieved 2007-03-21. Business Week summed up the general opinion by stating that it was "more impressed with what [the PlayStation 3] could do than with what it currently does".Cliff Edwards (2006-11-16). "PS3: Soon to Be a Great System". Business Week. . Retrieved 2007-07-10.Developers have also found the machine difficult to program for. In 2007, Gabe Newell of Valve CorporationValve said "The PS3 is a total disaster on so many levels, I think it's really clear that Sony lost track of what customers and what developers wanted". He continued "I'd say, even at this late date, they should just cancel it and do a do over. Just say, 'This was a horrible disaster and we're sorry and we're going to stop selling this and stop trying to convince people to develop for it'".Bishop, Stuart (2007-01-15). "Valve: PS3 a "total disaster on so many levels"". computerandvideogames.com. . Retrieved 2007-01-15. Doug Lombardi VP of Marketing for Valve has since stated that they are interested in developing for the console and are looking to hire talented PS3 programmers for future projects. "Valve open to work with PS3 development". Qj.net. "2009-07-05". . He later restated Valve's position, "Until we have the ability to get a PS3 team together, until we find the people who want to come to Valve or who are at Valve who want to work on that, I don't really see us moving to that platform".Guttridge, Luke (2009-07-03). "Valve dodge PS3 development". play.tm. . Retrieved 2009-07-13. At Sony's Electronic Entertainment Expo 2010E3 2010 press conference, Newell made a live appearance to recant his previous statements, citing Sony's move to make the system more developer friendly, and to announce that Valve would be developing Portal 2 for the system. He also claimed that the inclusion of Steamworks (Valve's system to automatically update their software independently) would help to make the PS3 version of Portal 2 the best console version on the market.Activision Blizzard CEO Bobby Kotick has criticized the PS3's high development costs and inferior attach rate and return to that of the Xbox 360 and Wii. He believes these factors are pushing developers away from working on the console. In an interview with The Times Kotick stated "I'm getting concerned about Sony; the PlayStation 3 is losing a bit of momentum and they don't make it easy for me to support the platform."Sabbagh, Dan (2009-06-19). "Sony should beware — Activision chief is not simply playing games". London: timesonline.co.uk. . Retrieved 2009-07-13. He continued, "It's expensive to develop for the console, and the Wii and the Xbox are just selling better. Games generate a better return on invested capital (ROIC) on the Xbox than on the PlayStation." Kotick also claimed that Activision Blizzard may stop supporting the system if the situation is not addressed. "[Sony has] to cut the [PS3's retail] price, because if they don't, the attach rates are likely to slow. If we are being realistic, we might have to stop supporting Sony."Crossley, Rob (2009-06-19). "Kotick slams PS3 for 'high' dev costs". develop-online.. . Retrieved 2009-07-13. Kotick received heavy criticism for the statement, notably from developer Bioware who questioned the wisdom of the threatened move, and referred to the statement as "silly." "BioWare: Activision PS3 comment "silly and biased towards what is popular."". .In an interview, Kazuo Hirai, chairman of Sony Computer Entertainment argued for the choice of a complex architecture. "Hirai: We're the "official" industry leader". Eurogamer. 2009-01-20. . Retrieved 2009-10-31. "We don't provide the easy to program for console that [developers] want, because easy to program for means that anybody will be able to take advantage of pretty much what the hardware can do, so then the question is what do you do for the rest of the nine-and-a-half years?"Despite the initial negative press, several websites have given the system very good reviews. CNET United Kingdom praised the system saying, "the PS3 is a versatile and impressive piece of home-entertainment equipment that lives up to the hype [...] the PS3 is well worth its hefty price tag."Carnoy, Dave (March 2007). "PS3 Reviewed". CNet.co.uk. Archived from the original on 2007-03-28. . Retrieved 2007-04-21. CNET awarded it a score of 8.8 out of 10 and voted it as its number one "must-have" gadget,CNET Staff (2006-11-21). "Top 10 Must-haves". CNET. . Retrieved 2007-02-22. praising its robust graphical capabilities and stylish exterior design while criticizing its limited selection of available games.Carnoy, David (2006-11-14). "Sony PlayStation 3 (60 GB) Reviews". CNET. . Retrieved 2007-02-03.In addition, both Home Theater Magazine and Ultimate AV have given the system's Blu-ray playback very favorable reviews, stating that the quality of playback exceeds that of many current standalone Blu-ray Disc players.Chiarella, Chris (December 2006). "Sony PlayStation 3". Home Theater Magazine. Primedia Magazines, Inc.. . Retrieved 2007-02-22.Buettner, Shane C. (December 2006). "PlayStation3 Blu-ray Disc Player". Primedia Magazines, Inc.. pp. 4 of 6. . Retrieved 2007-02-22.Hexus Gaming reviewed the PAL version and summed the review up by saying, "as the PlayStation 3 matures and developers start really pushing it, we'll see the PlayStation 3 emerge as the console of choice for gaming."Haywood, Nick (2007-03-23). "Review : Sony PlayStation 3". HEXUS.gaming. pp. 6 of 6. . Retrieved 2007-04-25. At Game Developers ConferenceGDC 2007, Shiny Entertainment founder David Perry (game developer)Dave Perry stated, "I think that Sony has made the best machine. It's the best piece of hardware, without question".Gibson, Ellie (2007-03-07). "GDC: What's Next for PS3?". GamesIndustry.biz. . Retrieved 2007-04-25. A second review of the PS3 by Ars Technica in June 2008 gave the console an overall mark of 9/10, while the original launch review marked only 6/10.Kuchera, Ben (2008-06-04). "They say it got smart: a 2008 review of the PS3". Ars Technica. . Retrieved 2008-06-10. In September 2009, IGN named the PlayStation 3 the 15th best gaming console of all time, behind both of its competitors: the Wii (10th) and Xbox 360 (6th). "PlayStation 3 is number 15". IGN Entertainment. . Retrieved 15-Oct-2009.Slim model and rebranding The PlayStation 3 Slim received extremely positive reviews as well as a boost in sales; less than 24 hours after its announcement the PS3 Slim took the number-one bestseller spot on Amazon.com in the video games section for fifteen consecutive days. It regained the number-one position again one day later. "PS3 Slim gets back Bestseller top spot on Amazon". qj.net. September 1, 2009. . Retrieved 3 September 2009. The PS3 Slim also received praise from PC World giving it a 90 out of 100 praising its new repackaging and the new value it brings at a lower price as well as praising its quietness and the reduction in its power consumption. This is in stark contrast to the original PS3's launch in which it was given position number-eight on their "The Top 21 Tech Screwups of 2006" list. "Priced right, Sony's slim-line PlayStation 3 (120GB) bundles the top-notch features of the earlier PlayStation 3 into a smarter, sleeker package". PC World (magazine)PC World. September 2, 2009. . Retrieved 3 September 2009.CNET awarded the PS3 Slim four out of five stars praising its Blu-ray capabilities, 120 GB hard drive, free online gaming service and more affordable pricing point, but complained about the lack of backward compatibility for PlayStation 2 games. "Sony PlayStation 3 Slim - 120GB Review". CNET. August 19, 2009. . Retrieved 3 September 2009. TechRadar gave the PS3 Slim four and a half stars out of five praising its new smaller size and summed up its review stating "Over all, the PS3 Slim is a phenomenal piece of kit. It's amazing that something so small can do so much". They criticized the exterior design of the PS3 Slim, calling it "ugly" and described the build quality as "cheap" compared to the original PS3. "Sony PlayStation 3 Slim 120GB review". TechRadar. September 1, 2009. . Retrieved 3 September 2009.PlayStation 3 supercomputer In

November 2010 the Air Force Research Laboratory (AFRL) created a powerful supercomputer by connecting together 1,760 Sony PS3s which include 168 separate graphical processing units and 84 coordinating servers in a parallel array capable of performing 500 trillion floating-point operations per second (500 TFLOPS). "AFRL to hold ribbon cutting for Condor supercomputer". www.wpafb.af.mil. The Official Web Site of Wright-Patterson Air Force Base. 17 November 2010, updated 29 November 2010. . Retrieved 2010-12-15. As built the Condor Cluster was the 33rd largest supercomputer in the world and would be used to analyse high definition satellite imagery.Stephen Koff (November 30, 2010). "Defense Department discusses new Sony PlayStation supercomputer". blog.cleveland.com. . Retrieved 2011-03-28.Content notesReferencesExternal linksOfficial websites Asia Australia Canada New Zealand United States United KingdomAuxiliary sites by Sony Hardware press images User's guideDirectories PlayStation 3 at the Open Directory Project

Xbox_360

Left: Xbox 360 Elite, Right: Xbox 360 S and new-style controller

Developer	Microsoft
Manufacturer	Flextronics, Wistron, Celestica, Foxconn[2] [3]
Product family	Xbox
Generation	Seventh generation
Retail availability	
Units sold	Worldwide: 57.6 million (as of September 30, 2011)[4] (details)
Media	DVD, Compact Disc, Download **Add-on:** HD DVD (*discontinued*)
CPU	3.2 GHz PowerPC Tri-Core Xenon
Storage capacity	
Memory	512 MB of GDDR3 RAM clocked at 700 MHz
Display	
Graphics	500 MHz ATI Xenos
Sound	• Analog stereo • Stereo LPCM (TOSLINK and HDMI) • Dolby Digital 5.1 (TOSLINK and HDMI) • Dolby Digital with WMA pro (TOSLINK and HDMI)
Controller input	
Connectivity	**Original models** 2.4 GHz wireless, 3 × USB 2.0, IR receiver, 100 Mbit Ethernet **Add-on:** Wifi 802.11a/b/g, Wifi 802.11a/b/g/n[5] **Revised "S" models** 2.4 GHz wireless, 5 × USB 2.0, IR receiver, 100 Mbit Ethernet, Wifi 802.11b/g/n, AUX port
Online services	Xbox Live
Best-selling game	*Halo 3* (14.5 million)[6]
Backward compatibility	478 original Xbox games[7] (requires hard drive and the latest update)

Predecessor	Xbox

The **Xbox 360** is the second video game console produced by Microsoft and the successor to the Xbox. The Xbox 360 competes with Sony's PlayStation 3 and Nintendo's Wii as part of the seventh generation of video game consoles. As of September 30, 2011, 57.6 million Xbox 360 consoles have been sold worldwide.[4]

The Xbox 360 was officially unveiled on MTV on May 12, 2005, with detailed launch and game information divulged later that month at the Electronic Entertainment Expo (E3). The console sold out completely upon release in all regions except in Japan.[8] [9] [10]

Several major features of the Xbox 360 are its integrated Xbox Live service that allows players to compete online, download arcade games, game demos, trailers, TV shows, music and movies and its Windows Media Center multimedia capabilities. The Xbox 360 also offers region specific access to third-party media streaming services such as Netflix and ESPN in the US or Sky Go in the UK.

At their E3 presentation on June 14, 2010, Microsoft announced a redesigned Xbox 360 that would ship on the same day.[11] The redesigned console is slimmer than the previous Xbox 360 model and features integrated 802.11 b/g/n Wi-Fi, TOSLINK S/PDIF optical audio output, 5 USB 2.0 ports (compared to the 3 from older versions) and a special AUX port.[12] Older models of the Xbox 360 have since been discontinued.[13] The first new console to be released features a 250 GB hard drive, while a later, less expensive SKU features 4 GB internal storage.

With the announcement of the Xbox 360 S, Microsoft have said that they believe that the console is only mid-way through its life-cycle and will continue through 2015.[14] [15] [16] However, unnamed sources have claimed that Microsoft's next video game console is being slated for release as early as the 2013 holiday season.[17] Development of the system has also been implied by updates Microsoft employees have made to their resumes, which state that they were working on the next Xbox console. [18]

Overview

Development

Known during development as **Xbox Next**, **Xenon**, **Xbox 2**, **Xbox FS** or **NextBox**,[19] the Xbox 360 was conceived in early 2003.[20] In February 2003, planning for the Xenon software platform began, and was headed by Microsoft's Vice President J Allard.[20] That month, Microsoft held an event for 400 developers in Bellevue, Washington to recruit support for the system.[20] Also that month, Peter Moore, former president of Sega of America, joined Microsoft. On August 12, 2003, ATI signed on to produce the graphic processing unit for the new console, a deal which was publicly announced two days later.[21] Before the launch of the Xbox 360, several Alpha development kits were spotted using Apple's Power Mac G5 hardware. This was because the system's PowerPC 970 processor running the same PowerPC architecture that the Xbox 360 would eventually run under IBM's Xenon processor. The cores of the Xenon processor were developed using a slightly modified version of the PlayStation 3's Cell Processor PPE architecture. According to David Shippy and Mickie Phipps, the IBM employees were "hiding" their work from Sony and Toshiba.[22] Jeff Minter created the music visualization program Neon which is included with the Xbox 360.[23]

Launch

The Xbox 360 was released on November 22, 2005, in the United States and Canada; December 2, 2005, in Europe and December 10, 2005, in Japan. It was later launched in Mexico, Brazil, Chile, Colombia, Hong Kong, Singapore, South Korea, Taiwan, Australia, New Zealand, South Africa, India, and Russia. In its first year on the market, the system launched in 36 countries, more countries than any other console has launched in a single year.[24]

Retail configurations

At launch, the Xbox 360 was available in two configurations: the "Xbox 360" package (unofficially known as the 20 GB Pro or Premium), priced at US$399 or GB£279.99, and the "Xbox 360 Core," priced at US$299 and GB£209.99. The original shipment of the Xbox 360 version included a cut-down version of the Media Remote as a promotion. The Elite package was launched later at US$479. The "Xbox 360 Core" was replaced by the "Xbox 360 Arcade" in October 2007[25] and a 60 GB version of the Xbox 360 Pro was released on August 1, 2008. The Pro package was discontinued and marked down to US$249 on August 28, 2009 to be sold until stock ran out, while the Elite was also marked down in price to US$299.[26] In June 2010, Microsoft announced a new, redesigned, model and the discontinuation of the Elite and Arcade models.[13]

Current models (Xbox 360 S)

Technically designated the **Xbox 360 S**[27] and marketed simply as the **Xbox 360**;[28] current Xbox 360 consoles are based around a redesign of the Xbox 360 hardware which was officially announced on June 14, 2010 during a press briefing prior to that year's E3.

Xbox 360 S console

It was speculated that a complete redesign of the Xbox 360 hardware was being produced after pictures of a possible new motherboard design surfaced on March 17, 2010.[29] Ads later surfaced on June 13, 2010 showing a slimmer Xbox 360 design, which was expected to include a 250 GB hard drive and integrated Wi-Fi functionality.[30]

Xbox 360 S consoles feature redesigned internal architecture with the Valhalla motherboard, which allows for around 30% more space than previous motherboards,[31] and the XCGPU, an integrated CPU/GPU/eDRAM chip using a 45 nm fabrication process.[32] This allows them to be both smaller and quieter than the previous versions of the Xbox 360. They also feature five standard USB 2.0 ports (two more than previous models) and an additional custom USB port for use with peripherals such as the Kinect sensor.[32] Unlike older models, 2.4 GHz 802.11 b/g/n Wi-Fi[33] and a TOSLINK S/PDIF optical audio connector are also built-in, allowing for digital audio and wireless networking without the need for external adapters.[32] The Memory Unit slots found on previous consoles have been removed in favor of the USB flash drive solution added in a previous system software update (released on April 6, 2010)[34] and the power and DVD drive eject 'buttons' are touch sensitive rather than the physical buttons found on previous models. The external hard disk drive connector has also been swapped for an internal bay for use with a proprietary hard drive. The hard drive bay is designed such that a specially formatted laptop hard drive may be loaded in. It has been noted that users can also open up the casing of the original model's hard drive and simply load it into the drive bay instead of purchasing a hard drive branded for use with the new model.[35] [36] Other notable hardware changes include the use of one larger fan compared to the previous Xbox 360 models (which used two smaller ones) and the design of the vents, which are similar to those used on the original Xbox. Unlike previous generations of the console which had names to distinguish different SKUs, the new models are to be marketed solely by the amount of included storage,[13] in a similar fashion to current models of its main competitor the PlayStation 3. When the first new models began to ship, remaining stock of the Elite package dropped in price to US$249.99[13] or A$349[37] and the Arcade dropped to US$149.99.[13]

Xbox 360 250 GB

The first Xbox 360 S SKU to be revealed includes a 250 GB hard drive[38] and its casing featured a glossy black finish. It was shipped to US retailers the same day it was announced () and went on sale later that week. It was released in Australia on ,[39] in New Zealand on [39] and in Europe on .[40] It retails at US$299.99,[39] £199.99,[39] A$449.00,[39] NZ$499.00[39] or €249.00,[41] replacing the Xbox 360 Elite at that price point.

In August 2011, Microsoft announced they will be streamlining their models by discontinuing the glossy finish and future 250 GB consoles will use the matte finish found on 4 GB models.[42] [43]

Xbox 360 4 GB

A second SKU which includes 4 GB of internal flash storage and has matte black casing (much like the Xbox 360 Elite)[44] was released on in the US[44] and in Europe.[45] It replaced the Xbox 360 Arcade and is priced at US$199.99, £149.99 or €199.99.[44] [45] [46] [47] Although this model has on-board storage, *Xbox Product Director* Aaron Greenberg confirmed that it does have a drive bay which Microsoft has "the opportunity to use in the future".[44] On , Microsoft announced a 250 GB stand-alone hard drive for use with Xbox 360 S models priced at US$129.99.[48]

Xbox 360 320 GB limited editions

Further information: #Xbox 360 S special editions

In June 2011, Microsoft announced a "Limited Collector's Edition" Xbox 360 S console to coincide with the launch of Gears of War 3, which featuring custom finish,[49] a 320 GB hard drive[49] and sounds from the Gears of War 3 game which are played when the console is switched on or the disc tray is opened.

Other 320 GB Xbox 360 S limited editions soon followed. Like the 250 GB "Super Elite" consoles, 320 GB Xbox 360 S consoles are only available as part of limited/special edition console bundles (as of September 2011), with stand-alone 320 GB hard drives also available for purchase.[50]

Discontinued models

Xbox 360 *Arcade* console with white wireless controller. The Arcade is cosmetically identical to the *Core* model.

Xbox 360 Core

The **Xbox 360 Core** was an entry level Xbox 360 which was later replaced with the "Arcade". Although available at launch in other regions, it was not available in Japan until November 2, 2006.[51] The *Core* system came bundled with composite video cables, capable of only SDTV resolutions. It was however capable of the same HDTV resolutions (up to 1080i) as the other models when connected to a separately sold component cable. In October 2006, 1080p support was added for all models in a system update,[52] including the "Core" using either the component cable, or the new VGA cable (although 1080p via component was not widely supported by televisions). It may also utilize a separately sold Xbox 360 hard drive. Unlike all other SKUs, it shipped with a wired version of the Xbox 360 controller, instead of the wireless version found in other SKUs.

Xbox 360 Pro / Xbox 360 Premium

The **Xbox 360 Pro**[53] (sometimes referred to as **Premium** and packaged as simply **Xbox 360** with the subheading "Go Pro") included all the features of the Xbox 360 Arcade and included a hybrid composite and component cable with optional optical out instead of a composite cable. This model also included a detachable Hard Disk Drive (initially 20 GB, while later models had 60 GB) to store downloaded content, provide compatibility with original Xbox games, and store game data. The included hard drive came with game demos, video clips and a free Live Arcade game, *Hexic HD*. In July 2007, this version of the Xbox 360 began appearing with the Zephyr motherboard (the motherboard used in the Elite) which features HDMI 1.2 output and an improved GPU heatsink. Although this package did include an HDMI output, it did not come with an HDMI cable.[54] [55] Starting at the end of September 2007, the newest systems were shipped with the new "Falcon" motherboard. This motherboard includes the new 65-nm CPUs, making them quieter and cooler than the older systems.[56] On August 1, 2008, the 20 GB version was discontinued and was replaced by a 60 GB HDD model at the same price. Holiday 2008 consoles were bundled with *Lego Indiana Jones: The Original Adventures* and *Kung Fu Panda*.[57] Price cuts that took effect on September 4, 2008 reduced the price from $349 to $299.[58] The Xbox 360 Pro, following its discontinuation, retailed for $249.99 until stocks were exhausted.

Xbox 360 Elite

The **Xbox 360 Elite** included a 120 GB hard drive and a matte black finish. The Elite retail package also included a controller and headset that match the system's black finish.[59] The initial release price was $479.99 USD,[60] C$549.99,[60] £299.99, and A$729.95. The Elite was released in North America on April 29, 2007,[61] Europe on August 24, 2007, and Australia on August 30, 2007. These Elites (and other Xbox 360 models using the Falcon) can be identified from earlier versions by a re-designed power connector and a power supply rated to 175 W. In 2009, Elite models using the Jasper chipset became available. These can also be identified by their power supply, which is rated at 150W and has a 12.1A 12v rail. Holiday 2008 consoles were bundled with *Lego Indiana Jones: The Original Adventures* and *Kung Fu Panda*.[57] The Elite's price tag was cut from $449 to $399 on September 4, 2008.[26] [62] With the announcement of the new Xbox 360 250 GB model, the Elite dropped in price to US$249.99 for remaining units until stocks were exhausted.[13]

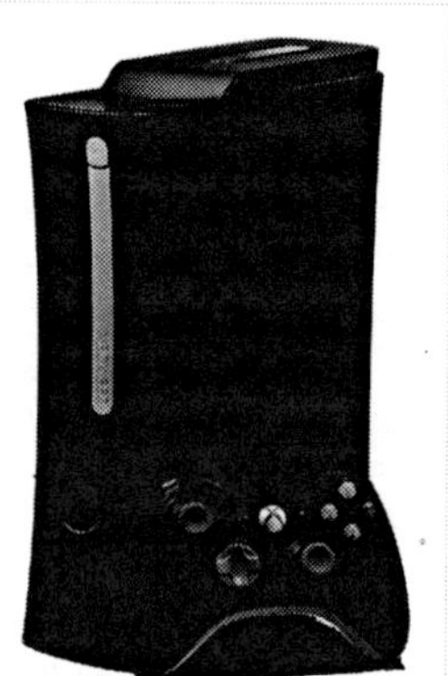

Xbox 360 *Elite* console with black wireless controller.

Xbox 360 Arcade

The **Xbox 360 Arcade**[63] replaced the "Xbox 360 Core" as the entry level Xbox 360 on , while retaining the Core's price of US$279.99.[64] It was publicly revealed by Microsoft's president of Entertainment Devices division Robbie Bach to the *Financial Times* on October 18, 2007,[65] and officially announced on October 22, 2007,[25] although it was available in stores far earlier.[63] It included a wireless controller, composite AV cable, HDMI 1.2 output, a 256 MB memory unit and 5 Xbox Live Arcade titles:[66] *Boom Boom Rocket*, *Feeding Frenzy*, *Luxor 2*, *Pac-Man Championship Edition*, and *Uno* on a single disk, which also included a "Welcome Video" and several game trailers/demos.[67] Like its predecessor the "Core", it did not include a hard disk drive, which is required for Xbox software backwards compatibility. In Autumn (Fall) 2008, with the introduction of the Jasper motherboard revision, the memory unit was removed from the package and replaced with an 256 MB internal memory chip.[68] This was later upgraded to a 512 MB chip in Summer 2009.[69] Holiday 2008 consoles were bundled with *Sega Superstars Tennis*.[57] With the price cuts on , the Arcade fell from US$279 to US$199 in the US.[58] In the UK, with the 2009

Elite price drop and discontinuation of the "Premium" Pro SKU, the Arcade price rose from GB£129.99 to GB£159.99.[70] With the unveiling of the Xbox 360 S redesign, the Arcade dropped in price to US$149.99 for remaining units until stocks are exhausted. The Arcade was replaced at the US$200 price tier by the 4 GB Xbox 360 S.

Xbox 360 Super Elite

The Xbox 360 Elite has also been configured with a 250 GB hard disk drive and 2 wireless controllers on special limited editions of the console. Also referred to as the **Xbox 360 Super Elite**,[71] the console retailed at US$399.99, GB£249.99 and A$599.00[72] as of November 10, 2009. A *Call of Duty: Modern Warfare 2* bundle included a special limited edition black console featuring Modern Warfare 2 branding. Other bundles included a standard Elite finish and either the game *Forza Motorsport 3*[73] or both Halo 3 and Halo 3: ODST.[74] On March 9, 2010, alongside the release of Final Fantasy XIII, Microsoft released another 250 GB bundle with the same extras as the Modern Warfare bundle except bundled with the Final Fantasy XIII game. Unlike other Super Elites, this console and its accessories featured a the same white color scheme as "Pro" models. This bundle also included exclusive downloadable items for use with Xbox 360 avatars and a specially branded 250 GB hard drive.[75] [76] In April 2010, a *Splinter Cell: Conviction* Super Elite bundle featuring special print on the console and the same extras as the other bundles was released along side the Splinter Cell: Conviction game.

Comparison of features

Information is based on current specifications for standard packages. Older or holiday packages may differ from current configurations.

Model	Storage	HDMI	Appearance	Accessories/Bundled items	Suggested retail price	First available	In production
S[13] [27]	320 GB HDD[a][49] [77]	Yes	Metallic red with black Gears of War 3 branding[49]	• 1 or 2 wireless controllers with "transforming d-pad"[i] [49] • Wired Headset • Composite AV Cable[b] For details of bundle specifics, see Xbox 360 S special editions below	US$399.99[49]	• NA [49]	Yes
			Matte blue and white R2-D2 artwork[77]		US$449.99,[77] GB£349.99[78]	• NA "Holiday" 2011[77]	Yes
			Matte grey and dark grey with white Modern Warfare 3 branding[79]		US$399[79]	• NA/AU/EMEA November 8, 2011[79]	Yes
	250 GB HDD	Yes	Matte black (glossy black prior to August 2011)	• Wireless controller • Wired Headset[80] • Composite AV Cable[b]	US$299.99, C$299.99, GB£199.99, €249.99,[81] A$449.00,[39] NZ$499.00[39]	• AU [39] • NZ [39] • EU [82]	Yes
	4 GB onboard[44] [46] [47]	Yes	Matte black	• Wireless controller • Composite AV Cable[b]	US$199.99,[47] GB£149.99,[47] C$199.99	• NA [44] • EU [45]	Yes

Elite	250 GB HDD[a]	Yes	Matte black Chrome disc drive	• Ethernet cable • Wired Headset • Composite AV Cable[b][g][83]	• 2 wireless controllers	US$399.99, C$399.99, GB£249.99, €329.99, A$599.00		No[84]
	120 GB HDD				• 1 wireless controller	US$299.99, C$299.99, GB£199.99, €299.99 ¥29,800, A$549.00[85] [86] [87]		No[13] [84]
Arcade	512 MB[69] onboard	Yes	Matte white Matte white disc drive	• Wireless controller • Composite AV Cable[b][88] • Xbox Live Arcade compilation disc[e] (not included with all units)		US$199.99, C$149.99, GB£159.99, €179.99, ¥19,800, A$299.00[70] [87]	June 2009	No[13] [84] [89]
	256 MB[68] onboard						December 2008	No
	256 MB memory unit							
Pro	60 GB HDD	Yes	Matte white Chrome disc drive	• Wireless controller • Ethernet cable • Wired Headset[f] • Component HD AV Cable[c][h]		US$249.99, C$299.99, GB£169.99, €239.99, ¥29,800, A$399.00[87]		No
	20 GB HDD	After 2007						No
Core	None	No	Matte white Matte white disc drive	• Wired controller • Composite AV Cable[b]		US$279.99, GB£199.99, ¥27,800		No

All Xbox 360s come with Xbox Live Free membership (Prior to October 2010, the free service was known as *Xbox Live Silver*)[90] and a one-month trial of Xbox Live Gold membership (only new accounts are eligible).
All Xbox 360s are backwards compatible with supported Xbox titles as long as they have an Xbox 360 HDD attached. This can be purchased separately for the Core/Arcade pack.
All hard drives (included with a console or bought separately) come with the Xbox Live Arcade game Hexic HD.
Included accessories match the color scheme of the console they are bundled with.
All European consoles also include a composite SCART adapter (*Advanced SCART AV Cable* sold separately).

a 250 GB Elite consoles and 320 GB Xbox 360 S consoles are/were only available as part of limited/special edition bundles (see below).

b The standard *Composite AV Cable* features three RCA connectors, for standard left and right channel audio and composite video, which supports an SD image (NTSC on NTSC consoles, PAL and PAL60 on PAL consoles). It also lacks the TOSLINK connector found on all other (pre-2010) AV cables (including the *Component HD AV Cable*).

c The *Component HD AV Cable* features six RCA connectors, for standard left and right channel audio, composite video and HD component (YP$_B$P$_R$) supporting up to 1080p image.[91] It also features a TOSLINK optical audio connector, which supports either 2 channel (stereo) LPCM or dolby digital 5.1.

d The audio dongle features two RCA connectors for left and right audio and a TOSLINK optical audio connector.

e Compilation disk includes Boom Boom Rocket, Feeding Frenzy, Luxor 2, Pac-Man Championship Edition, and Uno.

f Excluding Mexican and older Australian and New Zealand versions, where a Media Remote is bundled instead.

g *Component HD AV Cable,*[c][h] HDMI cable and *HDMI Audio Adapter*[d] were included with the 120 GB model prior to September 2009.

h *Component HD AV Cable* is replaced with a *D-Terminal HD AV Cable* (D HD AV) in Japan.
i "Transforming D-Pad" controllers feature a d-pad that can be rotated to switch between either a "plus" (4-way) or a "disk" (8-way) d-pad. These controllers also feature different concave analog stick tops than standard controllers.

Special editions

On a few occasions, Microsoft has produced special editions of the console, usually to coincide with the release of a major product. These special editions are typically custom-colored Xbox 360 models, and are produced in limited numbers.

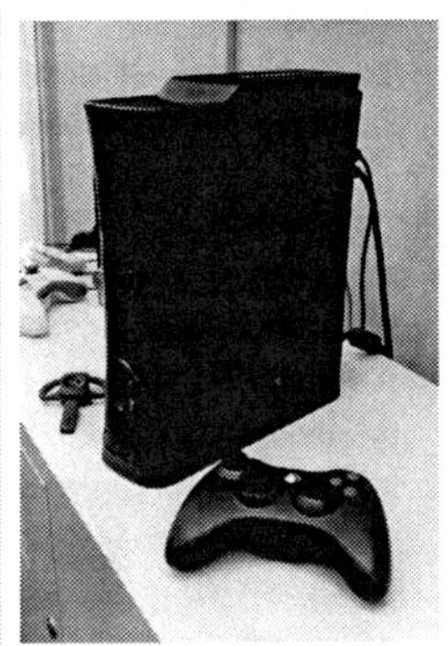

Halo 3 Special Edition

- At the E3 2007 press conference, Microsoft announced the *Halo 3* Special Edition console, released September 25, 2007. It sports a *Halo 3* theme on the console, a wired headset, a wireless controller, and a Play and Charge Kit. Other than the unique "Spartan green-and-gold" color scheme, exclusive dashboard theme and downloads, and an HDMI port,[92] its features were identical to those of the Xbox 360 system at the time.[93] It is priced at US$399.99 and £279.99 (the original price of the Xbox 360).[94]
- To promote *The Simpsons Movie*, Microsoft created a specially designed, yellow Xbox 360 console.[95] The configuration was based on the Xbox 360 package of the time, the only difference being the color scheme of the Xbox 360 console and wireless controller. The consoles were to be given out to winners of drawings taking place between July 18, 2007 and July 27, 2007, in which a name was randomly drawn each day in the "10 Days and 10 Chances to Win" sweepstakes.[96] 100 consoles were produced in total.[97]
- A *Resident Evil 5* bundle containing a red Xbox 360 Elite console was released on March 13, 2009.[98] The bundle also contains a red, wireless controller and a black, wired headset.[99]

Super Elite special editions

- On September 15, 2009 Microsoft announced a special 250 GB hard disk limited edition version of the Xbox 360 Elite for *Call of Duty: Modern Warfare 2*. The unit is highlighted by special game product branding and includes two wireless controllers and a headset. This was the first Xbox 360 announced to come with a 250 GB hard drive.[73]
- A *Final Fantasy XIII* 250 GB limited edition bundle of the Xbox 360 console was announced on February 11, 2010 and was released to coincide with the release of Final Fantasy XIII (March 9, 2010). The bundle includes an imprinted white 250 GB Xbox 360 Elite (Final Fantasy XIII imprinted where HDD size imprint goes), two wireless controllers, a copy of Final Fantasy XIII and exclusive downloadable avatar items. Other than the HDD imprint, this console is cosmetically identical to the discontinued Pro models.[100]
- In March 2010, Microsoft announced a special limited edition black Xbox 360 Elite console for *Tom Clancy's Splinter Cell: Conviction*. The unit includes a 250 GB hard drive, two black wireless controllers, a black wired headset, an Ethernet cable, a standard definition Composite A/V cable, and the standard edition version of the game.[101]

Xbox 360 S special editions

250 GB models

- A *Halo: Reach* special edition bundle was made available concurrently with the game on . It consists of a custom branded silver 250 GB Xbox 360 S console, two silver and black branded controllers, a black wired headset, a copy of the game and a download voucher for "Limited Edition Elite" armor and an episode from the Halo Legends anime. The press release in which it was announced also revealed that it "not only captures the look and feel of the game, but also features custom sound effects from the Halo universe". The bundle retailed for US$399. The branded controllers and headset were also sold as standalone products.[102] [103]

320 GB models

- In June 2011, Microsoft announced a "Limited Collector's Edition" (LCE) Xbox 360 S console to coincide with the launch of Gears of War 3.[49] The console features a metallic red finish with black Gears of War 3 Branding.[49] Unlike other Xbox 360 S consoles, the Gears of War 3 LCE features a 320 GB hard drive.[49] and sounds from the Gears of War 3 game are played when the console is switched on or the disc tray is opened.[104] It is bundled with two custom branded metallic red Gears of War 3 wireless controllers, a wired headset, a copy of the game and a download code for additional in-game content.[49] It is due for release on September 20, 2011.[49]
- On July 21, 2011, Microsoft announced the "Limited Edition Kinect Star Wars Bundle".[77] The bundle includes a 320 GB Xbox 360 S console, which features R2-D2-based artwork and custom R2-D2-themed power on/tray eject sounds, a white Kinect sensor, a gold and black C-3PO themed wireless controller, a copy of Star Wars Kinect and Kinect Adventures, and exclusive downloadable content.[77] It was subsequently delayed until 2012.[105]
- On September 2, 2011, Microsoft announced the "Limited Edition Call of Duty: Modern Warfare 3 Bundle". The bundle includes a 320 GB Xbox 360 S console, which features *Modern Warfare 3* based artwork and custom *Modern Warfare 3* themed power on/tray eject sounds, two *Modern Warfare 3* themed wireless controllers (with transforming d-pad), a black wired headset, and a copy of the game.[79] The console is due to launch on November 8, 2011 in North America, Australasia and the EMEA (Europe, Middle East and Africa) region.[79]

Xbox 360 Launch Team Edition

The Launch Team Console with controller

A white console with green accents was released in 2005 only to the Launch Team as a gift from Microsoft. The consoles came complete with a 20 GB HDD also in green to match the top and bottom sections that are typically grey in color. The HDD plate was also personalized and engraved with the team member's gamertag. Few examples have been seen with simply the console release date November 22, 2005 engraved on the HDD plate. The consoles came with a limited issue controller to accompany the console. The difference with the controller again is the grey part on the controller is replaced with a green molded plastic, not painted contrary to popular belief, to match the console. Each console came with a plain white faceplate. However as an additional gift, each team member was given an additional packaged faceplate with one of a kind art with the caption "I Made This" on the USB door of the faceplate. Very few examples have been sold off from original team member's collection. Larry Hryb, better known as "Major Nelson", is known to own one which he displays pictures of on his website having been a member of the Launch Team. The special edition Launch Team console, hard drive, controller and the special faceplate were never sold in stores or meant for the general public. It is unknown as of now how many of these very rare consoles exist today.

Holiday bundles

As with the original Xbox, Microsoft has continued bundling two video game titles in console retail packaging during the holiday season. During the holidays of 2007, the Xbox 360 Pro and Elite packages were bundled with *Forza Motorsport 2* and *Marvel: Ultimate Alliance*.[106] In the UK, Ireland and the Netherlands, *Forza Motorsport 2* was bundled with *Viva Piñata*.[107] Holiday 2008 Xbox 360 and Xbox 360 Elite packages were bundled with *Lego Indiana Jones: The Original Adventures* and *Kung Fu Panda*, while Arcade consoles were bundled with *Sega Superstars Tennis*.[57] Holiday 2009 packages included an Elite console, LEGO Batman and Pure.[108] Holiday 2010 bundles included a 250 GB Xbox 360 S console, Forza Motorsport 3 and a voucher to download Alan Wake from the Xbox Live Marketplace.[109]

Hardware and accessories

Hardware

The main unit of the Xbox 360 itself has slight double concavity in matte white or black. The official color of the white model is Arctic Chill. It features a port on the top when vertical (left side when horizontal) to which a custom-housed hard drive unit can be attached in sizes of either 20, 60, 120 or 250 GB. Inside, the Xbox 360 uses the triple-core IBM designed Xenon as its CPU, with each core capable of simultaneously processing two threads, and can therefore operate on up to six threads at once.[110] Graphics processing is handled by the ATI Xenos, which has 10 MB of eDRAM. Its main memory pool is 512 MB in size.

Xbox 360 Wireless Controller

Accessories

Many accessories are available for the console, including both wired and wireless controllers, faceplates for customization, headsets for chatting, a webcam for video chatting, dance mats and Gamercize for exercise, three sizes of memory units and four sizes of hard drives (20 GB, 60 GB, 120 GB and 250 GB (initially Japan only,[111] but later also available in elsewhere[112] [113]), among other items, all of which are styled to match the console.

Kinect

The *Kinect* sensor device

Kinect is a "controller-free gaming and entertainment experience" for the Xbox 360. It was first announced on June 1, 2009 at the Electronic Entertainment Expo, under the codename, *Project Natal*. The add-on peripheral enables users to control and interact with the Xbox 360 without a game controller by using gestures, spoken commands and presented objects and images. The Kinect accessory is compatible with all Xbox 360 models,[114] connecting to new models via a custom connector, and to older ones via a USB and mains power adapter.[115] During their CES 2010 keynote speech, Robbie Bach and Microsoft CEO Steve Ballmer went on to say that Kinect will be released during the holiday period (November–January) and it will work with every 360 console. Its name and release date of were officially announced on , prior to Microsoft's press conference at E3 2010.[116]

Software

Dashboard

The Xbox 360's original graphical user interface was the *Xbox 360 Dashboard*; a tabbed interface that featured five "Blades" (formerly four blades), and was designed by AKQA and Audiobrain. It could be launched automatically when the console booted without a disc in it, or when the disc tray was ejected, but the user had the option to select what the console does if a game is in the tray on start up, or if inserted when already on. A simplified version of it was also accessible at any time via the Xbox Guide button on the gamepad. This simplified version showed the user's gamercard, Xbox Live messages and friends list. It also allowed for personal and music settings, in addition to voice or video chats, or returning to the Xbox Dashboard from the game.

On November 19, 2008, the Xbox 360's dashboard was changed from the "Blade" interface, to a dashboard reminiscent of that present on the Zune and Windows Media Center, known as the "New Xbox Experience" or NXE.

Since the console's release, Microsoft has released several updates for the Dashboard software. These updates have included adding new features to the console, enhancing Xbox Live functionality and multimedia playback capabilities, adding compatibility for new accessories, and fixing bugs in the software. Such updates are mandatory for users wishing to use Xbox Live, as access to Xbox Live is disabled until the update is performed.

New Xbox Experience

Further information: Xbox Live, section "New Xbox Experience"

At E3 2008, at Microsoft's Show, Microsoft's Aaron Greenberg and Marc Whitten announced the new Xbox 360 interface called the "New Xbox Experience" (NXE). The update was intended to ease console menu navigation. Its GUI uses the Twist UI, previously used in Windows Media Center and the Zune. Its new Xbox Guide retains all Dashboard functionality (including the Marketplace browser and disk ejection) and the original "Blade" interface.

The NXE also provides many new features. Users can now install games from disc to the hard drive to play them with reduced load time and less disc drive noise, but each game's disc must remain in the system in order to run. A new, built-in Community system allows the creation of digitized Avatars that can be used for multiple activities, such as sharing photos or playing Arcade games like *1 vs. 100*. The update was released on November 19, 2008.[117]

While previous system updates have been stored on internal memory, the NXE was the first to require a storage device—at least a 128 MB memory card or a hard drive.

Microsoft is due to release a further update to the Xbox 360 Dashboard starting on December 6, 2011[118] . It will include a completely new user interface which utilizes Microsoft's Metro design language, and will add new features such as cloud storage for game saves and profiles, live television, Bing voice search,[119] access to YouTube videos, Skype integration, and better support for Kinect voice commands.[120]

Multimedia

The Xbox 360 supports videos in Windows Media Video (WMV) format (including high-definition and PlaysForSure videos), as well as H.264 and MPEG-4 media. The December 2007 dashboard update added support for the playback of MPEG-4 ASP format videos.[121] The console can also display pictures and perform slideshows of photo collections with various transition effects, and supports audio playback, with music player controls accessible through the Xbox 360 Guide button. Users may play back their own music while playing games or using the dashboard, and can play music with an interactive visual synthesizer.

Music, photos and videos can be played from standard USB mass storage devices, Xbox 360 proprietary storage devices (such as memory cards or Xbox 360 hard drives), and servers or computers with Windows Media Center or Windows XP with Service Pack 2 or higher within the local-area network in streaming mode.[122] [123] As the Xbox 360 uses a modified version of the UPnP AV protocol,[124] some alternative UPnP servers such as uShare (part of the

GeeXboX project) and MythTV can also stream media to the Xbox 360, allowing for similar functionality from non-Windows servers. This is possible with video files up to HD-resolution and with several codecs (MPEG-2, MPEG-4, WMV) and container formats (WMV, MOV, TS).[125]

As of 27 October 2009,[126] UK and Ireland users are also able to access live and on-demand streams of Sky television programming.[127]

Canadian users can access a streaming HDTV service from Telus.[128]

At the 2007, 2008, and 2009 Consumer Electronics Shows, Microsoft had announced that IPTV services would soon be made available to use through the Xbox 360. In 2007, Microsoft chairman Bill Gates stated that IPTV on Xbox 360 was expected to be available to consumers by the holiday season, using the Microsoft TV IPTV Edition platform.[129] In 2008, Gates and president of Entertainment & Devices Robbie Bach announced a partnership with BT in the United Kingdom, in which the BT Vision advanced TV service, using the newer Microsoft Mediaroom IPTV platform, would be accessible via Xbox 360, planned for the middle of the year.[130] BT Vision's DVR-based features would not be available on Xbox 360 due to limited hard drive capacity.[131] In 2010, while announcing version 2.0 of Microsoft Mediaroom, Microsoft CEO Steve Ballmer mentioned that AT&T's U-verse IPTV service would enable Xbox 360s to be used as set-top boxes later in the year.[132] As of January 2010, IPTV on Xbox 360 has yet to be deployed beyond limited trials.

XNA community

XNA Community is a feature whereby Xbox 360 owners can receive community-created games, made with Microsoft XNA Game Studio, from the XNA Creators Club . The games are written, published, and distributed through a community managed portal. XNA Community provides a channel for digital videogame delivery over Xbox Live that can be free of royalties, publishers and licenses.[133] [134] XNA game sales, however, did not meet original expectations.[135]

Games

The Xbox 360 launched with 14 games in North America and 13 in Europe. The console's best-selling game for 2005, *Call of Duty 2*, sold over a million copies.[136] Five other games sold over a million copies in the console's first year on the market: *Ghost Recon Advanced Warfighter*,[137] *The Elder Scrolls IV: Oblivion*,[138] *Dead or Alive 4*,[139] *Saints Row*,[140] and *Gears of War*.[141] *Gears of War* would become the best-selling game on the console with 3 million copies in 2006,[142] before being surpassed in 2007 by *Halo 3* with over 8 million copies.[143]

Six games were initially available in Japan, while eagerly anticipated titles such as *Dead or Alive 4* and *Enchanted Arms* were released in the weeks following the console's launch.[144] Games targeted specifically for the region, such as *Chromehounds*, *Ninety-Nine Nights*, and *Phantasy Star Universe*, were also released in the console's first year.[145] Microsoft also had the support of Japanese developer Mistwalker, founded by *Final Fantasy* creator Hironobu Sakaguchi. Mistwalker's first game, *Blue Dragon*, was released in 2006 and had a limited-edition bundle which sold out quickly with over 10,000 pre-orders.[146] *Blue Dragon* is one of three Xbox 360 games to surpass 200,000 units in Japan, along with *Tales of Vesperia* and *Star Ocean: The Last Hope*. Mistwalker's second game, *Lost Odyssey* also sold over 100,000 copies.[147]

The 2007 Game Critics Awards honored the Xbox 360 platform with 38 Nominations and 11 Wins.[148] By March 2008, the Xbox 360 had reached a software attach rate of 7.5 games per console—a record for any console in history, in EU the rate was 7.0 while its competitors were 3.8 (PS3) and 3.5 (Wii).[149]

Xbox Live

Upcoming Xbox Live Dashboard

When the Xbox 360 was released, Microsoft's online gaming service Xbox Live was shut down for 24 hours and underwent a major upgrade, adding a basic non-subscription service called Xbox Live Silver (later renamed Xbox Live Free) to its already established premium subscription-based service (which was renamed Gold). Xbox Live Free is included with all SKUs of the console. It allows users to create a user profile, join on message boards, and access Microsoft's Xbox Live Arcade and Marketplace and talk to other members.[90] [150] A Live Free account does not generally support multiplayer gaming; however, some games that have rather limited online functions already, (such as *Viva Piñata*) or games that feature their own subscription service (e.g. EA Sports games) can be played with a Free account. Xbox Live also supports voice and video communication, the latter a feature possible with the Xbox Live Vision; an add-on USB web camera designed specifically for the Xbox 360.[151]

Xbox Live Gold includes the same features as Free and includes integrated online game playing capabilities outside of third-party subscriptions. Microsoft has allowed previous Xbox Live subscribers to maintain their profile information, friends list, and games history when they make the transition to Xbox Live Gold. To transfer an Xbox Live account to the new system, users need to link a Windows Live ID to their gamertag on Xbox.com.[152] When users add an Xbox Live enabled profile to their console, they are required to provide the console with their passport account information and the last four digits of their credit card number, which is used for verification purposes and billing. An Xbox Live Gold account has an annual cost of US$49.99, C$59.99, NZ$90.00, GB£39.99, or €59.99. As of January 5, 2011, Xbox Live has over 30 million subscribers.[153]

Xbox Live Marketplace

Upcoming Xbox Live Marketplace

The Xbox Live Marketplace is a virtual market designed for the console that allows Xbox Live users to download purchased or promotional content. The service offers movie and game trailers, game demos, Xbox Live Arcade games, gamertag images, and Xbox 360 Dashboard themes as well as add-on game content (items, costumes, levels etc.). These features are available to both Free and Gold members on Xbox Live. A hard drive or memory unit is required to store products purchased from Xbox Live Marketplace.[154] In order to download priced content, users are required to purchase Microsoft Points for use as scrip;[155] though some products (such as trailers and demos) are free to download. Microsoft Points can be obtained through prepaid cards in 1,600 and 4,000 point denominations. Microsoft Points can also be purchased through Xbox Live with a credit card in 500, 1,000, 2,000, and 5,000 point denominations. Users are able to view items available to download on the service through a PC via the Xbox Live Marketplace website.[156] An estimated seventy percent of Xbox Live users have downloaded items from the Marketplace.[157]

Video store

On November 6, 2006, Microsoft announced the Xbox Video Marketplace, an exclusive video store accessible through the console. Launched in the United States on November 22, 2006, the first anniversary of the Xbox 360's launch, the service allows users in the United States to download high-definition and standard-definition television shows and movies onto an Xbox 360 console for viewing. With the exception of short clips, content is not currently available for streaming, and must be downloaded. Movies are also available for rental. They expire in 14 days after download or at the end of the first 24 hours after the movie has begun playing, whichever comes first. Television episodes can be purchased to own, and are transferable to an unlimited number of consoles. Downloaded files use 5.1 surround audio and are encoded using VC-1 for video at 720p, with a bitrate of 6.8 Mbit/s.[158] Television content is offered from MTV, VH1, Comedy Central, Turner Broadcasting, and CBS; and movie content is Warner Bros., Paramount, and Disney, along with other publishers.[159]

After the Spring 2007 update, the following video codecs are supported:

- H.264 video support: Up to 15 Mbit/s, Baseline, Main, and High (up to level 4.1) Profiles with 2 channel AAC LC and Main Profiles.
- MPEG-4 Part 2 video support: Up to 8 Mbit/s, Simple Profile with 2 channel AAC LC and Main Profiles.

As a late addition to the December Xbox 360 update, 25 movies were added to the European Xbox 360 video market place on the December 11, 2007 and cost 250 Microsoft points for the SD version on the movie and 380 Microsoft points for the HD version of the movie.[160] Xbox Live members in Canada featured the ability to go on the Xbox Live Marketplace as of December 10, 2007 with around 30 movies to be downloaded for the same amount of Microsoft Points.

On May 26, 2009, Microsoft announced the future release of the Zune HD (in the fall of 2009), the next addition to the Zune product range. This is of an impact on the Xbox Live Video Store as it was also announced that the Zune Video Marketplace and the Xbox Live Video Store will be merged to form the Zune Marketplace, which will be arriving on Xbox Live in 7 countries initially, the United Kingdom, the United States, France, Italy, Germany, Ireland and Spain.[161] Further details will be released at the Microsoft press conference at E3 2009.

Xbox Live Arcade

Xbox Live Arcade is an online service operated by Microsoft that is used to distribute downloadable video games to Xbox and Xbox 360 owners. In addition to classic arcade games such as *Ms. Pac-Man*, the service offers some new original games like *Assault Heroes*. The Xbox Live Arcade also features games from other consoles, such as the PlayStation game *Castlevania: Symphony of the Night* and PC games such as *Zuma*. The service was first launched on November 3, 2004,[162] using a DVD to load, and offered games for about US$5 to $15. Items are purchased using Microsoft Points, a proprietary currency used to reduce credit card transaction charges. On November 22, 2005, Xbox Live Arcade was re-launched with the release of the Xbox 360, in which it was now integrated with the Xbox 360's dashboard. The games are generally aimed toward more casual gamers; examples of the more popular titles are *Geometry Wars*, *Street Fighter II' Hyper Fighting*, and *Uno*.[163] On March 24, 2010, Microsoft introduced the Game Room to Xbox Live. Game Room is a gaming service for Xbox 360 and Microsoft Windows that lets players compete in classic arcade and console games in a virtual arcade.[164]

Reception and sales

Region	Units sold	First available
Canada	870,000 as of August 1, 2008[165]	November 22, 2005
United States	25.4 million as of March 31, 2011[166]	
EMEA region (Europe, Middle East and Africa)	13.7 million as of March 31, 2011[166] (Includes UK sales)	December 2, 2005
United Kingdom	3.9 million as of January 13, 2009[167]	
Japan	1.5 million as of June 17, 2011[168]	December 10, 2005
Australia & New Zealand	1 million as of April 19, 2010[169]	March 23, 2006

Worldwide	**57.6 million** as of September 30, 2011[4]	(more...)

The Xbox 360 began production only 69 days before launch.[170] [171] As a result, Microsoft was not able to supply enough systems to meet initial consumer demand in Europe or North America[172] and many potential customers were unable to purchase a console at launch. Forty thousand units appeared on eBay during the initial week of release, 10% of the total supply.[173] By year's end, Microsoft had shipped 1.5 million units, including 900,000 in North America, 500,000 in Europe, and 100,000 in Japan.[174]

At E3 in May 2006, Bill Gates announced that Microsoft would have a head start of 10 million units by the time Sony and Nintendo entered the market.[175] Microsoft later specified that goal and estimated shipments of 10 million units by the end of 2006.[176]

Cumulative sales from the system's launch until June 30, 2007 were predicted to reach 12 million units, down from 13 to 15 million units estimated earlier. Being released one year ahead of its competitors, the Xbox 360 was the market leader throughout the first half of 2007; however, on September 12, 2007, it was reported by the *Financial Times* that the Xbox 360 had been surpassed by the Wii in terms of worldwide home console sales.[177]

On October 4, 2007, after the launch of *Halo 3*, Microsoft stated that Xbox 360 sales had more than doubled compared to the previous average.[178] According to the NPD Group, the Xbox 360 outsold the Wii for the month of September in the United States, helped in part by the spike in sales seen after the launch of *Halo 3*, which sold 3.3 million copies in the US in a 12-day period.[179] The Xbox had sales of 528,000 units for September, while the Wii had sales of 501,000 units.[180]

Despite the relatively strong sales figures, Microsoft's gaming division was losing money. Through 2005, the Xbox gaming division had lost over $4 billion,[181] However, Microsoft expected the console to start making money in 2008.[182] due to a loss leader market strategy of selling consoles below cost in order to obtain market saturation and gain profits on software and peripherals with a much higher profit margin.[183] [184] Additionally, Microsoft took a charge of $1 billion dollars on its June 2007 Income Statement to account for the cost of replacing bricked Xbox 360s.[185]

In Europe, the Xbox 360 sold 2 million units in 2006 and 1.9 million in 2007 according to estimates by Electronic Arts.[186] [187] In 2007, the Xbox 360 sold 4.62 million units in the US according to the NPD Group,[188] [189] [190] and 257,841 in Japan according to Enterbrain.[191] [192] [193] In 2008, the Xbox 360 sold 317,859 units in Japan, according to Enterbrain.[193] [194]

On February 13, 2008, Microsoft announced that the Xbox 360 suffered shortages in the US in January 2008, possibly continued into February.[195] [196] [197] Prior to the release of the NPD Group's video game statistics for January 2008, the Xbox 360 was in second place behind the Wii in US sales for most months since the Wii and PS3's release.[196] On May 14, 2008, Microsoft announced that 10 million Xbox 360s had been sold and that it was the "first current generation gaming console" to surpass the 10 million figure in the US.[198] In the US, the Xbox 360 was the leader in current-generation home console sales until June 2008, when it was surpassed by the Wii.[199] [200] [201]

In the first seven months of 2008, the Xbox 360 has sold 154,000 units in Canada, being outsold by the PlayStation 3 and Wii;[165] the Xbox 360 has sold a total of 870,000 units in Canada as of August 1, 2008.[165] In Europe, the Xbox 360 has sold seven million units as of November 20, 2008, according to Microsoft.[202] In Japan, the Xbox 360 has sold 866,167 units as of December 28, 2008, according to Enterbrain.[194] [203] In the United Kingdom, the Xbox 360 has sold 3.9 million units as of June 27, 2009, according to GfK Chart-Track.[167]

On May 13, 2008, Microsoft stated that there will not be a release of a newer version of the Xbox 360 in 2009. Microsoft representative says, "While we [Microsoft] don't normally comment on rumors like this [release of newer and smaller Xbox 360], we can tell you that we have no plans to release a new console in 2009".[204]

On May 28, 2009, Microsoft announced that sales have surpassed the 30-million unit-mark, stating that 2008 was Microsoft's "biggest year in history" also due to the increase of Xbox Live users.[205]

On January 6, 2011, Microsoft stated that the Xbox 360 sold over 50 million units worldwide during their CES 2011 conference. In addition, they stated that the Kinect sold about 8 million units, 3 million more than initially predicted, during the product's first 60 days on the market. It was also announced that Xbox live had more than 30 million users, making 2010 the best year to date for the online service.[4]

Edge magazine reported in August 2011 that initially lacklustre and subsequently falling sales in Japan, where Microsoft had been unable to make serious inroads into the dominance of domestic rivals Sony and Nintendo, had led to retailers scaling back and in some cases discontinuing sales of the Xbox 360 completely.[206]

In 2009, IGN named the Xbox 360 the sixth greatest video game console of all time, out of a field of 25.[207]

Timeline of hardware models

United States

October 22, 2005

- Launch of Xbox 360 Premium (20 GB) – $399.99
- Launch of Xbox 360 Core – $299.99

April 29, 2007

- Launch Xbox 360 Elite (120 GB) – $479.99[208]

August 6, 2007[94]

- Price Cut on Xbox 360 Premium (20 GB) – $349.99
- Price Cut on Xbox 360 Core – $279.99
- Price Cut on Xbox 360 Elite – $449.99

October 27, 2007

- Launch of Xbox 360 Arcade – $279.99[209]
- Discontinuation of Xbox 360 Core

July 13, 2008

- Discontinuation of Xbox 360 (20 GB) (Price cut to $299.99 for remaining stock)[210]

August 1, 2008[210]

- Launch of Xbox 360 Premium (60 GB) – $349.99

September 5, 2008[211]

- Price Cut on Xbox 360 Elite – $399.99
- Price Cut on Xbox 360 (60 GB) – $299.99
- Price Cut on Xbox 360 Arcade – $199.99

August 28, 2009

- Discontinuation of Xbox 360 (60 GB) (Price cut to $249.99 for remaining stock)
- Price Cut on Xbox 360 Elite – $299.99

June 19, 2010[13]

- Launch of Xbox 360 S 250 GB – $299.99
- Discontinuation of Xbox 360 Elite (Price Cut to $249.99 for remaining stock)
- Discontinuation of Xbox 360 Arcade (Price Cut to $149.99 for remaining stock)

August 3, 2010

- Launch of Xbox 360 S 4 GB – $199.99

Technical problems

Three red lights on the Xbox 360's ring indicator representing a "General Error" requiring service of the Console or Power Adapter, nicknamed the "Red Ring of Death", "Red Ring of Fire" or "Red Lights of Death."

The Xbox 360 has been subject to a number of technical problems. Since the console's release in 2005, users have reported concerns over its reliability and failure rate.[212] [213] [214]

To aid customers with defective consoles, Microsoft extended the Xbox 360's manufacturer's warranty to three years for hardware failure problems that generate a "General Hardware Failure" error report. A "General Hardware Failure" is recognized on all models released before the Xbox 360 S by three quadrants of the ring around the power button flashing red. This error is often known as the "Red Ring of Death".[215] In April 2009 the warranty was extended to also cover failures related to the E74 error code.[216] The warranty extension is not granted for any other types of failures that do not generate these specific error codes.

Since these problems surfaced, Microsoft has attempted to modify the console to improve its reliability. Modifications include a reduction in the number, size, and placement of components, the addition of dabs of epoxy on the corners and edges of the CPU and GPU as glue to prevent movement relative to the board during heat expansion,[217] and a second GPU heatsink to dissipate more heat.[218] With the release of the redesigned Xbox 360 S, the warranty for the newer models do not include the three-year extended coverage for "General Hardware Failures".[219] The newer Xbox 360 S model indicates system overheating when the console's power button begins to flash red, unlike previous models where the first and third quadrant of the ring would light up red around the power button if overheating occurred. The system will then warn the user of imminent system shutdown until the system has cooled,[220] whereas a flashing power button that alternates between green and red is an indication of a "General Hardware Failure" unlike older models where three of the quadrants would light up red.[221]

References

[1] "Google translation of Xbox.com/JA-JP" (http://translate.google.co.uk/translate?hl=en&sl=ja&u=http://www.xbox.com/JA-JP/&ei=HWaaS8aPC6X20wT1s8HdAQ&sa=X&oi=translate&ct=result&resnum=1&ved=0CBIQ7gEwAA&prev=/search?q=http://www.xbox.com/ja-JP/&num=50&hl=en&safe=off&client=firefox-a&hs=9Fn&rls=org.mozilla:en-GB:official). microsoft. . Retrieved 2010-12-03.

[2] Radd, David (2005-08-16). "Xbox 360 Manufacturers Revealed" (http://www.businessweek.com/innovate/content/aug2005/id20050816_595546.htm). Bloomberg Businessweek. . Retrieved 2010-08-19.

[3] Ashcraft, Brian (2010-11-19). "Report: Foxconn Denies Protest Over Pay | Kotaku Australia" (http://www.kotaku.com.au/2010/11/report-foxconn-workers-protest-over-pay-foxconn-denies-protest/#more-424340). Kotaku.com.au. . Retrieved 2011-10-23.

[4] "Earnings Release FY12 Q1" (http://www.microsoft.com/investor/EarningsAndFinancials/Earnings/Kpi/fy12/Q1/detail.aspx). Microsoft. .

[5] "Microsoft confirms Xbox 360 802.11n adapter" (http://www.joystiq.com/2009/09/21/microsoft-confirms-xbox-360-802-11n-adapter/). joystiq. 2009-09-21. . Retrieved 2009-11-23.

[6] "Halo 3 - Halo Official Site" (http://halo.xbox.com/en-us/intel/titles/halo3). Microsoft. 2011-10-23. . Retrieved 2011-10-23.

[7] "Original Xbox Games Playable on Xbox 360" (http://www.xbox.com/en-US/games/backwardcompatibilitygameslist.htm). . Retrieved 17 November 2009.

[8] "Shortages spoil Xbox 360 launch" (http://news.bbc.co.uk/2/hi/technology/4462744.stm). *BBC News*. 2005-11-23. . Retrieved 2009-12-31.

[9] "Xbox 360 Launch Takes Europe by Storm" (http://www.xbox365.com/news.cgi?id=GGPPLLiddi12060836). .

[10] "Retailers Sell Out of Initial Xbox 360 Shipments" (http://www.tgdaily.com/content/view/21759/). . Retrieved 2008–10–14.

[11] "New Xbox 360 Shipping Today, Costs $299" (http://kotaku.com/5563154/new-xbox-360-shipping-today-costs-299). Kotaku.com. 2010-06-14. . Retrieved 2010-11-11.

[12] Dean Takahashi, VentureBeat. " Microsoft to Launch Smaller XBOX 360 Game Console (http://venturebeat.com/2010/06/14/microsoft-to-launch-smaller-xbox-360-game-console/)." June 14, 2010.

[13] Thorsen, Tor (14 June 2010). "New $200 Xbox 360 planned, Elite & Arcade getting $50 price cut" (http://e3.gamespot.com/story/6265465/new-200-xbox-360-planned-elite-and-arcade-getting-50-price-cut). *GameSpot*. CBS Interactive. . Retrieved 2010-06-17. "Moore said that Microsoft is currently working on a second new Xbox 360, which will be offered at the arcade's price point of $200. He declined to say what functionalities the cheaper model would or would not have. [...] One likely scenario is the cheaper model will essentially be a slim arcade with no Wi-Fi capabilities or a hard drive. [...] Moore also said that going forward, all future models of the Xbox 360 would not have names and would only be designated by their memory capacity."

[14] Robinson, Martin (2009-06-04). "E3 2009: 360 to Stick Around Until 2015 - Xbox360 News at IGN" (http://uk.xbox360.ign.com/articles/991/991485p1.html). Uk.xbox360.ign.com. . Retrieved 2010-11-11.

[15] Bunker, Adam (2010-06-21). "Xbox 720 set to launch at the E3 expo in June 2015 | T3 magazine" (http://www.t3.com/news/xbox-720-set-to-launch-at-the-e3-expo-in-june-2015?=46467). T3.com. . Retrieved 2010-11-11.

[16] "New Xbox 360 (Xbox 360 'Slim') Delays Xbox 720 Till 2015 [Xbox 360 Presently 'Mid-Cycle' Says Microsoft: "We've Got at Least Another Five Years of this Generation"] » TFTS – Technology, Gadgets & Curiosities" (http://nexus404.com/Blog/2010/06/21/new-xbox-360-xbox-360-slim-delays-xbox-720-till-2015-xbox-360-presently-mid-cycle-says-microsoft-weve-got-at-least-another-five-years-of-this-generation/). Nexus404.com. . Retrieved 2010-11-11.

[17] "Rumor: Next Xbox Won't Appear Until E3 2013" (http://www.tomsguide.com/us/xbox-next-xbox-720-console-jeff-faulkner-Fable-Next,news-12958.html). Tom's Hardware. .

[18] "Next generation Xbox pops up in Microsoft resumes" (http://www.joystiq.com/2011/10/10/next-generation-xbox-pops-up-in-microsoft-resumes/). joystiq. .

[19] Smith, David (June 2004). "Microsoft Narrowing Down Xbox Successor's Name?" (http://web.archive.org/web/20070517222211/http://findarticles.com/p/articles/mi_zd1up/is_200406/ai_n9520284). Ziff Davis Media. Archived from the original (http://findarticles.com/p/articles/mi_zd1up/is_200406/ai_n9520284) on 2007-05-17. . Retrieved 2008-07-26.

[20] Takahashi, Dean (2006-05-01). "Chronology of Xbox History, 2003" (http://blogs.mercurynews.com/aei/2006/04/24/the_xbox_360_un/). Mercury News. . Retrieved 2006-11-02.

[21] Becker, David (2003-08-14). "ATI wins bid for next Xbox" (http://news.cnet.com/2100-1043-5063661.html). CNET. . Retrieved 2008-07-26.

[22] Jonathan V. Last (December 30, 2008). "Playing the Fool" (http://online.wsj.com/article/SB123069467545545011.html). Wall Street Journal. .

[23] "Jeff Minter vs Xbox 360: how Microsoft bought the light synth vision" (http://www.guardian.co.uk/technology/gamesblog/2005/jul/05/jeffmintervs). The Guardian. July 5, 2005. .

[24] Perry, Douglass (2006-10-17). "Microsoft Spreads 360 Love" (http://xbox360.ign.com/articles/739/739880p1.html). IGN. . Retrieved 2006-10-18.

[25] "Microsoft Launches New Xbox 360 Console for Families" (http://www.microsoft.com/Presspass/press/2007/oct07/10-22ArcadeConsolePR.mspx). Microsoft orporation. 2007-10-22. . Retrieved 2008-07-26.

[26] "Microsoft drops price of 120 GB Xbox 360 system - Game Hunters: In search of video games and interactive awesomeness" (http://content.usatoday.com/communities/gamehunters/post/2009/08/microsoft-drops-price-of-120gb-xbox-360-system-/1). Content.usatoday.com. 2009-08-27. . Retrieved 2009-09-10.

[27] "New Xbox 360 is Actually Called the Xbox 360 'S'" (http://www.tomsguide.com/us/xbox-360-slim-s-360-s-xbox-slim,news-7133.html). *Tom's Guide*. Bestofmedia. . Retrieved 2010-06-22. "Even though Microsoft is still calling it the same name as old, the sticker on the back shows clearly that it's the Xbox 360 **S** – perhaps taking a page out of Porsche naming conventions."

[28] "Xbox 360 Transforms the Way Everyone Has Fun" (http://www.microsoft.com/Presspass/press/2010/jun10/06-14E3UmbrellaPR.mspx). Microsoft. June 14, 2010. . Retrieved 2010-06-16. "The New Xbox 360"

[29] "Xbox 360 Slim in the Works?" (http://www.industrygamers.com/news/xbox-360-slim-in-the-works/). . Retrieved May 1, 2010.

[30] "So What's Inside The New Xbox 360?" (http://kotaku.com/5562412/so-whats-inside-the-new-xbox-360). Kotaku.com. 2010-06-13. . Retrieved 2010-11-11.

[31] Anand Lal Shimpi (2010-06-18). "Inside the New Xbox 360" (http://www.anandtech.com/show/3774/welcome-to-valhalla-inside-the-new-250gb-xbox-360-slim). AnandTech. . Retrieved 2010-06-19.

[32] Patel, Nilay (2010-06-14). "New Xbox 360 looks angular and Ominous" (http://www.engadget.com/2010/06/14/slim-xbox-360-gets-official-at-299-shipping-today-looks-angul/). Engadget. . Retrieved 2010-06-14.

[33] "New Xbox 360 guide: Microsoft's slim console explained" (http://www.engadget.com/2010/06/15/new-xbox-360-guide-microsofts-slim-console-explained/3). Engadget. 2010-06-15. . Retrieved 2010-07-01. "*Biggest news on the connectivity front is 802.11n WiFi built-in. Interestingly the system detects it as if it were connected externally, but it's right in there, meaning no dongles for you (unless if you want 5GHz 802.11n, this one is single band 2.4GHz)*"

[34] "USB Flash Drive Memory Support Coming to the 360 on April 6th" (http://www.xboxpotato.com/2010/03/usb-flash-drive-support-coming/). . Retrieved 2010-03-26.

[35] "Picking up an Xbox 360 Slim? Read this FIRST | Game Hacks | a Chron.com blog" (http://blog.chron.com/gamehacks/2010/07/picking-up-an-xbox-360-slim-read-this-first/). Blog.chron.com. 2010-07-22. . Retrieved 2011-07-24.

[36] "How To Replace the Hard Drive in an Xbox 360 Slim 4GB Arcade" (http://flicktitio.us/how-to-replace-the-hard-drive-in-an-xbox-360-slim-4gb-arcade). Flicktitio.us. 2011-07-20. . Retrieved 2011-07-24.

[37] Dudley-Nicholson, Jennifer (2010-06-15). "Microsoft reveals Kinect and slimmer Xbox 360" (http://www.news.com.au/technology/microsoft-launches-kinect-and-slimmer-xbox-360/story-e6frfro0-1225879783329). *News Limited newspapers*. .

[38] "Xbox.com - Hardware Catalog - Xbox 360 250 GB Console" (http://mktplassets.xbox.com/NR/exeres/3EE4C7AF-0C7A-4B0D-92DC-83961F1B4F44.htm). *Xbox.com*. Microsoft. . Retrieved 2010-06-16.

[39] var authorId = "47061140" by Levi Buchanan (2010-06-14). "E3 2010: Xbox 360 Slims Down Today - Xbox360 News at IGN" (http://uk.xbox360.ign.com/articles/109/1097102p1.html). Uk.xbox360.ign.com. . Retrieved 2010-11-11.

[40] "Twitter / Suspended" (http://twitter.com/Microsoft_Xbox/status/16170140427). Twitter.com. . Retrieved 2010-11-11.

[41] Beavis, Gareth. "Xbox 360 Slim announced by Microsoft | News | TechRadar UK" (http://www.techradar.com/news/gaming/consoles/microsoft-announces-new-slim-xbox-360-696530). Techradar.com. . Retrieved 2010-11-11.

[42] "Xbox 360 S Gets a Little Less Shiny" (http://news.teamxbox.com/xbox/24403/Xbox-360-S-Gets-a-Little-Less-Shiny/). TeamXbox. . Retrieved 2011-08-06.

[43] "The official word on the glossy Xbox 360" (http://majornelson.com/2011/08/12/the-official-word-on-the-glossy-xbox-360/). MajorNelson.com. . Retrieved 2011-08-15.

[44] "Kinect bundled with slim 4GB Xbox 360 Arcade for $300, new console for $200 in August" (http://www.joystiq.com/2010/07/20/kinect-bundled-with-slim-4gb-xbox-360-arcade-for-300-new-conso/). Joystiq. 2010-07-20. . Retrieved 2010-07-20. "*Though there are currently no Xbox 360 hard drives on the market that support the new "slim" form factor, Greenberg told us that this 4GB unit "absolutely does have that expansion bay that we have the opportunity to use in the future.""*

[45] "Microsoft confirms Kinect price" (http://www.eurogamer.net/articles/microsoft-confirms-kinect-price). Eurogamer. 2010-07-20. . Retrieved 2010-07-21.

[46] "Microsoft Announces Kinect Bundle And Xbox 360 4GB Console" (http://zunited.net/2010/07/microsoft-announces-kinect-bundle-and-xbox-360-4gb-console/). [zunited.net Zunited Microsoft Community]. 2010-07-20. . Retrieved 2010-07-20.

[47] "New Xbox 360 Model Hits Next Month, Kinect Bundle Confirmed" (http://kotaku.com/5591653/new-xbox-360-model-hits-next-month-kinect-bundle-confirmed?utm_source=feedburner&utm_medium=feed&utm_campaign=Feed:+kotaku/full+(Kotaku)). Kotaku. 2010-07-20. . Retrieved 2010-07-20.

[48] Miller, Paul (2010-08-20). "Xbox 360 S 250GB hard drive sneaks into stores, Microsoft cackles as you fall into its trap (update: official)" (http://www.engadget.com/2010/08/20/xbox-360-s-250gb-hard-drive-sneaks-into-stores-microsoft-cackle/). Joystiq. . Retrieved 2010-08-21.

[49] "Gears 3 LCE Console and Controller Announced" (http://gearsofwar.xbox.com/en/news/gallery/2011-06-controller-console-announce). Microsoft. .

[50] "MS announces 320GB Xbox 360 hard-drive News - Xbox 360 - Page 1 | Eurogamer.net" (http://www.eurogamer.net/articles/2011-10-11-ms-announces-320gb-xbox-360-hard-drive). *eurogamer.net*. Eurogamer. .

[51] Kietzmann, Ludwig (2006-09-07). "Microsoft to launch Xbox 360 Core in Japan" (http://www.joystiq.com/2006/09/07/microsoft-to-launch-xbox-360-core-in-japan/). Weblogs, Inc.. . Retrieved 2008-07-26.

[52] "Latest update adds 1080p native support, Live Arcade upgrades and much, much more!" (http://web.archive.org/web/20080731071201/http://www.xbox.com/en-GB/community/news/systemupdate.htm). Archived from the original (http://www.xbox.com/en-GB/community/news/systemupdate.htm) on 2008-07-31. . Retrieved 2009-10-30.

[53] "Xbox 360 System" (http://web.archive.org/web/20080731143424/http://www.xbox.com/en-US/hardware/x/xbox360prosystem/default.htm). Microsoft Corporation. Archived from the original (http://www.xbox.com/en-US/hardware/x/xbox360prosystem/default.htm) on 2008-07-31. . Retrieved 2008-07-31.

[54] "HDMI Premium Xbox 360 caught on camera" (http://www.joystiq.com/2007/08/08/hdmi-premium-xbox-360-potentially-caught-on-film/). . Retrieved 2008-07-26.

[55] "Microsoft officially adds HDMI to Xbox 360 Premium" (http://www.engadget.com/2007/08/08/microsoft-officially-adds-hdmi-to-xbox-360-premium/). . Retrieved 2008-07-26.

[56] Block, Gerry (2007-10-01). "Xbox 360 Falcon Mother Board and 65-nm CPU Arrive" (http://gear.ign.com/articles/824/824183p1.html). IGN. . Retrieved 2009-07-01.

[57] Dumitrescu, Andrei (2008-10-10). "Xbox 360 Bundles LEGO Indiana Jones and Kung Fu Panda" (http://news.softpedia.com/news/Xbox-360-Bundles-LEGO-Indiana-Jones-and-Kung-Fu-Panda-95385.shtml). Softpedia. . Retrieved 2008-11-16.

[58] McDougall, Paul (2008-09-04). "Xbox 360 Price Cuts Take Effect Friday" (http://www.informationweek.com/news/internet/b2c/showArticle.jhtml?articleID=210300528). InformationWeek. . Retrieved 2009-09-10.

[59] "Introducing Xbox 360 Elite" (http://web.archive.org/web/20070510235323/http://www.xbox.com/en-US/hardware/xbox360elite/default.htm). Microsoft Corporation. Archived from the original (http://www.xbox.com/en-US/hardware/xbox360elite/default.htm) on 2007-05-10. . Retrieved 2008-07-26.

[60] "Microsoft Unveils Xbox 360 Elite" (http://web.archive.org/web/20070405185941/http://www.xbox.com/en-US/community/news/2007/0327-microsoftunveilsxbox360elite.htm). Microsoft Corporation. Archived from the original (http://www.xbox.com/en-US/community/news/2007/0327-microsoftunveilsxbox360elite.htm) on 2007-04-05. . Retrieved 2008-07-26.

[61] "Happy Xbox 360 Elite launch day!" (http://xbox.joystiq.com/2007/04/29/happy-xbox-360-elite-launch-day/). Joystiq. . Retrieved 2009-11-25.

[62] "2nd UPDATE: Microsoft Cuts High-End Xbox 360 Price By $100" (http://online.wsj.com/article/SB125136429316263083.html). Online.wsj.com. 2009-08-27. . Retrieved 2009-09-10.

[63] Paul Miller (2007-10-12). "Xbox 360 Arcade finds a shelf in the wild" (http://www.engadget.com/2007/10/12/xbox-360-arcade-finds-a-shelf-in-the-wild/). Engadget. . Retrieved 2008-07-26.

[64] Carlos Bergfeld (2007-10-02). "New Xbox 360 SKU Shows Up Online" (http://www.shacknews.com/onearticle.x/49262). Shacknews. . Retrieved 2008-07-26.

[65] Magrino, Tom (2007-10-18). "$279 360 Arcade official" (http://www.gamespot.com/news/6181267.html). Gamespot. . Retrieved 2008-07-26.

[66] Ben Kuchera (2007-10-11). "First actual pictures of Xbox 360 Arcade bundle leaked, with packing list" (http://arstechnica.com/gaming/news/2007/10/first-actual-pictures-of-xbox-360-arcade-bundle-leaked-with-packing-list.ars). Ars Technica. . Retrieved 2008-07-26.

[67] Conrad Quilty-Harper (2007-10-14). "Xbox 360 Arcade unboxed" (http://www.engadget.com/2007/10/14/xbox-360-arcade-unboxed/). Weblogs, Inc. . Retrieved 2008-07-26.

[68] "Microsoft confirms 256 MB internal memory" (http://xbox.joystiq.com/2008/12/02/microsoft-confirms-256mb-internal-storage-in-xbox-360-arcade-uni/). Joystiq. . Retrieved 2009-11-07.

[69] "Xbox 360 Jasper units with 512mb internal memory spotted in the wild" (http://www.joystiq.com/2009/06/18/confirmed-new-xbox-360-arcade-systems-have-512-mb-onboard-memor/). Joystiq. . Retrieved 2009-11-07.

[70] "MS increases Arcade 360 price in the UK news" (http://www.eurogamer.net/articles/ms-increases-arcade-360-price-in-uk). Eurogamer. . Retrieved 2009-11-11.

[71] "MS unveils 250GB 'Super Elite' 360" (http://www.computerandvideogames.com/223527/news/ms-unveils-250gb-super-elite-360/). Computer and Video Games. . Retrieved 2011-07-18.

[72] "That page is missing! - EB Games Australia" (http://ebgames.com.au/xbox/product.cfm?id=15737&refer=consoles). Ebgames.com.au. . Retrieved 2010-11-11.

[73] "Limited Edition Console for *Call of Duty: Modern Warfare 2*" (http://web.archive.org/web/20100107131653/http://www.xbox.com/en-US/press/2009/0915-mw2.htm). Microsoft. 2009-09-15. Archived from the original (http://www.xbox.com/en-US/press/2009/0915-mw2.htm) on 2010-01-07. . Retrieved 2009-10-01.

[74] "Microsoft Announces Canadian Exclusive Xbox 360 Elite Special 250 GB Edition including Halo 3 and Halo 3 ODST" (http://web.archive.org/web/20100109080528/http://www.xbox.com/en-CA/community/news/2009/1201-xbox360elitespecial.htm). Web.archive.org. 2010-01-09. . Retrieved 2011-08-24.

[75] "Final Fantasy XIII Special Edition Xbox 360 250GB Bundle Revealed" (http://www.cinemablend.com/games/Final-Fantasy-XIII-Special-Edition-Xbox-360-250GB-Bundle-Revealed-22836.html?catid=6&id=22836&tid=11743). Cinemablend.com. . Retrieved 2010-11-11.

[76] "Xbox.com - Final Fantasy XIII" (http://marketplace.xbox.com/es-ES/Product/FINAL-FANTASY-XIII/66acd000-77fe-1000-9115-d802535107e4) (in Spanish). *Xbox.com*. Microsoft. . Retrieved 2010-06-16. "¡Hazte ahora con uno de los packs exclusivos de Final Fantasy XIII para Xbox 360! [...] Comprar [...] Xbox 360 Super Elite"

[77] "Xbox Live's Major Nelson » Limited Edition Star Wars Console" (http://majornelson.com/2011/07/21/limited-edition-star-wars-console/). Microsoft. .

[78] "320GB Star Wars Xbox 360 costs £350 News - Xbox 360 - Page 1 | Eurogamer.net" (http://www.eurogamer.net/articles/2011-07-25-320gb-star-wars-xbox-360-costs-gbp350). Eurogamer. .

[79] "Xbox Live's Major Nelson » Xbox 360 Limited Edition Call of Duty: Modern Warfare 3 Console, Wireless Headset with Bluetooth and Wireless Controller" (http://majornelson.com/2011/09/02/xbox-360-limited-edition-call-of-duty-modern-warfare-3-console-wireless-headset-with-bluetooth-and-wireless-controller/). *majornelson.com*. Microsoft. .

[80] (http://www.xbox.com/en-CA/hardware/x/xbox360250GBconsole/)

[81] "New Xbox 360 priced at €249.99 in Europe" (http://www.engadget.com/2010/06/15/new-xbox-360-priced-at-249-99-in-europe/). Engadget. . Retrieved 2010-11-11.

[82] Tom Bramwell (2010-06-14). "New 360 in US this week, July in Europe Xbox 360 News - Page 1" (http://www.eurogamer.net/articles/new-xbox-360-launching-this-week). Eurogamer.net. . Retrieved 2010-11-11.

[83] "Xbox 360 Elite System" (http://web.archive.org/web/20090602022639/http://www.xbox.com/zh-SG/hardware/x/xbox360elitesystem/default.htm). Web.archive.org. 2009-06-02. . Retrieved 2011-08-24.

[84] Nelson, Randy (2010-06-14). "Microsoft no longer making current Xbox 360 console" (http://www.joystiq.com/2010/06/14/microsoft-no-longer-making-current-360-console/). Joystiq. . Retrieved 2010-11-11.

[85] "Microsoft cuts Japan price for Xbox" (http://www.digitaltrends.com/gaming/microsoft-cuts-japan-price-for-xbox/). .

[86] "Xbox elite now GB199.99" (http://web.archive.org/web/20090502052920/http://www.xbox.com/en-GB/hardware/compare101.htm). Archived from the original (http://www.xbox.com/en-GB/hardware/compare101.htm) on 2009-05-02. .

[87] Hill, Jason (2008-11-11). "Xbox price falls again" (http://blogs.smh.com.au/screenplay/archives/020533.html). *Screen Play*. The Sydney Morning Herald. . Retrieved 2008-11-16.

[88] "Xbox 360 Component AV Cable" (http://web.archive.org/web/20080212052249/http://www.xbox.com/en-US/support/systemsetup/xbox360/accessories/avcable.htm). Archived from the original (http://support.xbox.com/support/en/us/xbox360/hardware/accessories/av/compositecable.aspx) on 2008-02-12. . Retrieved 2008-03-31.

[89] "Microsoft confirms: $199 new Xbox 360 coming soon, as well as Kinect bundles" (http://www.engadget.com/2010/06/19/microsoft-confirms-199-for-new-xbox-360-kinect-bundles-coming/). Engadget. . Retrieved 2010-11-11.

[90] Kyle Orland. "Microsoft Renames Xbox Live Silver to 'Xbox Live Free'" (http://www.gamasutra.com/view/news/31092/Microsoft_Renames_Xbox_Live_Silver_to_Xbox_Live_Free.php). .

[91] "Xbox.com | Catalog - Xbox 360™ Component HD AV Cable" (http://replay.waybackmachine.org/20090601193425/http://www.xbox.com/en-US/hardware/x/xbox360componenthdcable/). Microsoft. Archived from the original (http://www.xbox.com/en-US/hardware/x/xbox360componenthdcable/).. .

[92] "Halo 3 Console Announced" (http://web.archive.org/web/20070713224019/http://www.xbox.com/en-US/community/events/e32007/articles/halo3console.htm). Microsoft. 2007-07-10. Archived from the original (http://www.xbox.com/en-US/community/events/e32007/articles/halo3console.htm) on 2007-07-13. . Retrieved 2008-07-26.

[93] "Halo 3 Special Edition Console" (http://web.archive.org/web/20080725042904/http://www.xbox.com/en-US/hardware/x/xbox360halo3console/default.htm). Microsoft. Archived from the original (http://www.xbox.com/en-US/hardware/x/xbox360halo3console/default.htm) on 2008-07-25. . Retrieved 2008-07-26.

[94] "Microsoft Lowers Xbox 360 Price" (http://web.archive.org/web/20070820211136/http://www.xbox.com/en-US/community/news/2007/0806-consoleprice.htm). Microsoft Corporation. Archived from the original (http://www.xbox.com/en-US/community/news/2007/0806-consoleprice.htm) on 2007-08-20. . Retrieved 2008-07-26.

[95] "Win A Limited Edition Xbox 360" (http://web.archive.org/web/20080619023538/http://www.xbox.com/en-US/community/events/simpsonsmovie/winxbox360.htm). Microsoft. Archived from the original (http://www.xbox.com/en-US/community/events/simpsonsmovie/winxbox360.htm) on 2008-06-19. . Retrieved 2008-07-26.

[96] "Limited Edition Simpsons Movie Xbox 360 Sweepstakes" (http://www.thesimpsons.com/sweepstakes/movie/index.htm). Fox Broadcasting Company. . Retrieved 2008-01-15.

[97] "Xbox 360 Has Movie Fans Seeing Stars" (http://www.prnewswire.com/cgi-bin/stories.pl?ACCT=104&STORY=/www/story/05-10-2007/0004585255&EDATE=). PR Newswire Association LLC. . Retrieved 2009-07-22.

[98] Hinkle, David. "Major Nelson unboxes the red Resident Evil 5 Xbox 360 bundle [update (http://www.joystiq.com/2009/02/25/major-nelson-unboxes-the-red-resident-evil-5-xbox-360-bundle?icid=sphere_blogsmith_inpage_joystiq)"]. Joystiq. . Retrieved 2009-02-26.

[99] Fletcher, JC. "Red Resident Evil 5 Xbox 360 bundle: it's real!" (http://www.joystiq.com/2009/02/24/red-resident-evil-5-xbox-360-bundle-its-real/). Joystiq. . Retrieved 2009-02-26.

[100] "Get Ready for the Biggest Year in Xbox 360 Gaming History" (http://www.xbox.com/en-US/Press/archive/2010/0212-x10). Microsoft. 2010-02-11. . Retrieved 2010-03-20.

[101] "Xbox 360 Splinter Cell Conviction Special Edition Announced" (http://web.archive.org/web/20100307160806/http://www.xbox.com/en-US/press/2010/0302-sccspecialedition.htm). Microsoft. 2010-03-02. Archived from the original (http://www.xbox.com/en-US/press/2010/0302-sccspecialedition.htm) on 2010-03-07. . Retrieved 2010-03-20.

[102] "Halo Reach Xbox 360 bundle available September 14 for $399" (http://www.joystiq.com/2010/07/22/halo-reach-xbox-360-bundle-available-september-14-for-399/). Joystiq. 2010-07-22. . Retrieved 2010-07-23.

[103] Ross Miller (2010-07-22). "New Xbox 360 250GB Limited Edition Halo: Reach bundle revealed" (http://www.engadget.com/2010/07/22/new-xbox-360-250gb-limited-edition-halo-reach-bundle-revealed/). Engadget. . Retrieved 2010-07-24.

[104] "Announcing the Xbox 360 Gears of War 3 Limited Edition Console" (http://majornelson.com/2011/06/06/announcing-the-xbox-360-gears-of-war-3-limited-edition-console/). Major Nelson.com. . Retrieved 25 August 2011.

[105] "Star Wars Kinect and Console Bundle Reportedly Delayed" (http://news.teamxbox.com/xbox/24562/Star-Wars-Kinect-and-Console-Bundle-Reportedly-Delayed/). Team Xbox. . Retrieved 25 August 2011.

[106] "Holiday Consoles Include Two Free Games" (http://web.archive.org/web/20071011012329/http://www.xbox.com/en-US/community/news/2007/1008-holidayconsoles.htm). Microsoft. Archived from the original (http://www.xbox.com/en-US/community/news/2007/1008-holidayconsoles.htm) on 2007-10-11. . Retrieved 2008-01-22.

[107] "Wireless Entertainment Pack 2 for Xbox 360 (Includes Forza 2, Viva Pinata and Wireless Controller) - Xbox 360 Console Sold Separately" (http://www.amazon.co.uk/dp/B000YIMXD0). Amazon.co.uk. . Retrieved 2009-09-10.

[108] Reisinger, Don (2009-09-29). "New Xbox 360 holiday bundle announced" (http://news.cnet.com/8301-13506_3-10363475-17.html). *CNET*. . Retrieved 2011-09-06.

[109] Mitchell, Richard (2010-11-02). "Xbox 360 bundle with Alan Wake, Forza 3 confirmed for US" (http://www.joystiq.com/2010/11/02/xbox-360-bundle-with-alan-wake-forza-3-confirmed-for-us/). *Joystiq*. . Retrieved 2011-09-06.

[110] "Xbox 360 Technical Specifications" (http://web.archive.org/web/20080822024003/http://www.xbox.com/en-AU/support/xbox360/manuals/xbox360specs.htm). Xbox.com. Archived from the original (http://www.xbox.com/en-AU/support/xbox360/manuals/xbox360specs.htm) on 2008-08-22. . Retrieved 2009-09-10.

[111] "Xbox 360 Hard Drive accessory bumped to 250GB in Japan" (http://www.engadget.com/2010/02/01/xbox-360-hard-drive-accessory-bumped-to-250gb-in-japan). . Retrieved 2010-02-01.

[112] "250GB Xbox 360 hard drive now available for $129.99" (http://www.engadget.com/2010/03/23/250gb-xbox-360-hard-drive-now-available-for-129-99/). 2010-03-23. . Retrieved 2010-03-23.

[113] "MS unveils standalone 250GB HDD" (http://www.eurogamer.net/articles/ms-unveils-standalone-250gb-hdd). Eurogamer. 2010-03-24. . Retrieved 2010-03-25.

[114] "Xbox 360 Fat & Slim Comparison" (http://wireninja.com/xbox-360-fat-slim-comparison/). Wireninja.com. 2010-06-21. . Retrieved 2010-11-11.

[115] "Project Natal & Its Instruction Manual(!) In The Wild" (http://kotaku.com/5520812/project-natal--its-instruction-manual-in-the-wild). Kotaku. 2010-04-20. . Retrieved 2010-05-27.

[116] Snider, Mike (2010-06-14). "Microsoft Kinect gets into motion as E3 confab kicks off" (http://www.usatoday.com/tech/gaming/2010-06-14-vidgame14_ST_N.htm). *USA Today*. .

[117] Tom Bramwell (October 9, 2008). "New Xbox Experience" (http://www.eurogamer.net/articles/new-xbox-experience-dated). Eurogamer. . Retrieved 2008-10-14.

[118] {{cite web|url=http://www.microsoft.com/Presspass/press/2011/dec11/12-04Xbox360TV.mspx|accessdate=

[119] "Microsoft's new Xbox 360 UI has Bing voice search across Hulu, Netflix, YouTube, and live TV -- Engadget" (http://www.engadget.com/2011/06/06/xbox-com-leaks-e3-secrets-early-voice-search-and-a-new-look-f/). *Engadget*. AOL. .

[120] "Xbox Live Fall 2011 Dashboard update preview: Bing search, voice control, and a Metro overhaul -- Engadget" (http://www.engadget.com/2011/09/09/xbox-live-fall-2011-dashboard-update-preview-bing-search-voice/). *Engadget*. AOL. .

[121] "Xbox 360 System Software - Fall Update Summary" (http://web.archive.org/web/20071202193403/http://www.xbox.com/en-US/community/personality/elle/20071130-systemupdate.htm). Microsoft. 2007-11-30. Archived from the original (http://www.xbox.com/en-US/community/personality/elle/20071130-systemupdate.htm) on 2007-12-02. . Retrieved 2008-07-26.

[122] "Enjoy Windows Media Center on your Xbox 360" (http://web.archive.org/web/20071223104154/http://www.xbox.com/en-US/pcsetup/MCEVista.htm). Microsoft. Archived from the original (http://www.xbox.com/en-US/pcsetup/MCEVista.htm) on 2007-12-23. . Retrieved 2008-01-01.

[123] "Windows Connect Now Technology" (http://www.microsoft.com/windowsxp/using/networking/getstarted/windowsconnectnow.mspx). Microsoft. 2004-12-21. . Retrieved 2008-07-26.

[124] "Why do I hate DLNA protocol so much ? " Ben's Lost World - Diary of a GeeXboX developer" (http://gxben.wordpress.com/2008/08/24/why-do-i-hate-dlna-protocol-so-much/). . Retrieved 2008-11-24.

[125] Joystiq (2007-07-27). *HD-Streaming: Guide For Streaming HD To The Xbox360* (http://www.joystiq.com/2006/11/07/how-to-transcode-and-stream-videos-on-xbox-360/). . Retrieved 2008-07-26.

[126] "Xbox 360 Sky Player" (http://web.archive.org/web/20100107182127/http://www.xbox.com/en-GB/skyplayer/). Web.archive.org. 2010-01-07. . Retrieved 2011-08-24.

[127] "Xbox 360 to show Sky live and on-demand programming | Industries | Technology, Media & Telecommunications | Reuters" (http://www.reuters.com/article/rbssTechMediaTelecomNews/idUSLS97781220090528). Reuters<!. 2009-05-28. . Retrieved 2009-09-10.

[128] "Hardware | The service & hardware | Optik TV | TV | TELUS" (http://www.telus.com/content/tv/optik/services-hardware/hardware.jsp). Telus. .

[129] "Microsoft Integrates IPTV Software Platform With Xbox 360, Creating All-in-One Digital Entertainment Device" (http://www.microsoft.com/presspass/press/2007/jan07/01-08IPTVXboxPR.mspx) (Press release). Microsoft. 7 January 2007. . Retrieved 2010-01-07. "IPTV on Xbox 360 is expected to be available to consumers by holiday season 2007 and will be offered by providers that are deploying TV services over broadband networks based on the IPTV Edition software platform."

[130] "BT and Microsoft Announce Partnership to Deliver Powerful, First-of-its-Kind Entertainment Experience to Consumers Through Xbox 360" (http://www.microsoft.com/presspass/press/2008/jan08/01-06MSBTXboxPR.mspx) (Press release). Microsoft. 8 January 2008. . Retrieved 2010-01-07. "All BT broadband customers will have the unique opportunity to receive the best in high-definition gaming, television and movies through an Xbox 360 console, in a powerful, all-in-one entertainment experience. BT Vision, which is powered by the Microsoft Mediaroom Internet Protocol TV (IPTV) platform, plans to make this first-of-its-kind device and service offering available to customers in the middle of 2008."

[131] "Xbox will host BT's TV service" (http://news.bbc.co.uk/1/hi/technology/7178661.stm). BBC News. 2008-01-09. . Retrieved 2008-01-22.

[132] "Steve Ballmer and Robbie Bach Keynote: 2010 International Consumer Electronics Show" (http://www.microsoft.com/presspass/exec/steve/2010/01-06CES.mspx). Microsoft. 6 January 2010. . Retrieved 2010-01-07. "J.D. Power and Associates have recently rated AT&T Uverse TV the best-rated, highest-rated residential TV satisfaction in the United States Southern and Western Regions for the second year in a row. I'm happy to announce that later this year AT&T will also enable subscribers to U-Verse to enjoy that service on their Xbox 360 consoles."

[133] "XNA Community" (http://www.gamasutra.com/view/feature/3545/sponsored_feature_democratizing_.php). .

[134] "XNA Creators Club" (http://creators.xna.com/). .

[135] McElroy, Griffin (2009-03-30). "GamerBytes study shows disappointing sales for XNA Community Games" (http://www.joystiq.com/2009/03/30/gamerbytes-study-shows-disappointing-sales-for-xna-community-gam/). Joystiq.com. . Retrieved 2009-09-10.

[136] Snider, Mike (2006-07-11). "WWII shows no battle fatigue" (http://www.usatoday.com/tech/gaming/2006-07-10-world-war-II_x.htm). USA Today. . Retrieved 2008-07-26.

[137] "Ubisoft Entertainment Reports Revenue For Fiscal Year 2005–2006" (http://www.ubisoftgroup.com/gallery_files/site/270/416/418/926.pdf) (PDF). Ubisoft. 2006-04-27. . Retrieved 2008-07-26.

[138] Thorsen, Tor (2006-04-10). "Oblivion enjoying epic sales" (http://www.gamespot.com/news/6147359.html). *GameSpot News*. GameSpot. . Retrieved 2008-07-26.

[139] Calonne, Stéphane (2006-07-11). "Dead or Alive 4 passe le million" (http://www.gamekyo.com/news18066_dead-or-alive-4-passe-le-million.html) (in French). JeuxFrance. . Retrieved 2006-11-22.

[140] Kris Graft (2006-11-03). "Strong Quarter for THQ" (http://www.next-gen.biz/news/strong-quarter-thq). Next-Gen.biz. . Retrieved 2008-07-26.

[141] Berardini, César (2006-11-21). "Gears of War Hits 1 Million Units Sold" (http://news.teamxbox.com/xbox/12306/Gears-of-War-Hits-1-Million-Units-Sold/). TeamXbox. . Retrieved 2008-07-26.

[142] "Gears of War Juggernauts sells three million units" (http://web.archive.org/web/20070202034555/http://www.xbox.com/en-US/community/news/2007/0119-gearsofwarjuggernautcontinues.htm). Xbox.com. 2007-01-19. Archived from the original (http://www.xbox.com/en-US/community/news/2007/0119-gearsofwarjuggernautcontinues.htm) on 2007-02-02. . Retrieved 2008-03-19.

[143] Brendan Sinclair (2008-01-03). "MS: 17.7 million 360s sold" (http://www.gamespot.com/news/6184291.html). GameSpot. . Retrieved 2008-01-04.

[144] "Dead or Alive 4 to miss Japanese 360 launch" (http://www.gamespot.com/xbox360/action/deadoralive4/news.html?sid=6140447&mode=all). GameSpot. 2005-11-28. . Retrieved 2008-07-26.

[145] Douglas J. Parry (2006-06-27). "The Future of Japanese Gaming" (http://xbox360.ign.com/articles/715/715308p1.html). IGN. . Retrieved 2008-07-26.

[146] Ashcraft, Brian (2006-10-18). "Blue Dragon Sells Well in Japan" (http://kotaku.com/gaming/top/blue-dragon-pre+orders-selling-out-in-japan-no-really-208601.php). Kotaku. . Retrieved 2008-07-26.

[147] "Microsoft Xbox 360 Japanese Ranking" (http://web.archive.org/web/20080318173956/http://www.japan-gamecharts.com/360.php). Japan Game Charts. Archived from the original (http://www.japan-gamecharts.com/360.php) on 2008-03-18. . Retrieved 2008-03-19.

[148] "2007 Winners" (http://www.gamecriticsawards.com/2007winners.html). GameCriticsAwards.com. .

[149] "Xbox 360 Comes In 2nd For March Sales" (http://www.thebitbag.com/2008/04/17/xbox-360-comes-in-2nd-for-march-sales/). The Bit Bag. 2008-04-17. . Retrieved 2011-01-05.

[150] Barker, Ben (2005-09-19). "Xbox Live The Silver Age" (http://web.archive.org/web/20090204212851/http://www.xbox.com/NR/exeres/F925FE0C-E310-491A-82E0-7A604A353BFB.htm). Microsoft. Archived from the original (http://www.xbox.com/NR/exeres/F925FE0C-E310-491A-82E0-7A604A353BFB.htm) on 2009-02-04. . Retrieved 2006-09-30.

[151] Gettys, Jim W. (2005-09-17). "More than Games" (http://web.archive.org/web/20060908092939/http://www.xbox.com/en-AU/hardware/xbox360/morethangames.htm). Microsoft. Archived from the original (http://www.xbox.com/en-AU/hardware/xbox360/morethangames.htm) on September 8, 2006. . Retrieved 2006-09-30.

[152] "Xbox Live Account Migration to Xbox 360" (http://web.archive.org/web/20081024111920/http://support.xbox.com/support/en/us/xbox360/xboxlive/myaccount/managemyaccount/accountmanagement.aspx). Microsoft. Archived from the original (http://support.xbox.com/support/en/us/xbox360/xboxlive/myaccount/managemyaccount/accountmanagement.aspx) on 2008-10-24. . Retrieved 2006-10-07.

[153] "2010: A Year Filled with New Experiences for Consumers" (http://www.microsoft.com/Presspass/Features/2011/jan11/01-05CESBallmer.mspx). Microsoft. .

[154] Gettys, Jim W. (2005-08-19). "To Hard Drive or Not?" (http://web.archive.org/web/20070427045612/http://www.xbox.com/en-US/hardware/x/xbox360harddrive/harddriveornot.htm). Microsoft. Archived from the original (http://www.xbox.com/en-US/hardware/x/xbox360harddrive/harddriveornot.htm) on April 27, 2007. . Retrieved 2006-09-30.

[155] Barker, Ben (2005-09-20). "Xbox 360 Get the Points" (http://www.xbox.com/en-US/live/microsoftpoints.htm). Microsoft. . Retrieved 2006-09-30.

[156] Berardini, César (2006-10-12). "Browse Live Marketplace from your PC" (http://news.teamxbox.com/xbox/11982/Xbox-Live-Pipeline-Browse-Live-Marketplace-from-your-PC/). TeamXbox. . Retrieved 2006-10-13.

[157] Berardini, César (2006-10-18). "Latest Xbox Live Facts and Stats" (http://news.teamxbox.com/xbox/12028/Latest-Xbox-Live-Facts-and-Stats/). TeamXbox. . Retrieved 2006-10-19.

[158] Block, Ryan (2006-11-07). "HDTV and HD movie downloads for your 360" (http://www.engadget.com/2006/11/06/microsofts-xbox-live-video-hdtv-and-hd-movie-downloads-for-you/). Endaget. . Retrieved 2006-11-11.

[159] "Movies and TV on your Xbox" (http://web.archive.org/web/20061211062527/http://www.xbox.com/en-US/community/news/2006/1106-moviestv.htm) (Press release). Microsoft. 2006-11-06. Archived from the original (http://www.xbox.com/en-US/community/news/2006/1106-moviestv.htm) on 2006-12-11. . Retrieved 2008-07-26.

[160] "Movies and TV on your Xbox 360 in Europe" (http://web.archive.org/web/20080620173136/http://www.xbox.com/en-GB/news-features/news/Video-Marketplace-announcement-041207.htm) (Press release). Microsoft. 2007-07-12. Archived from the original (http://www.xbox.com/en-GB/news-features/news/Video-Marketplace-announcement-041207.htm) on 2008-06-20. . Retrieved 2008-07-26.

[161] "Zune Marketplace" (http://tech.uk.msn.com/news/articles.aspx?cp-documentid=149641612) (Press release). Microsoft. 2009-05-27. . Retrieved 2009-05-27.

[162] Morris, Chris (2004-10-14). "Xbox Live Arcade to launch Nov. 3 – Oct. 14, 2004" (http://money.cnn.com/2004/10/13/commentary/game_over/column_gaming/index.htm). CNN. . Retrieved 2009-03-23.

[163] Hryb, Larry (2006-12-30). "Top Xbox Live Games of 2006" (http://web.archive.org/web/20080530191824/http://majornelson.com/archive/2006/12/30/2006-top-xbox-live-games.aspx). MajorNelson. Archived from the original (http://majornelson.com/archive/2006/

12/30/2006-top-xbox-live-games.aspx) on 2008-05-30. . Retrieved 2008-07-26.

[164] Fried, Ina (2010-03-24). "Xbox Game Room launches, but with issues" (http://news.cnet.com/8301-13860_3-20001073-56.html). CNET. . Retrieved 2010-03-25.

[165] Neil Davidson (2008-08-26). "Nintendo Wii surpasses mark of one million consoles sold in Canada" (http://www.fftimes.com/node/214646). *The Canadian Press*. Yahoo!. . Retrieved 2008-09-08.

[166] Leigh Alexander (March 31, 2011). "GameStop Details Europe, U.S. Installed Base For Consoles" (http://www.gamasutra.com/view/news/33842/GameStop_Details_Europe_US_Installed_Base_For_Consoles.php). Gamasutra. . Retrieved March 31, 2011.

[167] "360 sells more than Wii and DS in UK's 2009" (http://www.gamezine.co.uk/news-story/2009/7/2/360-sells-more-than-wii-and-ds-in-uk-s-2009-u). Gamezine. 2009-01-13. . Retrieved 2009-07-02.

[168] "Xbox 360 sells 1.5 million in Japan" (http://www.eurogamer.net/articles/2011-06-17-xbox-360-sells-1-5-million-in-japan). Eurogamer. .

[169] "This Is A Nice, Nice Xbox 360" (http://kotaku.com/5520115/maybe-the-best-xbox-360-weve-ever-seen?utm_source=feedburner&utm_medium=feed&utm_campaign=Feed:+kotaku/full+(Kotaku)). Kotaku. 2010-04-19. . Retrieved 2010-04-19.

[170] Morris, Chris (2006-07-05). "Nintendo's Wii may get early launch" (http://money.cnn.com/2006/07/05/commentary/column_gaming/index.htm?section=money_latest). CNN Money. . Retrieved 2009-03-23.

[171] "Microsoft cooking 360 figures?" (http://www.gamespot.com/news/6235933.html). GameSpot. 2007-04-16. . Retrieved 2008-07-26.

[172] "Xbox 360 sells out within hours" (http://news.bbc.co.uk/2/hi/technology/4491804.stm). BBC. 2005-12-02. . Retrieved 2008-07-26.

[173] Svensson, Christian (2005-12-02). "40,000 Xbox 360s Sold on eBay" (http://www.next-gen.biz/news/40000-xbox-360s-sold-ebay). Future Network USA. . Retrieved 2008-07-26.

[174] "Microsoft FY2006 Quarter 2 report" (http://web.archive.org/web/20060225181443/http://www.microsoft.com/msft/download/FY06/MSFTQ2_06.ppt#14) (Powerpoint). Microsoft. 2006-01-26. Archived from the original (http://www.microsoft.com/msft/download/FY06/MSFTQ2_06.ppt#14) on 2006-02-25. . Retrieved 2008-07-26.

[175] "Bill Gates Declares 10 Million-Unit Head Start for Xbox 360 and Outlines uture of Interactive Entertainment" (http://www.microsoft.com/presspass/press/2006/may06/05-09E32006BriefingPR.mspx). Microsoft Corporation. 2006-05-09. . Retrieved 2008-07-26.

[176] Berardini, César (2006-10-27). "Microsoft Confident to Meet Xbox 360 Shipping Target" (http://news.teamxbox.com/xbox/12113/Microsoft-Confident-to-Meet-Xbox-360-Shipping-Target/). TeamXbox. . Retrieved 2008-07-26.

[177] Mariko Sanchanta (September 12, 2007). "Nintendo's Wii takes console lead" (http://www.ft.com/cms/s/0/51df0c84-6154-11dc-bf25-0000779fd2ac.html?nclick_check=1). Financial Times. . Retrieved 2007-09-12.

[178] Hillis, Scott (2007-10-04). "Microsoft says "Halo" 1st-week sales were $300 mln" (http://uk.reuters.com/article/2007/10/05/tech-microsoft-halo3-dc-idUKN0438777720071005). Reuters. . Retrieved 2008-10-05.

[179] Geddes, Ryan (2007-10-18). "Halo 3 sells over 3 million units in 12 days" (http://xbox360.ign.com/articles/828/828504p1.html). IGN. . Retrieved 2008-07-26.

[180] Mattice, Bradley (2007-10-19). "Bragging Rights: Microsoft's Xbox finally outsells the Wii" (http://techland.blogs.fortune.cnn.com/2007/10/19/game-on-xbox-outsells-the-wii/). Fortune Magazine. . Retrieved 2008-07-26.

[181] Murphy, Victoria (2005-09-13). "Microsoft's Midlife Crisis" (http://www.forbes.com/2005/09/12/microsoft-management-software_cz_vm_0913microsoft.html). Forbes.com. . Retrieved 2008-07-26.

[182] "Bach: Xbox Business Profitable Next Year" (http://web.archive.org/web/20080629192037/http://www.gamedaily.com/articles/features/bach-xbox-business-profitable-next-year/70371/?biz=1). GameDaily BIZ. 2007-05-04. Archived from the original (http://www.gamedaily.com/articles/features/bach-xbox-business-profitable-next-year/70371/?biz=1) on 2008-06-29. . Retrieved 2008-07-26.

[183] Tony Smith (2005-11-24). "Xbox 360 costs third more to make than it sells for" (http://www.theregister.co.uk/2005/11/24/xbox360_component_breakdown/). The Register. . Retrieved 2008-07-26.

[184] Blake Snow (2005-12-28). "Xbox 360 costs $715 to make" (http://www.joystiq.com/2005/12/28/xbox-360-costs-715-to-make/). Weblogs, Inc. . Retrieved 2008-07-26.

[185] Topolsky, Joshua (2007-07-05). "Xbox 360 warranty extended to three years" (http://www.engadget.com/2007/07/05/xbox-360-warranty-extended-to-three-years). Engadget. . Retrieved 2008-07-26.

[186] Electronic Arts (2008-01-31). "Supplemental Segment Information" (http://media.corporate-ir.net/media_files/IROL/88/88189/Q3FY08SupSeg.pdf#page=4) (PDF). *Thomson Financial*. p. 4. . Retrieved 2008-02-09.

[187] David Jenkins (2008-02-01). "EA Reveals European Hardware Estimates" (http://www.gamasutra.com/php-bin/news_index.php?story=17206). *Gamasutra*. CMP Media. . Retrieved 2008-02-09.

[188] Michael McWhertor (2008-01-18). "Who's Winning The Console War In The US?" (http://kotaku.com/346301/whos-winning-the-console-war-in-the-us). Kotaku. . Retrieved 2008-05-24.

[189] James Brightman (2008-01-17). "NPD: U.S. Video Game Industry Totals $17.94 billion, Halo 3 Tops All" (http://web.archive.org/web/20080119014303/http://www.gamedaily.com/articles/news/npd-us-video-game-industry-totals-1794-billion-halo-3-tops-all/19119/?biz=1). GameDaily. Archived from the original (http://www.gamedaily.com/articles/news/npd-us-video-game-industry-totals-1794-billion-halo-3-tops-all/19119/?biz=1) on 2008-01-19. . Retrieved 2008-05-24.

[190] Brandon Boyer (2008-01-18). "NPD: 2007 U.S. Game Industry Growth Up 43% To $17.9 billion" (http://www.gamasutra.com/php-bin/news_index.php?story=17006). Gamasutra. . Retrieved 2008-05-24.

[191] Kiyoshi Takenaka and Hugh Lawson (2008-01-07). "Nintendo Wii outsells PS3 3-to-1 in Japan" (http://www.reuters.com/article/2008/01/07/us-sony-nintendo-idUSTFA00293420080107?sp=true). *Reuters* (Reuters). . Retrieved 2008-05-25.

[192] David Jenkins (2008-01-11). "Wii Sports Named Best Selling Game Of 2007 In Japan" (http://www.gamasutra.com/php-bin/news_index.php?story=16914). Gamasutra. . Retrieved 2008-05-25.

[193] "Japanese 2008 Market Report" (http://www.mcvuk.com/news/read/japanese-2008-market-report). Market for Home Computing and Video Games. 2009-01-09. . Retrieved 2009-01-15.

[194] "200858261000" (http://www.famitsu.com/game/news/1221045_1124.html) (in Japanese). *Famitsu*. Enterbrain. 2009-01-05. . Retrieved 2009-01-15.

[195] Scott Hillis, Andre Grenon (2008-02-13). "Microsoft seeing Xbox shortages in U.S." (http://www.reuters.com/article/2008/02/13/us-microsoft-xbox-idUSN1337448620080213). *Reuters*. . Retrieved 2008-02-15.

[196] P-I STAFF (2008-02-14). "Supply shortages hurt Xbox sales last month" (http://www.seattlepi.com/default/article/Supply-shortages-hurt-Xbox-sales-last-month-1264396.php). *Seattle Post-Intelligencer*. Hearst Corporation. . Retrieved 2008-02-15.

[197] David Jenkins, Staff (2008-02-14). "Stock Low For Xbox 360 Hardware As NPD Approaches" (http://www.gamasutra.com/php-bin/news_index.php?story=17381). *Gamasutra*. CMP Media. . Retrieved 2008-02-15.

[198] "Xbox 360 First to Reach Ten Million Console Sales in U.S." (http://web.archive.org/web/20080517113549/http://www.xbox.com/en-US/community/news/2008/0514-10million.htm). Microsoft. 2008-05-14. Archived from the original (http://www.xbox.com/en-US/community/news/2008/0514-10million.htm) on 2008-05-17. . Retrieved 2008-05-14.

[199] Tom Magrino (2008-07-17). "NPD: PS3 sales spike on MGS4" (http://www.gamespot.com/news/6194691.html). GameSpot. . Retrieved 2008-11-22.

[200] Joe Keiser (2008-07-17). "NPD: Wii Overtakes 360 in US" (http://www.next-gen.biz/news/npd-wii-overtakes-360-us). *Edge*. . Retrieved 2008-11-22.

[201] Ryan Kim (2008-07-17). "E3: Nintendo Wii pulls ahead of Xbox 360 in console sales" (http://www.sfgate.com/cgi-bin/blogs/techchron/detail?entry_id=28286). San Francisco Chronicle. . Retrieved 2008-11-22.

[202] Dan Howdle (2009-11-13). "Xbox 360 European Sales Top 10m" (http://www.nowgamer.com/news/914963/xbox_360_european_sales_top_10m.html). Microsoft. . Retrieved 2010-03-02.

[203] Brian Ashcraft (2009-01-05). "Last Year, Japanese Game Market Experienced Shrinkage" (http://kotaku.com/5123288/last-year-japanese-game-market-experienced-shrinkage). Kotaku. . Retrieved 2009-01-15.

[204] Microsoft says no new Xbox 360s in 2009 (http://www.engadget.com/2008/05/13/microsoft-says-no-new-xbox-360s-in-2009/), retrieved on 2009-10-08

[205] Xbox 360 Reaches 30 Million Units Globally, 20 Million Active XBL Users (http://www.gamasutra.com/php-bin/news_index.php?story=23796) - gamasutra.com 05-29-2009

[206] Nathan Brown. "Xbox 360: dead in Japan? - Edge Magazine" (http://www.next-gen.biz/news/xbox-360-dead-japan). Next-gen.biz. . Retrieved 2011-08-24.

[207] "Xbox 360 is number 6" (http://www.ign.com/top-25-consoles/6.html). IGN. . Retrieved 2010-11-11.

[208] "Xbox.com | Xbox News - Microsoft Unveils Xbox 360 Elite" (http://web.archive.org/web/20070405185941/http://www.xbox.com/en-US/community/news/2007/0327-microsoftunveilsxbox360elite.htm). Web.archive.org. Archived from the original (http://www.xbox.com/en-US/community/news/2007/0327-microsoftunveilsxbox360elite.htm) on 2007-04-05. . Retrieved 2010-11-11.

[209] "Xbox.com | Xbox News - New Xbox 360 Family Console Released" (http://web.archive.org/web/20071025003931/http://www.xbox.com/en-US/community/news/2007/1022-familyconsole.htm). Web.archive.org. 2007-10-25. Archived from the original (http://www.xbox.com/en-US/community/news/2007/1022-familyconsole.htm) on 2007-10-25. . Retrieved 2010-11-11.

[210] "E3 2008: 60GB Xbox 360 official, 20GB discounted and discontinued" (http://uk.gamespot.com/news/6193719.html). .

[211] "Xbox News - Xbox 360 Invites Everyone to Play" (http://web.archive.org/web/20080905184538/http://www.xbox.com/en-US/community/news/2008/xbox360pricedrop.htm). Web.archive.org. 2008-09-05. . Retrieved 2011-08-24.

[212] "Microsoft admit Xbox problems" (http://news.bbc.co.uk/cbbcnews/hi/newsid_6270000/newsid_6277700/6277716.stm). BBC. 2007-07-06. . Retrieved 2009-02-13.

[213] "Rings of Red" (http://www.gamesindustry.biz/articles/rings-of-red). GamesIndustry.biz. . Retrieved 2008-07-26.

[214] Philip Kollar (2007-06-26). "A Tale of 11 Broken Xbox 360s" (http://www.1up.com/news/tale-11-broken-360s). Ziff Davis Publishing Holdings Inc. . Retrieved 2008-07-26.

[215] Microsoft (2007-06-05). "Xbox 360 Warranty Coverage Expanded" (http://web.archive.org/web/20070605083902/http://www.xbox.com/en-US/support/systemuse/xbox360/resources/warrantyupdate.htm). Archived from the original (http://support.xbox.com/support/en/us/xbox360/hardware/warranty/warrantyupdate/WarrantyUpdate.aspx) on 2007-06-05. . Retrieved 2008-07-26.

[216] "Xbox 360 Warranty extended for E74" (http://majornelson.com/2009/04/14/xbox-360-warranty-extended-for-e74/). majornelson.com. . Retrieved 2009-04-17.

[217] "Inside the Xbox 360 Elite" (http://www.llamma.com/xbox360/news/inside_the_xbox_360_elite.htm). The Llamma's Adventures. 2007-04-19. . Retrieved 2008-07-26.

[218] XanTium (2007-06-13). "New Pictures of 2nd GPU HeatSink on Xbox 360" (http://www.xbox-scene.com/xbox1data/sep/EElEZukFEleqkRrcnN.php). Xbox-Scene. . Retrieved 2008-07-26.

[219] "Product Warranty and Software License: New Xbox 360 Console" (http://support.xbox.com/en-us/pages/xbox-360/troubleshoot/warranty/xbox-360-new-warranty.aspx). Microsoft. . Retrieved 2011-01-10. ""Warranty Period" for Xbox 360 S, means 1 year from the date You purchased the Xbox 360 S from an authorized retailer"

[220] Microsoft (2010-07-23). "Flashing red light on the Xbox 360 S console power button" (http://support.xbox.com/en-us/Pages/xbox-360/troubleshoot/kb/flashing-red-light-on-the-xbox-360-s-console-power-button.aspx?kbid=2122791). Microsoft. . Retrieved 2011-09-02.

[221] Plunkett, Luke. "The New Xbox 360 Won't "Red Ring"" (http://kotaku.com/5566600/the-new-xbox-360-wont-red-ring). Kotaku. . Retrieved 1 November 2010.

External links

- Official Xbox 360 website (http://www.xbox.com/)
- Microsoft Support for Xbox 360 (http://support.xbox.com/en-us/pages/xbox-360/troubleshoot/default.aspx)
- Xbox development team blog (http://blogs.msdn.com/xboxteam/default.aspx)
- Xbox 360 (http://www.dmoz.org/Games/Video_Games/Console_Platforms/Microsoft/Xbox_360/) at the Open Directory Project

Artificial_intelligence

TOPIO, a humanoid robot, played table tennis at Tokyo International Robot Exhibition (IREX) 2009.[1]

Artificial intelligence (AI) is the intelligence of machines and the branch of computer science that aims to create it. AI textbooks define the field as "the study and design of intelligent agents"[2] where an intelligent agent is a system that makes people smarter perceives its environment and takes actions that maximize its chances of success.[3] John McCarthy, who coined the term in 1956,[4] defines it as "the science and engineering of making intelligent machines."[5]

The field was founded on the claim that a central property of humans, intelligence—the sapience of *Homo sapiens*—can be so precisely described that it can be simulated by a machine.[6] This raises philosophical issues about the nature of the mind and the ethics of creating artificial beings, issues which have been addressed by myth, fiction and philosophy since antiquity.[7] Artificial intelligence has been the subject of optimism,[8] but has also suffered setbacks[9] and, today, has become an essential part of the technology industry, providing the heavy lifting for many of the most difficult problems in computer science.[10]

AI research is highly technical and specialized, *deeply* divided into subfields that often fail in the task of communicating with each other.[11] Subfields have grown up around particular institutions, the work of individual researchers, the solution of specific problems, longstanding differences of opinion about how AI should be done and the application of widely differing tools. The central problems of AI include such traits as reasoning, knowledge, planning, learning, communication, perception and the ability to move and manipulate objects.[12] General intelligence (or "strong AI") is still among the field's long term goals.[13]

History

Thinking machines and artificial beings appear in Greek myths, such as Talos of Crete, the bronze robot of Hephaestus, and Pygmalion's Galatea.[14] Human likenesses believed to have intelligence were built in every major civilization: animated cult images were worshipped in Egypt and Greece[15] and humanoid automatons were built by Yan Shi, Hero of Alexandria and Al-Jazari.[16] It was also widely believed that artificial beings had been created by Jābir ibn Hayyān, Judah Loew and Paracelsus.[17] By the 19th and 20th centuries, artificial beings had become a common feature in fiction, as in Mary Shelley's *Frankenstein* or Karel Čapek's *R.U.R. (Rossum's Universal Robots)*.[18] Pamela McCorduck argues that all of these are examples of an ancient urge, as she describes it, "to forge the gods".[7] Stories of these creatures and their fates discuss many of the same hopes, fears and ethical concerns that are presented by artificial intelligence.

Mechanical or "formal" reasoning has been developed by philosophers and mathematicians since antiquity. The study of logic led directly to the invention of the programmable digital electronic computer, based on the work of mathematician Alan Turing and others. Turing's theory of computation suggested that a machine, by shuffling symbols as simple as "0" and "1", could simulate any conceivable act of mathematical deduction.[19] [20] This, along with concurrent discoveries in neurology, information theory and cybernetics, inspired a small group of researchers to begin to seriously consider the possibility of building an electronic brain.[21]

The field of AI research was founded at a conference on the campus of Dartmouth College in the summer of 1956.[22] The attendees, including John McCarthy, Marvin Minsky, Allen Newell and Herbert Simon, became the leaders of AI research for many decades.[23] They and their students wrote programs that were, to most people, simply astonishing:[24] Computers were solving word problems in algebra, proving logical theorems and speaking English.[25] By the middle of the 1960s, research in the U.S. was heavily funded by the Department of Defense[26] and laboratories had been established around the world.[27] AI's founders were profoundly optimistic about the future of the new field: Herbert Simon predicted that "machines will be capable, within twenty years, of doing any work a man can do" and Marvin Minsky agreed, writing that "within a generation ... the problem of creating 'artificial intelligence' will substantially be solved".[28]

They had failed to recognize the difficulty of some of the problems they faced.[29] In 1974, in response to the criticism of Sir James Lighthill and ongoing pressure from the US Congress to fund more productive projects, both the U.S. and British governments cut off all undirected exploratory research in AI. The next few years, when funding for projects was hard to find, would later be called the "AI winter".[30]

In the early 1980s, AI research was revived by the commercial success of expert systems,[31] a form of AI program that simulated the knowledge and analytical skills of one or more human experts. By 1985 the market for AI had reached over a billion dollars. At the same time, Japan's fifth generation computer project inspired the U.S and British governments to restore funding for academic research in the field.[32] However, beginning with the collapse of the Lisp Machine market in 1987, AI once again fell into disrepute, and a second, longer lasting AI winter began.[33]

In the 1990s and early 21st century, AI achieved its greatest successes, albeit somewhat behind the scenes. Artificial intelligence is used for logistics, data mining, medical diagnosis and many other areas throughout the technology industry.[10] The success was due to several factors: the increasing computational power of computers (see Moore's law), a greater emphasis on solving specific subproblems, the creation of new ties between AI and other fields working on similar problems, and a new commitment by researchers to solid mathematical methods and rigorous scientific standards.[34]

On 11 May 1997, Deep Blue became the first computer chess-playing system to beat a reigning world chess champion, Garry Kasparov.[35] In 2005, a Stanford robot won the DARPA Grand Challenge by driving autonomously for 131 miles along an unrehearsed desert trail.[36] Two years later, a team from CMU won the DARPA Urban Challenge by autonomously navigating 55 miles in an Urban environment while adhering to traffic hazards and all traffic laws.[37] In February 2011, in a Jeopardy! quiz show exhibition match, IBM's question answering system, Watson, defeated the two greatest Jeopardy! champions, Brad Rutter and Ken Jennings, by a significant margin.[38]

The leading-edge definition of artificial intelligence research is changing over time. One pragmatic definition is: "AI research is that which computing scientists do not know how to do cost-effectively today." For example, in 1956 optical character recognition (OCR) was considered AI, but today, sophisticated OCR software with a context-sensitive spell checker and grammar checker software comes for free with most image scanners. No one would any longer consider already-solved computing science problems like OCR "artificial intelligence" today.

Low-cost entertaining chess-playing software is commonly available for tablet computers. DARPA no longer provides significant funding for chess-playing computing system development. The Kinect which provides a 3D body–motion interface for the Xbox 360 uses algorithms that emerged from lengthy AI research,[39] but few consumers realize the technology source.

AI applications are no longer the exclusive domain of Department of defense R&D, but are now common place consumer items and inexpensive intelligent toys.

In common usage, the term "AI" no longer seems to apply to off-the-shelf solved computing-science problems, which may have originally emerged out of years of AI research.

Problems

"Can a machine act intelligently?" is still an open problem. Taking "A machine can act intelligently" as a working hypothesis, many researchers have attempted to build such a machine.

The general problem of simulating (or creating) intelligence has been broken down into a number of specific sub-problems. These consist of particular traits or capabilities that researchers would like an intelligent system to display. The traits described below have received the most attention.[12]

Deduction, reasoning, problem solving

Early AI researchers developed algorithms that imitated the step-by-step reasoning that humans use when they solve puzzles or make logical deductions.[40] By the late 1980s and '90s, AI research had also developed highly successful methods for dealing with uncertain or incomplete information, employing concepts from probability and economics.[41]

For difficult problems, most of these algorithms can require enormous computational resources — most experience a "combinatorial explosion": the amount of memory or computer time required becomes astronomical when the problem goes beyond a certain size. The search for more efficient problem-solving algorithms is a high priority for AI research.[42]

Human beings solve most of their problems using fast, intuitive judgments rather than the conscious, step-by-step deduction that early AI research was able to model.[43] AI has made some progress at imitating this kind of "sub-symbolic" problem solving: embodied agent approaches emphasize the importance of sensorimotor skills to higher reasoning; neural net research attempts to simulate the structures inside human and animal brains that give rise to this skill.

Knowledge representation

Knowledge representation[44] and knowledge engineering[45] are central to AI research. Many of the problems machines are expected to solve will require extensive knowledge about the world. Among the things that AI needs to represent are: objects, properties, categories and relations between objects;[46] situations, events, states and time;[47] causes and effects;[48] knowledge about knowledge (what we know about what other people know);[49] and many other, less well researched domains. A representation of "what exists" is an ontology (borrowing a word from traditional philosophy), of which the most general are called upper ontologies.[50]

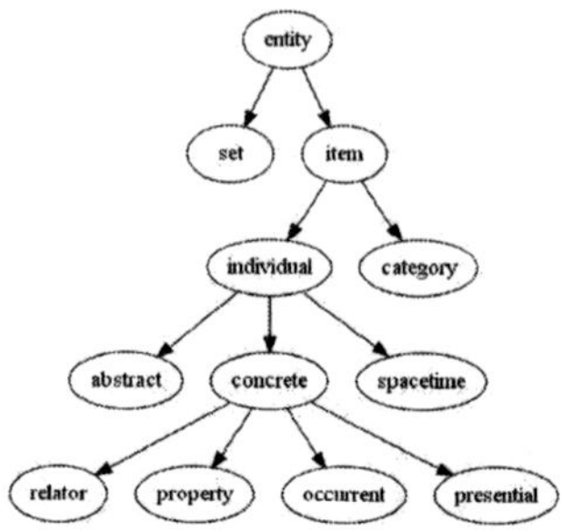

An ontology represents knowledge as a set of concepts within a domain and the relationships between those concepts.

Among the most difficult problems in knowledge representation are:

Default reasoning and the qualification problem

Many of the things people know take the form of "working assumptions." For example, if a bird comes up in conversation, people typically picture an animal that is fist sized, sings, and flies. None of these things are true about all birds. John McCarthy identified this problem in 1969[51] as the qualification problem: for any commonsense rule that AI researchers care to represent, there tend to be a huge number of exceptions. Almost nothing is simply true or false in the way that abstract logic requires. AI research has explored a number of solutions to this problem.[52]

The breadth of commonsense knowledge

The number of atomic facts that the average person knows is astronomical. Research projects that attempt to build a complete knowledge base of commonsense knowledge (e.g., Cyc) require enormous amounts of laborious ontological engineering — they must be built, by hand, one complicated concept at a time.[53] A major goal is to have the computer understand enough concepts to be able to learn by reading from sources like the internet, and thus be able to add to its own ontology.

The subsymbolic form of some commonsense knowledge

Much of what people know is not represented as "facts" or "statements" that they could express verbally. For example, a chess master will avoid a particular chess position because it "feels too exposed"[54] or an art critic can take one look at a statue and instantly realize that it is a fake.[55] These are intuitions or tendencies that are represented in the brain non-consciously and sub-symbolically.[56] Knowledge like this informs, supports and provides a context for symbolic, conscious knowledge. As with the related problem of sub-symbolic reasoning, it is hoped that situated AI or computational intelligence will provide ways to represent this kind of knowledge.[56]

Planning

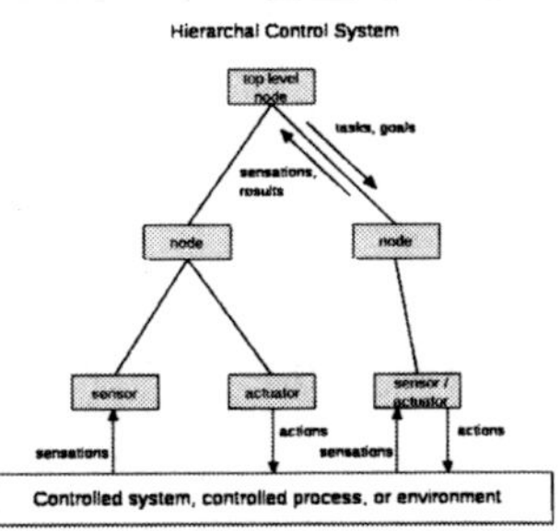

A hierarchical control system is a form of control system in which a set of devices and governing software is arranged in a hierarchy.

Intelligent agents must be able to set goals and achieve them.[57] They need a way to visualize the future (they must have a representation of the state of the world and be able to make predictions about how their actions will change it) and be able to make choices that maximize the utility (or "value") of the available choices.[58]

In classical planning problems, the agent can assume that it is the only thing acting on the world and it can be certain what the consequences of its actions may be.[59] However, if this is not true, it must periodically check if the world matches its predictions and it must change its plan as this becomes necessary, requiring the agent to reason under uncertainty.[60]

Multi-agent planning uses the cooperation and competition of many agents to achieve a given goal. Emergent behavior such as this is used by evolutionary algorithms and swarm intelligence.[61]

Learning

Machine learning[62] has been central to AI research from the beginning.[63] In 1956, at the original Dartmouth AI summer conference, Ray Solomonoff wrote a report on unsupervised probabilistic machine learning: "An Inductive Inference Machine".[64] Unsupervised learning is the ability to find patterns in a stream of input. Supervised learning includes both classification and numerical regression. Classification is used to determine what category something belongs in, after seeing a number of examples of things from several categories. Regression is the attempt to produce a function that describes the relationship between inputs and outputs and predicts how the outputs should change as the inputs change. In reinforcement learning[65] the agent is rewarded for good responses and punished for bad ones. These can be analyzed in terms of decision theory, using concepts like utility. The mathematical analysis of machine learning algorithms and their performance is a branch of theoretical computer science known as computational learning theory.[66]

Natural language processing

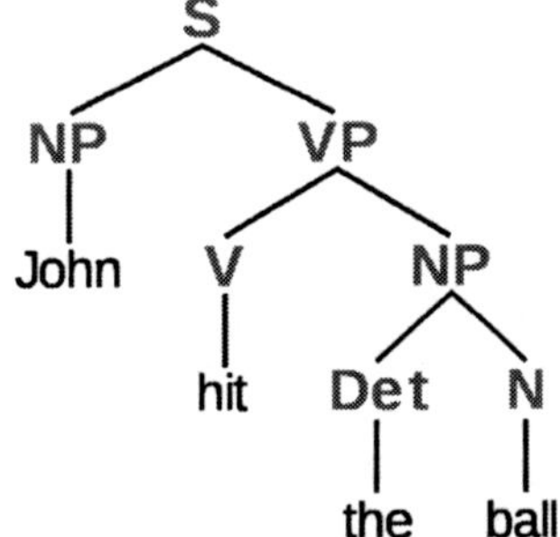

A parse tree represents the syntactic structure of a sentence according to some formal grammar.

Natural language processing[67] gives machines the ability to read and understand the languages that humans speak. Many researchers hope that a sufficiently powerful natural language processing system would be able to acquire knowledge on its own, by reading the existing text available over the internet. Some straightforward applications of natural language processing include information retrieval (or text mining) and machine translation.[68]

Motion and manipulation

The field of robotics[69] is closely related to AI. Intelligence is required for robots to be able to handle such tasks as object manipulation[70] and navigation, with sub-problems of localization (knowing where you are), mapping (learning what is around you) and motion planning (figuring out how to get there).[71]

Perception

Machine perception[72] is the ability to use input from sensors (such as cameras, microphones, sonar and others more exotic) to deduce aspects of the world. Computer vision[73] is the ability to analyze visual input. A few selected subproblems are speech recognition,[74] facial recognition and object recognition.[75]

Social intelligence

Emotion and social skills[76] play two roles for an intelligent agent. First, it must be able to predict the actions of others, by understanding their motives and emotional states. (This involves elements of game theory, decision theory, as well as the ability to model human emotions and the perceptual skills to detect emotions.) Also, for good human-computer interaction, an intelligent machine also needs to *display* emotions. At the very least it must appear polite and sensitive to the humans it interacts with. At best, it should have normal emotions itself.

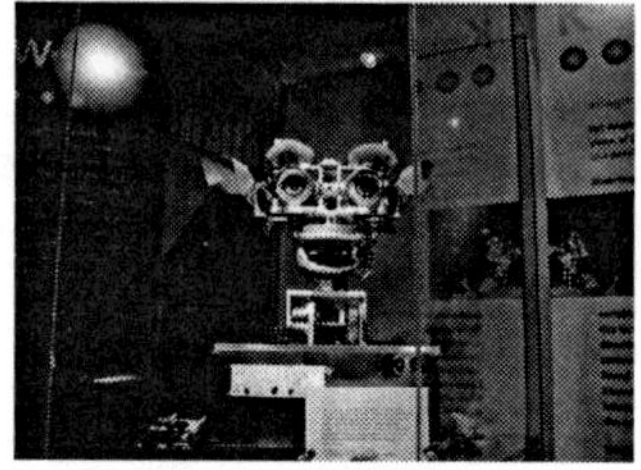

Kismet, a robot with rudimentary social skills

Creativity

A sub-field of AI addresses creativity both theoretically (from a philosophical and psychological perspective) and practically (via specific implementations of systems that generate outputs that can be considered creative, or systems that identify and assess creativity). A related area of computational research is Artificial intuition and Artificial imagination.

General intelligence

Most researchers hope that their work will eventually be incorporated into a machine with *general* intelligence (known as strong AI), combining all the skills above and exceeding human abilities at most or all of them.[13] A few believe that anthropomorphic features like artificial consciousness or an artificial brain may be required for such a project.[77] [78]

Many of the problems above are considered AI-complete: to solve one problem, you must solve them all. For example, even a straightforward, specific task like machine translation requires that the machine follow the author's argument (reason), know what is being talked about (knowledge), and faithfully reproduce the author's intention (social intelligence). Machine translation, therefore, is believed to be AI-complete: it may require strong AI to be done as well as humans can do it.[79]

Approaches

There is no established unifying theory or paradigm that guides AI research. Researchers disagree about many issues.[80] A few of the most long standing questions that have remained unanswered are these: should artificial intelligence simulate natural intelligence by studying psychology or neurology? Or is human biology as irrelevant to AI research as bird biology is to aeronautical engineering?[81] Can intelligent behavior be described using simple, elegant principles (such as logic or optimization)? Or does it necessarily require solving a large number of completely unrelated problems?[82] Can intelligence be reproduced using high-level symbols, similar to words and ideas? Or does it require "sub-symbolic" processing?[83] John Haugeland, who coined the term GOFAI (Good Old-Fashioned Artificial Intelligence), also proposed that AI should more properly be referred to as synthetic intelligence,[84] a term which has since been adopted by some non-GOFAI researchers.[85] [86]

Cybernetics and brain simulation

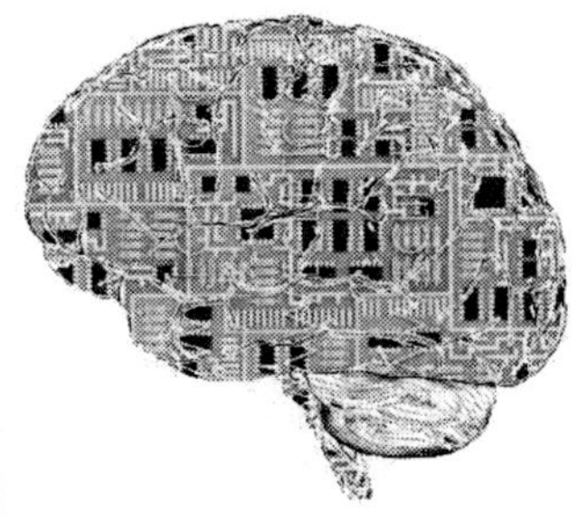
There is currently no consensus on how closely the brain should be simulated.

In the 1940s and 1950s, a number of researchers explored the connection between neurology, information theory, and cybernetics. Some of them built machines that used electronic networks to exhibit rudimentary intelligence, such as W. Grey Walter's turtles and the Johns Hopkins Beast. Many of these researchers gathered for meetings of the Teleological Society at Princeton University and the Ratio Club in England.[21] By 1960, this approach was largely abandoned, although elements of it would be revived in the 1980s.

Symbolic

When access to digital computers became possible in the middle 1950s, AI research began to explore the possibility that human intelligence could be reduced to symbol manipulation. The research was centered in three institutions: CMU, Stanford and MIT, and each one developed its own style of research. John Haugeland named these approaches to AI "good old fashioned AI" or "GOFAI".[87]

Cognitive simulation

Economist Herbert Simon and Allen Newell studied human problem-solving skills and attempted to formalize them, and their work laid the foundations of the field of artificial intelligence, as well as cognitive science, operations research and management science. Their research team used the results of psychological experiments to develop programs that simulated the techniques that people used to solve problems. This tradition, centered at Carnegie Mellon University would eventually culminate in the development of the Soar architecture in the middle 80s.[88] [89]

Logic-based

Unlike Newell and Simon, John McCarthy felt that machines did not need to simulate human thought, but should instead try to find the essence of abstract reasoning and problem solving, regardless of whether people used the same algorithms.[81] His laboratory at Stanford (SAIL) focused on using formal logic to solve a wide variety of problems, including knowledge representation, planning and learning.[90] Logic was also focus of the work at the University of Edinburgh and elsewhere in Europe which led to the development of the programming language Prolog and the science of logic programming.[91]

"Anti-logic" or "scruffy"

Researchers at MIT (such as Marvin Minsky and Seymour Papert)[92] found that solving difficult problems in vision and natural language processing required ad-hoc solutions – they argued that there was no simple and general principle (like logic) that would capture all the aspects of intelligent behavior. Roger Schank described their "anti-logic" approaches as "scruffy" (as opposed to the "neat" paradigms at CMU and Stanford).[82] Commonsense knowledge bases (such as Doug Lenat's Cyc) are an example of "scruffy" AI, since they must be built by hand, one complicated concept at a time.[93]

Knowledge-based

When computers with large memories became available around 1970, researchers from all three traditions began to build knowledge into AI applications.[94] This "knowledge revolution" led to the development and deployment of expert systems (introduced by Edward Feigenbaum), the first truly successful form of AI software.[31] The knowledge revolution was also driven by the realization that enormous amounts of knowledge would be required by many simple AI applications.

Sub-symbolic

During the 1960s, symbolic approaches had achieved great success at simulating high-level thinking in small demonstration programs. Approaches based on cybernetics or neural networks were abandoned or pushed into the background.[95] By the 1980s, however, progress in symbolic AI seemed to stall and many believed that symbolic systems would never be able to imitate all the processes of human cognition, especially perception, robotics, learning and pattern recognition. A number of researchers began to look into "sub-symbolic" approaches to specific AI problems.[83]

Bottom-up, embodied, situated, behavior-based or nouvelle AI

Researchers from the related field of robotics, such as Rodney Brooks, rejected symbolic AI and focused on the basic engineering problems that would allow robots to move and survive.[96] Their work revived the non-symbolic viewpoint of the early cybernetics researchers of the 50s and reintroduced the use of control theory in AI. This coincided with the development of the embodied mind thesis in the related field of cognitive science: the idea that aspects of the body (such as movement, perception and visualization) are required for higher intelligence.

Computational Intelligence

Interest in neural networks and "connectionism" was revived by David Rumelhart and others in the middle 1980s.[97] These and other sub-symbolic approaches, such as fuzzy systems and evolutionary computation, are now studied collectively by the emerging discipline of computational intelligence.[98]

Statistical

In the 1990s, AI researchers developed sophisticated mathematical tools to solve specific subproblems. These tools are truly scientific, in the sense that their results are both measurable and verifiable, and they have been responsible for many of AI's recent successes. The shared mathematical language has also permitted a high level of collaboration with more established fields (like mathematics, economics or operations research). Stuart Russell and Peter Norvig describe this movement as nothing less than a "revolution" and "the victory of the neats."[34] Critiques argue that these techniques are too focussed on particular problems and have failed to address the long term goal of general intelligence.

Integrating the approaches

Intelligent agent paradigm

An intelligent agent is a system that perceives its environment and takes actions which maximize its chances of success. The simplest intelligent agents are programs that solve specific problems. More complicated agents include human beings and organizations of human beings (such as firms). The paradigm gives researchers license to study isolated problems and find solutions that are both verifiable and useful, without agreeing on one single approach. An agent that solves a specific problem can use any approach that works — some agents are symbolic and logical, some are sub-symbolic neural networks and others may use new approaches. The paradigm also gives researchers a common language to communicate with other fields—such as decision theory and economics—that also use concepts of abstract agents. The intelligent agent paradigm became widely accepted during the 1990s.[3]

Agent architectures and cognitive architectures

Researchers have designed systems to build intelligent systems out of interacting intelligent agents in a multi-agent system.[99] A system with both symbolic and sub-symbolic components is a hybrid intelligent system, and the study of such systems is artificial intelligence systems integration. A hierarchical control system provides a bridge between sub-symbolic AI at its lowest, reactive levels and traditional symbolic AI at its highest levels, where relaxed time constraints permit planning and world modelling.[100] Rodney Brooks' subsumption architecture was an early proposal for such a hierarchical system.[101]

Tools

In the course of 50 years of research, AI has developed a large number of tools to solve the most difficult problems in computer science. A few of the most general of these methods are discussed below.

Search and optimization

Many problems in AI can be solved in theory by intelligently searching through many possible solutions:[102] Reasoning can be reduced to performing a search. For example, logical proof can be viewed as searching for a path that leads from premises to conclusions, where each step is the application of an inference rule.[103] Planning algorithms search through trees of goals and subgoals, attempting to find a path to a target goal, a process called means-ends analysis.[104] Robotics algorithms for moving limbs and grasping objects use local searches in configuration space.[70] Many learning algorithms use search algorithms based on optimization.

Simple exhaustive searches[105] are rarely sufficient for most real world problems: the search space (the number of places to search) quickly grows to astronomical numbers. The result is a search that is too slow or never completes. The solution, for many problems, is to use "heuristics" or "rules of thumb" that eliminate choices that are unlikely to lead to the goal (called "pruning the search tree"). Heuristics supply the program with a "best guess" for the path on which the solution lies.[106]

A very different kind of search came to prominence in the 1990s, based on the mathematical theory of optimization. For many problems, it is possible to begin the search with some form of a guess and then refine the guess incrementally until no more refinements can be made. These algorithms can be visualized as blind hill climbing: we begin the search at a random point on the landscape, and then, by jumps or steps, we keep moving our guess uphill, until we reach the top. Other optimization algorithms are simulated annealing, beam search and random optimization.[107]

Evolutionary computation uses a form of optimization search. For example, they may begin with a population of organisms (the guesses) and then allow them to mutate and recombine, selecting only the fittest to survive each generation (refining the guesses). Forms of evolutionary computation include swarm intelligence algorithms (such as ant colony or particle swarm optimization)[108] and evolutionary algorithms (such as genetic algorithms and genetic programming).[109]

Logic

Logic[110] is used for knowledge representation and problem solving, but it can be applied to other problems as well. For example, the satplan algorithm uses logic for planning[111] and inductive logic programming is a method for learning.[112]

Several different forms of logic are used in AI research. Propositional or sentential logic[113] is the logic of statements which can be true or false. First-order logic[114] also allows the use of quantifiers and predicates, and can express facts about objects, their properties, and their relations with each other. Fuzzy logic,[115] is a version of first-order logic which allows the truth of a statement to be represented as a value between 0 and 1, rather than simply True (1) or False (0). Fuzzy systems can be used for uncertain reasoning and have been widely used in modern industrial and consumer product control systems. Subjective logic[116] models uncertainty in a different and more explicit manner than fuzzy-logic: a given binomial opinion satisfies belief + disbelief + uncertainty = 1 within a Beta distribution. By this method, ignorance can be distinguished from probabilistic statements that an agent makes with high confidence.

Default logics, non-monotonic logics and circumscription[52] are forms of logic designed to help with default reasoning and the qualification problem. Several extensions of logic have been designed to handle specific domains of knowledge, such as: description logics;[46] situation calculus, event calculus and fluent calculus (for representing events and time);[47] causal calculus;[48] belief calculus; and modal logics.[49]

Probabilistic methods for uncertain reasoning

Many problems in AI (in reasoning, planning, learning, perception and robotics) require the agent to operate with incomplete or uncertain information. AI researchers have devised a number of powerful tools to solve these problems using methods from probability theory and economics.[117]

Bayesian networks[118] are a very general tool that can be used for a large number of problems: reasoning (using the Bayesian inference algorithm),[119] learning (using the expectation-maximization algorithm),[120] planning (using decision networks)[121] and perception (using dynamic Bayesian networks).[122] Probabilistic algorithms can also be used for filtering, prediction, smoothing and finding explanations for streams of data, helping perception systems to analyze processes that occur over time (e.g., hidden Markov models or Kalman filters).[122]

A key concept from the science of economics is "utility": a measure of how valuable something is to an intelligent agent. Precise mathematical tools have been developed that analyze how an agent can make choices and plan, using decision theory, decision analysis,[123] information value theory.[58] These tools include models such as Markov decision processes,[124] dynamic decision networks,[122] game theory and mechanism design.[125]

Classifiers and statistical learning methods

The simplest AI applications can be divided into two types: classifiers ("if shiny then diamond") and controllers ("if shiny then pick up"). Controllers do however also classify conditions before inferring actions, and therefore classification forms a central part of many AI systems. Classifiers are functions that use pattern matching to determine a closest match. They can be tuned according to examples, making them very attractive for use in AI. These examples are known as observations or patterns. In supervised learning, each pattern belongs to a certain predefined class. A class can be seen as a decision that has to be made. All the observations combined with their class labels are known as a data set. When a new observation is received, that observation is classified based on previous experience.[126]

A classifier can be trained in various ways; there are many statistical and machine learning approaches. The most widely used classifiers are the neural network,[127] kernel methods such as the support vector machine,[128] k-nearest neighbor algorithm,[129] Gaussian mixture model,[130] naive Bayes classifier,[131] and decision tree.[132] The performance of these classifiers have been compared over a wide range of tasks. Classifier performance depends greatly on the characteristics of the data to be classified. There is no single classifier that works best on all given problems; this is also referred to as the "no free lunch" theorem. Determining a suitable classifier for a given problem is still more an art than science.[133]

Neural networks

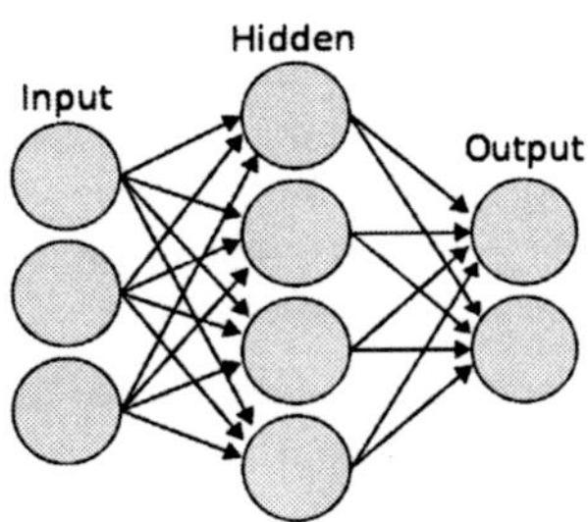

A neural network is an interconnected group of nodes, akin to the vast network of neurons in the human brain.

The study of artificial neural networks[127] began in the decade before the field AI research was founded, in the work of Walter Pitts and Warren McCullough. Other important early researchers were Frank Rosenblatt, who invented the perceptron and Paul Werbos who developed the backpropagation algorithm.[134]

The main categories of networks are acyclic or feedforward neural networks (where the signal passes in only one direction) and recurrent neural networks (which allow feedback). Among the most popular feedforward networks are perceptrons, multi-layer perceptrons and radial basis networks.[135] Among recurrent networks, the most famous is the Hopfield net, a form of attractor network, which was first described by John Hopfield in 1982.[136] Neural networks can be applied to the problem of intelligent control (for robotics) or learning, using such techniques as Hebbian learning and competitive learning.[137]

Hierarchical temporal memory is an approach that models some of the structural and algorithmic properties of the neocortex.[138]

Control theory

Control theory, the grandchild of cybernetics, has many important applications, especially in robotics.[139]

Languages

AI researchers have developed several specialized languages for AI research, including Lisp[140] and Prolog.[141]

Evaluating progress

In 1950, Alan Turing proposed a general procedure to test the intelligence of an agent now known as the Turing test. This procedure allows almost all the major problems of artificial intelligence to be tested. However, it is a very difficult challenge and at present all agents fail.[142]

Artificial intelligence can also be evaluated on specific problems such as small problems in chemistry, hand-writing recognition and game-playing. Such tests have been termed subject matter expert Turing tests. Smaller problems provide more achievable goals and there are an ever-increasing number of positive results.[143]

The broad classes of outcome for an AI test are: (1) Optimal: it is not possible to perform better. (2) Strong super-human: performs better than all humans. (3) Super-human: performs better than most humans. (4) Sub-human: performs worse than most humans. For example, performance at draughts is optimal,[144] performance at chess is super-human and nearing strong super-human (see Computer chess#Computers versus humans) and performance at many everyday tasks (such as recognizing a face or crossing a room without bumping into something) is sub-human.

A quite different approach measures machine intelligence through tests which are developed from *mathematical* definitions of intelligence. Examples of these kinds of tests start in the late nineties devising intelligence tests using notions from Kolmogorov complexity and data compression.[145] Two major advantages of mathematical definitions are their applicability to nonhuman intelligences and their absence of a requirement for human testers.

Applications

Artificial intelligence techniques are pervasive and are too numerous to list. Frequently, when a technique reaches mainstream use, it is no longer considered artificial intelligence; this phenomenon is described as the AI effect.[146]

Competitions and prizes

There are a number of competitions and prizes to promote research in artificial intelligence. The main areas promoted are: general machine intelligence, conversational behavior, data-mining, driverless cars, robot soccer and games.

Platforms

A platform (or "computing platform") is defined as "some sort of hardware architecture or software framework (including application frameworks), that allows software to run." As Rodney Brooks[147] pointed out many years ago, it is not just the artificial intelligence software that defines the AI features of the platform, but rather the actual platform itself that affects the AI that results, i.e., we need to be working out AI problems on real-world platforms rather than in isolation.

An automated online assistant providing customer service on a web page - one of many applications of artificial intelligence.

A wide variety of platforms has allowed different aspects of AI to develop, ranging from expert systems, albeit PC-based but still an entire real-world system, to various robot platforms such as the widely available Roomba with open interface.[148]

Philosophy

Artificial intelligence, by claiming to be able to recreate the capabilities of the human mind, is both a challenge and an inspiration for philosophy. Are there limits to how intelligent machines can be? Is there an essential difference between human intelligence and artificial intelligence? Can a machine have a mind and consciousness? A few of the most influential answers to these questions are given below.[149]

Turing's "polite convention": We need not decide if a machine can "think"; we need only decide if a machine can act as intelligently as a human being. This approach to the philosophical problems associated with artificial intelligence forms the basis of the Turing test.[142]

The Dartmouth proposal: "Every aspect of learning or any other feature of intelligence can be so precisely described that a machine can be made to simulate it." This conjecture was printed in the proposal for the Dartmouth Conference of 1956, and represents the position of most working AI researchers.[150]

Newell and Simon's physical symbol system hypothesis: "A physical symbol system has the necessary and sufficient means of general intelligent action." Newell and Simon argue that intelligences consist of formal operations on symbols.[151] Hubert Dreyfus argued that, on the contrary, human expertise depends on unconscious instinct rather than conscious symbol manipulation and on having a "feel" for the situation rather than explicit symbolic knowledge. (See Dreyfus' critique of AI.)[152] [153]

Gödel's incompleteness theorem: A formal system (such as a computer program) cannot prove all true statements.[154] Roger Penrose is among those who claim that Gödel's theorem limits what machines can do. (See *The Emperor's New Mind*.)[155]

Searle's strong AI hypothesis: "The appropriately programmed computer with the right inputs and outputs would thereby have a mind in exactly the same sense human beings have minds."[156] John Searle counters this assertion with his Chinese room argument, which asks us to look *inside* the computer and try to find where the "mind" might be.[157]

The artificial brain argument: The brain can be simulated. Hans Moravec, Ray Kurzweil and others have argued that it is technologically feasible to copy the brain directly into hardware and software, and that such a simulation will be essentially identical to the original.[78]

Predictions and ethics

Artificial Intelligence is a common topic in both science fiction and projections about the future of technology and society. The existence of an artificial intelligence that rivals human intelligence raises difficult ethical issues, and the potential power of the technology inspires both hopes and fears.

In fiction, Artificial Intelligence has appeared fulfilling many roles, including a servant (R2D2 in *Star Wars*), a law enforcer (K.I.T.T. "Knight Rider"), a comrade (Lt. Commander Data in *Star Trek: The Next Generation*), a conqueror/overlord (*The Matrix*), a dictator (*With Folded Hands*), a benevolent provider/de facto ruler (*The Culture*), an assassin (*Terminator*), a sentient race (*Battlestar Galactica*/Transformers), an extension to human abilities (*Ghost in the Shell*) and the savior of the human race (R. Daneel Olivaw in the *Asimov's Robot Series*).

Mary Shelley's *Frankenstein* considers a key issue in the ethics of artificial intelligence: if a machine can be created that has intelligence, could it also *feel*? If it can feel, does it have the same rights as a human? The idea also appears in modern science fiction, including the films *I Robot*, *Blade Runner* and *A.I.: Artificial Intelligence*, in which humanoid machines have the ability to feel human emotions. This issue, now known as "robot rights", is currently being considered by, for example, California's Institute for the Future, although many critics believe that the discussion is premature.[158] The subject is profoundly discussed in the 2010 documentary film *Plug & Pray*.[159]

Martin Ford, author of *The Lights in the Tunnel: Automation, Accelerating Technology and the Economy of the Future*,[160] and others argue that specialized artificial intelligence applications, robotics and other forms of automation will ultimately result in significant unemployment as machines begin to match and exceed the capability of workers to perform most routine and repetitive jobs. Ford predicts that many knowledge-based occupations—and in particular entry level jobs—will be increasingly susceptible to automation via expert systems, machine learning[161] and other AI-enhanced applications. AI-based applications may also be used to amplify the capabilities of low-wage offshore workers, making it more feasible to outsource knowledge work.[162]

Joseph Weizenbaum wrote that AI applications can not, by definition, successfully simulate genuine human empathy and that the use of AI technology in fields such as customer service or psychotherapy[163] was deeply misguided. Weizenbaum was also bothered that AI researchers (and some philosophers) were willing to view the human mind as nothing more than a computer program (a position now known as computationalism). To Weizenbaum these points suggest that AI research devalues human life.[164]

Many futurists believe that artificial intelligence will ultimately transcend the limits of progress. Ray Kurzweil has used Moore's law (which describes the relentless exponential improvement in digital technology) to calculate that desktop computers will have the same processing power as human brains by the year 2029. He also predicts that by 2045 artificial intelligence will reach a point where it is able to improve *itself* at a rate that far exceeds anything conceivable in the past, a scenario that science fiction writer Vernor Vinge named the "singularity".[165]

Robot designer Hans Moravec, cyberneticist Kevin Warwick and inventor Ray Kurzweil have predicted that humans and machines will merge in the future into cyborgs that are more capable and powerful than either.[166] This idea, called transhumanism, which has roots in Aldous Huxley and Robert Ettinger, has been illustrated in fiction as well, for example in the manga *Ghost in the Shell* and the science-fiction series *Dune*.

Edward Fredkin argues that "artificial intelligence is the next stage in evolution," an idea first proposed by Samuel Butler's "Darwin among the Machines" (1863), and expanded upon by George Dyson in his book of the same name in 1998.[167]

Pamela McCorduck writes that all these scenarios are expressions of the ancient human desire to, as she calls it, "forge the gods".[7]

See also

- AI-complete
- Artificial intelligence in fiction
- Artificial Intelligence (journal)
- Artificial intelligence (video games)
- Synthetic intelligence
- Cognitive sciences
- Human Cognome Project
- Friendly artificial intelligence
- List of basic artificial intelligence topics
- List of AI researchers
- List of important AI publications
- List of AI projects
- List of emerging technologies
- List of scientific journals
- Philosophy of mind
- Technological singularity
- Never-Ending Language Learning

References

Notes

[1] TOPIO:

- "A Ping-Pong-Playing Terminator" (http://www.popsci.com/technology/article/2010-02/ping-pong-playing-terminator). Popular Science. .
- "Best robot 2009" (http://www.gadgetrivia.com/8164-best_robot_international_robot_exhibition). www.gadgetrivia.com. .

[2] Definition of AI as the study of intelligent agents:

- Poole, Mackworth & Goebel 1998, p. 1 (http://people.cs.ubc.ca/~poole/ci/ch1.pdf), which provides the version that is used in this article. Note that they use the term "computational intelligence" as a synonym for artificial intelligence.
- Russell & Norvig (2003) (who prefer the term "rational agent") and write "The whole-agent view is now widely accepted in the field" (Russell & Norvig 2003, p. 55).
- Nilsson 1998

[3] The intelligent agent paradigm:

- Russell & Norvig 2003, pp. 27, 32–58, 968–972
- Poole, Mackworth & Goebel 1998, pp. 7–21
- Luger & Stubblefield 2004, pp. 235–240

The definition used in this article, in terms of goals, actions, perception and environment, is due to Russell & Norvig (2003). Other definitions also include knowledge and learning as additional criteria.

[4] Although there is some controversy on this point (see Crevier (1993, p. 50)), McCarthy states unequivocally "I came up with the term" in a c|net interview. (Skillings 2006)

[5] McCarthy's definition of AI:

- McCarthy 2007

[6] See the Dartmouth proposal, under Philosophy, below.

[7] This is a central idea of Pamela McCorduck's *Machines That Think*. She writes: "I like to think of artificial intelligence as the scientific apotheosis of a venerable cultural tradition." (McCorduck 2004, p. 34) "Artificial intelligence in one form or another is an idea that has pervaded Western intellectual history, a dream in urgent need of being realized." (McCorduck 2004, p. xviii) "Our history is full of attempts—nutty, eerie, comical, earnest, legendary and real—to make artificial intelligences, to reproduce what is the essential us—bypassing the ordinary means. Back and forth between myth and reality, our imaginations supplying what our workshops couldn't, we have engaged for a long time in this odd form of self-reproduction." (McCorduck 2004, p. 3) She traces the desire back to its Hellenistic roots and calls it the urge to "forge the Gods." (McCorduck 2004, pp. 340–400)

[8] The optimism referred to includes the predictions of early AI researchers (see optimism in the history of AI) as well as the ideas of modern transhumanists such as Ray Kurzweil.

[9] The "setbacks" referred to include the ALPAC report of 1966, the abandonment of perceptrons in 1970, the Lighthill Report of 1973 and the collapse of the lisp machine market in 1987.

[10] AI applications widely used behind the scenes:

- Russell & Norvig 2003, p. 28
- Kurzweil 2005, p. 265
- NRC 1999, pp. 216–222

[11] Pamela McCorduck (2004, pp. 424) writes of "the rough shattering of AI in subfields—vision, natural language, decision theory, genetic algorithms, robotics ... and these with own sub-subfield—that would hardly have anything to say to each other."

[12] This list of intelligent traits is based on the topics covered by the major AI textbooks, including:

- Russell & Norvig 2003
- Luger & Stubblefield 2004
- Poole, Mackworth & Goebel 1998
- Nilsson 1998

[13] General intelligence (strong AI) is discussed in popular introductions to AI:

- Kurzweil 1999 and Kurzweil 2005

[14] AI in myth:

- McCorduck 2004, pp. 4–5
- Russell & Norvig 2003, p. 939

[15] Cult images as artificial intelligence:

- Crevier (1993, p. 1) (statue of Amun)
- McCorduck (2004, pp. 6–9)

These were the first machines to be believed to have true intelligence and consciousness. Hermes Trismegistus expressed the common belief that with these statues, craftsman had reproduced "the true nature of the gods", their *sensus* and *spiritus*. McCorduck makes the connection between sacred automatons and Mosaic law (developed around the same time), which expressly forbids the worship of robots (McCorduck 2004, pp. 6–9)

[16] Humanoid automata:

Yan Shi:

- Needham 1986, p. 53

Hero of Alexandria:

- McCorduck 2004, p. 6

Al-Jazari:

- "A Thirteenth Century Programmable Robot" (http://www.shef.ac.uk/marcoms/eview/articles58/robot.html). Shef.ac.uk. . Retrieved 2009-04-25.

Wolfgang von Kempelen:

- McCorduck 2004, p. 17

[17] Artificial beings:

Jābir ibn Hayyān's Takwin:

- O'Connor, Kathleen Malone (1994). *The alchemical creation of life (takwin) and other concepts of Genesis in medieval Islam* (http://repository.upenn.edu/dissertations/AAI9503804). University of Pennsylvania. . Retrieved 2007-01-10.

Judah Loew's Golem:

- McCorduck 2004, pp. 15–16
- Buchanan 2005, p. 50

Paracelsus' Homunculus:

- McCorduck 2004, pp. 13–14

[18] AI in early science fiction.

- McCorduck 2004, pp. 17–25

[19] This insight, that digital computers can simulate any process of formal reasoning, is known as the Church–Turing thesis.

[20] Formal reasoning:

- Berlinski, David (2000). *The Advent of the Algorithm*. Harcourt Books. ISBN 0-15-601391-6. OCLC 46890682.

[21] AI's immediate precursors:

- McCorduck 2004, pp. 51–107
- Crevier 1993, pp. 27–32
- Russell & Norvig 2003, pp. 15, 940
- Moravec 1988, p. 3

See also Cybernetics and early neural networks (in History of artificial intelligence). Among the researchers who laid the foundations of AI were Alan Turing, John Von Neumann, Norbert Wiener, Claude Shannon, Warren McCullough, Walter Pitts and Donald Hebb.

[22] Dartmouth conference:

- McCorduck 2004, pp. 111–136
- Crevier 1993, pp. 47–49, who writes "the conference is generally recognized as the official birthdate of the new science."
- Russell & Norvig 2003, p. 17, who call the conference "the birth of artificial intelligence."
- NRC 1999, pp. 200–201

[23] Hegemony of the Dartmouth conference attendees:

- Russell & Norvig 2003, p. 17, who write "for the next 20 years the field would be dominated by these people and their students."
- McCorduck 2004, pp. 129–130

[24] Russell and Norvig write "it was astonishing whenever a computer did anything kind of smartish." Russell & Norvig 2003, p. 18

[25] "Golden years" of AI (successful symbolic reasoning programs 1956-1973):

- McCorduck 2004, pp. 243–252
- Crevier 1993, pp. 52–107
- Moravec 1988, p. 9
- Russell & Norvig 2003, pp. 18–21

The programs described are Daniel Bobrow's STUDENT, Newell and Simon's Logic Theorist and Terry Winograd's SHRDLU.

[26] DARPA pours money into undirected pure research into AI during the 1960s:

- McCorduck 2004, pp. 131
- Crevier 1993, pp. 51, 64–65
- NRC 1999, pp. 204–205

[27] AI in England:

- Howe 1994

[28] Optimism of early AI:

- Herbert Simon quote: Simon 1965, p. 96 quoted in Crevier 1993, p. 109.
- Marvin Minsky quote: Minsky 1967, p. 2 quoted in Crevier 1993, p. 109.

[29] See The problems (in History of artificial intelligence)

[30] First AI Winter, Mansfield Amendment, Lighthill report

- Crevier 1993, pp. 115–117
- Russell & Norvig 2003, p. 22
- NRC 1999, pp. 212–213
- Howe 1994

[31] Expert systems:

- ACM 1998, I.2.1,
- Russell & Norvig 2003, pp. 22–24
- Luger & Stubblefield 2004, pp. 227–331,
- Nilsson 1998, chpt. 17.4
- McCorduck 2004, pp. 327–335, 434–435
- Crevier 1993, pp. 145–62, 197–203

[32] Boom of the 1980s: rise of expert systems, Fifth Generation Project, Alvey, MCC, SCI:

- McCorduck 2004, pp. 426–441
- Crevier 1993, pp. 161–162,197–203, 211, 240
- Russell & Norvig 2003, p. 24
- NRC 1999, pp. 210–211

[33] Second AI winter:

- McCorduck 2004, pp. 430–435
- Crevier 1993, pp. 209–210
- NRC 1999, pp. 214–216

[34] Formal methods are now preferred ("Victory of the neats"):

- Russell & Norvig 2003, pp. 25–26
- McCorduck 2004, pp. 486–487

[35] McCorduck 2004, pp. 480–483

[36] DARPA Grand Challenge – home page (http://www.darpa.mil/grandchallenge/)

[37] "Welcome" (http://archive.darpa.mil/grandchallenge/). Archive.darpa.mil. . Retrieved 2011-10-31.

[38] Markoff, John (16 February 2011). "On 'Jeopardy!' Watson Win Is All but Trivial" (http://www.nytimes.com/2011/02/17/science/17jeopardy-watson.html). *The New York Times*. .

[39] Kinect's AI breakthrough explained (http://www.i-programmer.info/news/105-artificial-intelligence/2176-kinects-ai-breakthrough-explained.html)

[40] Problem solving, puzzle solving, game playing and deduction:

- Russell & Norvig 2003, chpt. 3-9,
- Poole, Mackworth & Goebel 1998, chpt. 2,3,7,9,
- Luger & Stubblefield 2004, chpt. 3,4,6,8,
- Nilsson 1998, chpt. 7-12

[41] Uncertain reasoning:

- Russell & Norvig 2003, pp. 452–644,
- Poole, Mackworth & Goebel 1998, pp. 345–395,
- Luger & Stubblefield 2004, pp. 333–381,
- Nilsson 1998, chpt. 19

[42] Intractability and efficiency and the combinatorial explosion:

- Russell & Norvig 2003, pp. 9, 21–22

[43] Psychological evidence of sub-symbolic reasoning:

- Wason & Shapiro (1966) showed that people do poorly on completely abstract problems, but if the problem is restated to allow the use of intuitive social intelligence, performance dramatically improves. (See Wason selection task)
- Kahneman, Slovic & Tversky (1982) have shown that people are terrible at elementary problems that involve uncertain reasoning. (See list of cognitive biases for several examples).
- Lakoff & Núñez (2000) have controversially argued that even our skills at mathematics depend on knowledge and skills that come from "the body", i.e. sensorimotor and perceptual skills. (See Where Mathematics Comes From)

[44] Knowledge representation:

- ACM 1998, I.2.4,
- Russell & Norvig 2003, pp. 320–363,
- Poole, Mackworth & Goebel 1998, pp. 23–46, 69–81, 169–196, 235–277, 281–298, 319–345,
- Luger & Stubblefield 2004, pp. 227–243,
- Nilsson 1998, chpt. 18

[45] Knowledge engineering:

- Russell & Norvig 2003, pp. 260–266,
- Poole, Mackworth & Goebel 1998, pp. 199–233,
- Nilsson 1998, chpt. ~17.1-17.4

[46] Representing categories and relations: Semantic networks, description logics, inheritance (including frames and scripts):

- Russell & Norvig 2003, pp. 349–354,
- Poole, Mackworth & Goebel 1998, pp. 174–177,
- Luger & Stubblefield 2004, pp. 248–258,
- Nilsson 1998, chpt. 18.3

[47] Representing events and time:Situation calculus, event calculus, fluent calculus (including solving the frame problem):

- Russell & Norvig 2003, pp. 328–341,
- Poole, Mackworth & Goebel 1998, pp. 281–298,
- Nilsson 1998, chpt. 18.2

[48] Causal calculus:

- Poole, Mackworth & Goebel 1998, pp. 335–337

[49] Representing knowledge about knowledge: Belief calculus, modal logics:

- Russell & Norvig 2003, pp. 341–344,
- Poole, Mackworth & Goebel 1998, pp. 275–277

[50] Ontology:

- Russell & Norvig 2003, pp. 320–328

[51] Qualification problem:

- McCarthy & Hayes 1969

- Russell & Norvig 2003

While McCarthy was primarily concerned with issues in the logical representation of actions, Russell & Norvig 2003 apply the term to the more general issue of default reasoning in the vast network of assumptions underlying all our commonsense knowledge.

[52] Default reasoning and default logic, non-monotonic logics, circumscription, closed world assumption, abduction (Poole *et al.* places abduction under "default reasoning". Luger *et al.* places this under "uncertain reasoning"):

- Russell & Norvig 2003, pp. 354–360,
- Poole, Mackworth & Goebel 1998, pp. 248–256, 323–335,
- Luger & Stubblefield 2004, pp. 335–363,
- Nilsson 1998, ~18.3.3

[53] Breadth of commonsense knowledge:

- Russell & Norvig 2003, p. 21,
- Crevier 1993, pp. 113–114,
- Moravec 1988, p. 13,
- Lenat & Guha 1989 (Introduction)

[54] Dreyfus & Dreyfus 1986

[55] Gladwell 2005

[56] Expert knowledge as embodied intuition:

- Dreyfus & Dreyfus 1986 (Hubert Dreyfus is a philosopher and critic of AI who was among the first to argue that most useful human knowledge was encoded sub-symbolically. See Dreyfus' critique of AI)
- Gladwell 2005 (Gladwell's *Blink* is a popular introduction to sub-symbolic reasoning and knowledge.)
- Hawkins & Blakeslee 2005 (Hawkins argues that sub-symbolic knowledge should be the primary focus of AI research.)

[57] Planning:

- ACM 1998, ~I.2.8,
- Russell & Norvig 2003, pp. 375–459,
- Poole, Mackworth & Goebel 1998, pp. 281–316,
- Luger & Stubblefield 2004, pp. 314–329,
- Nilsson 1998, chpt. 10.1-2, 22

[58] Information value theory:

- Russell & Norvig 2003, pp. 600–604

[59] Classical planning:

- Russell & Norvig 2003, pp. 375–430,
- Poole, Mackworth & Goebel 1998, pp. 281–315,
- Luger & Stubblefield 2004, pp. 314–329,
- Nilsson 1998, chpt. 10.1-2, 22

[60] Planning and acting in non-deterministic domains: conditional planning, execution monitoring, replanning and continuous planning:

- Russell & Norvig 2003, pp. 430–449

[61] Multi-agent planning and emergent behavior:

- Russell & Norvig 2003, pp. 449–455

[62] Learning:

- ACM 1998, I.2.6,
- Russell & Norvig 2003, pp. 649–788,
- Poole, Mackworth & Goebel 1998, pp. 397–438,
- Luger & Stubblefield 2004, pp. 385–542,
- Nilsson 1998, chpt. 3.3 , 10.3, 17.5, 20

[63] Alan Turing discussed the centrality of learning as early as 1950, in his classic paper Computing Machinery and Intelligence. (Turing 1950)

[64] (pdf scanned copy of the original) (http://world.std.com/~rjs/indinf56.pdf) (version published in 1957, An Inductive Inference Machine," IRE Convention Record, Section on Information Theory, Part 2, pp. 56-62)

[65] Reinforcement learning:

- Russell & Norvig 2003, pp. 763–788
- Luger & Stubblefield 2004, pp. 442–449

[66] Computational learning theory:

- CITATION IN PROGRESS.

[67] Natural language processing:

- ACM 1998, I.2.7
- Russell & Norvig 2003, pp. 790–831
- Poole, Mackworth & Goebel 1998, pp. 91–104

- Luger & Stubblefield 2004, pp. 591–632

[68] Applications of natural language processing, including information retrieval (i.e. text mining) and machine translation:

- Russell & Norvig 2003, pp. 840–857,
- Luger & Stubblefield 2004, pp. 623–630

[69] Robotics:

- ACM 1998, I.2.9,
- Russell & Norvig 2003, pp. 901–942,
- Poole, Mackworth & Goebel 1998, pp. 443–460

[70] Moving and configuration space:

- Russell & Norvig 2003, pp. 916–932

[71] Robotic mapping (localization, etc):

- Russell & Norvig 2003, pp. 908–915

[72] Machine perception:

- Russell & Norvig 2003, pp. 537–581, 863–898
- Nilsson 1998, ~chpt. 6

[73] Computer vision:

- ACM 1998, I.2.10
- Russell & Norvig 2003, pp. 863–898
- Nilsson 1998, chpt. 6

[74] Speech recognition:

- ACM 1998, ~I.2.7
- Russell & Norvig 2003, pp. 568–578

[75] Object recognition:

- Russell & Norvig 2003, pp. 885–892

[76] Emotion and affective computing:

- Minsky 2006

[77] Gerald Edelman, Igor Aleksander and others have both argued that artificial consciousness is required for strong AI. (Aleksander 1995; Edelman 2007)

[78] Artificial brain arguments: AI requires a simulation of the operation of the human brain

- Russell & Norvig 2003, p. 957
- Crevier 1993, pp. 271 and 279

A few of the people who make some form of the argument:

- Moravec 1988
- Kurzweil 2005, p. 262
- Hawkins & Blakeslee 2005

The most extreme form of this argument (the brain replacement scenario) was put forward by Clark Glymour in the mid-70s and was touched on by Zenon Pylyshyn and John Searle in 1980.

[79] AI complete: Shapiro 1992, p. 9

[80] Nils Nilsson writes: "Simply put, there is wide disagreement in the field about what AI is all about" (Nilsson 1983, p. 10).

[81] Biological intelligence vs. intelligence in general:

- Russell & Norvig 2003, pp. 2–3, who make the analogy with aeronautical engineering.
- McCorduck 2004, pp. 100–101, who writes that there are "two major branches of artificial intelligence: one aimed at producing intelligent behavior regardless of how it was accomplioshed, and the other aimed at modeling intelligent processes found in nature, particularly human ones."
- Kolata 1982, a paper in *Science*, which describes McCathy's indifference to biological models. Kolata quotes McCarthy as writing: "This is AI, so we don't care if it's psychologically real" (http://books.google.com/books?id=PEkqAAAAMAAJ&q="we+don't+care+if+it's+psychologically+real"&dq="we+don't+care+if+it's+psychologically+real"&output=html&pgis=1). McCarthy recently reiterated his position at the AI@50 conference where he said "Artificial intelligence is not, by definition, simulation of human intelligence" (Maker 2006).

[82] Neats vs. scruffies:

- McCorduck 2004, pp. 421–424, 486–489
- Crevier 1993, pp. 168
- Nilsson 1983, pp. 10–11

[83] Symbolic vs. sub-symbolic AI:

- Nilsson (1998, p. 7), who uses the term "sub-symbolic".

[84] Haugeland 1985, p. 255.

[85] http://citeseerx.ist.psu.edu/viewdoc/download?doi=10.1.1.38.8384&rep=rep1&type=pdf

[86] Pei Wang (2008). *Artificial general intelligence, 2008: proceedings of the First AGI Conference* (http://books.google.com/books?id=a_ZR81Z25z0C&pg=PA63). IOS Press. p. 63. ISBN 978-1-58603-833-5. . Retrieved 31 October 2011.

[87] Haugeland 1985, pp. 112–117

[88] Cognitive simulation, Newell and Simon, AI at CMU (then called Carnegie Tech):

- McCorduck 2004, pp. 139–179, 245–250, 322–323 (EPAM)
- Crevier 1993, pp. 145–149

[89] Soar (history):

- McCorduck 2004, pp. 450–451
- Crevier 1993, pp. 258–263

[90] McCarthy and AI research at SAIL and SRI International:

- McCorduck 2004, pp. 251–259
- Crevier 1993

[91] AI research at Edinburgh and in France, birth of Prolog:

- Crevier 1993, pp. 193–196
- Howe 1994

[92] AI at MIT under Marvin Minsky in the 1960s :

- McCorduck 2004, pp. 259–305
- Crevier 1993, pp. 83–102, 163–176
- Russell & Norvig 2003, p. 19

[93] Cyc:

- McCorduck 2004, p. 489, who calls it "a determinedly scruffy enterprise"
- Crevier 1993, pp. 239–243
- Russell & Norvig 2003, p. 363–365
- Lenat & Guha 1989

[94] Knowledge revolution:

- McCorduck 2004, pp. 266–276, 298–300, 314, 421
- Russell & Norvig 2003, pp. 22–23

[95] The most dramatic case of sub-symbolic AI being pushed into the background was the devastating critique of perceptrons by Marvin Minsky and Seymour Papert in 1969. See History of AI, AI winter, or Frank Rosenblatt.

[96] Embodied approaches to AI:

- McCorduck 2004, pp. 454–462
- Brooks 1990
- Moravec 1988

[97] Revival of connectionism:

- Crevier 1993, pp. 214–215
- Russell & Norvig 2003, p. 25

[98] Computational intelligence

- IEEE Computational Intelligence Society (http://www.ieee-cis.org/)

[99] Agent architectures, hybrid intelligent systems:

- Russell & Norvig (2003, pp. 27, 932, 970–972)
- Nilsson (1998, chpt. 25)

[100] Hierarchical control system:

- Albus, J. S. 4-D/RCS reference model architecture for unmanned ground vehicles. (http://www.isd.mel.nist.gov/documents/albus/4DRCS.pdf) In G Gerhart, R Gunderson, and C Shoemaker, editors, Proceedings of the SPIE AeroSense Session on Unmanned Ground Vehicle Technology, volume 3693, pages 11—20

[101] Subsumption architecture:

- CITATION IN PROGRESS.

[102] Search algorithms:

- Russell & Norvig 2003, pp. 59–189
- Poole, Mackworth & Goebel 1998, pp. 113–163
- Luger & Stubblefield 2004, pp. 79–164, 193–219
- Nilsson 1998, chpt. 7-12

[103] Forward chaining, backward chaining, Horn clauses, and logical deduction as search:

- Russell & Norvig 2003, pp. 217–225, 280–294

- Poole, Mackworth & Goebel 1998, pp. ~46–52
- Luger & Stubblefield 2004, pp. 62–73
- Nilsson 1998, chpt. 4.2, 7.2

[104] State space search and planning:

- Russell & Norvig 2003, pp. 382–387
- Poole, Mackworth & Goebel 1998, pp. 298–305
- Nilsson 1998, chpt. 10.1-2

[105] Uninformed searches (breadth first search, depth first search and general state space search):

- Russell & Norvig 2003, pp. 59–93
- Poole, Mackworth & Goebel 1998, pp. 113–132
- Luger & Stubblefield 2004, pp. 79–121
- Nilsson 1998, chpt. 8

[106] Heuristic or informed searches (e.g., greedy best first and A*):

- Russell & Norvig 2003, pp. 94–109,
- Poole, Mackworth & Goebel 1998, pp. pp. 132–147,
- Luger & Stubblefield 2004, pp. 133–150,
- Nilsson 1998, chpt. 9

[107] Optimization searches:

- Russell & Norvig 2003, pp. 110–116,120–129
- Poole, Mackworth & Goebel 1998, pp. 56–163
- Luger & Stubblefield 2004, pp. 127–133

[108] Artificial life and society based learning:

- Luger & Stubblefield 2004, pp. 530–541

[109] Genetic programming and genetic algorithms:

- Luger & Stubblefield 2004, pp. 509–530,
- Nilsson 1998, chpt. 4.2.
- Holland, John H. (1975). *Adaptation in Natural and Artificial Systems*. University of Michigan Press. ISBN 0262581116.
- Koza, John R. (1992). *Genetic Programming*. MIT Press. ISBN 0262111705.
- Poli, R., Langdon, W. B., McPhee, N. F. (2008). *A Field Guide to Genetic Programming*. Lulu.com, freely available from http://www.gp-field-guide.org.uk/. ISBN 978-1-4092-0073-4.

[110] Logic:

- ACM 1998, ~I.2.3,
- Russell & Norvig 2003, pp. 194–310,
- Luger & Stubblefield 2004, pp. 35–77,
- Nilsson 1998, chpt. 13-16

[111] Satplan:

- Russell & Norvig 2003, pp. 402–407,
- Poole, Mackworth & Goebel 1998, pp. 300–301,
- Nilsson 1998, chpt. 21

[112] Explanation based learning, relevance based learning, inductive logic programming, case based reasoning:

- Russell & Norvig 2003, pp. 678–710,
- Poole, Mackworth & Goebel 1998, pp. 414–416,
- Luger & Stubblefield 2004, pp. ~422–442,
- Nilsson 1998, chpt. 10.3, 17.5

[113] Propositional logic:

- Russell & Norvig 2003, pp. 204–233,
- Luger & Stubblefield 2004, pp. 45–50
- Nilsson 1998, chpt. 13

[114] First-order logic and features such as equality:

- ACM 1998, ~I.2.4,
- Russell & Norvig 2003, pp. 240–310,
- Poole, Mackworth & Goebel 1998, pp. 268–275,
- Luger & Stubblefield 2004, pp. 50–62,
- Nilsson 1998, chpt. 15

[115] Fuzzy logic:

- Russell & Norvig 2003, pp. 526–527

[116] Subjective logic:

- CITATION IN PROGRESS.

[117] Stochastic methods for uncertain reasoning:

- ACM 1998, ~I.2.3,
- Russell & Norvig 2003, pp. 462–644,
- Poole, Mackworth & Goebel 1998, pp. 345–395,
- Luger & Stubblefield 2004, pp. 165–191, 333–381,
- Nilsson 1998, chpt. 19

[118] Bayesian networks:

- Russell & Norvig 2003, pp. 492–523,
- Poole, Mackworth & Goebel 1998, pp. 361–381,
- Luger & Stubblefield 2004, pp. ~182–190, ~363–379,
- Nilsson 1998, chpt. 19.3-4

[119] Bayesian inference algorithm:

- Russell & Norvig 2003, pp. 504–519,
- Poole, Mackworth & Goebel 1998, pp. 361–381,
- Luger & Stubblefield 2004, pp. ~363–379,
- Nilsson 1998, chpt. 19.4 & 7

[120] Bayesian learning and the expectation-maximization algorithm:

- Russell & Norvig 2003, pp. 712–724,
- Poole, Mackworth & Goebel 1998, pp. 424–433,
- Nilsson 1998, chpt. 20

[121] Bayesian decision theory and Bayesian decision networks:

- Russell & Norvig 2003, pp. 597–600

[122] Stochastic temporal models:

- Russell & Norvig 2003, pp. 537–581

Dynamic Bayesian networks:

- Russell & Norvig 2003, pp. 551–557

Hidden Markov model:

- (Russell & Norvig 2003, pp. 549–551)

Kalman filters:

- Russell & Norvig 2003, pp. 551–557

[123] decision theory and decision analysis:

- Russell & Norvig 2003, pp. 584–597,
- Poole, Mackworth & Goebel 1998, pp. 381–394

[124] Markov decision processes and dynamic decision networks:

- Russell & Norvig 2003, pp. 613–631

[125] Game theory and mechanism design:

- Russell & Norvig 2003, pp. 631–643

[126] Statistical learning methods and classifiers:

- Russell & Norvig 2003, pp. 712–754,
- Luger & Stubblefield 2004, pp. 453–541

[127] Neural networks and connectionism:

- Russell & Norvig 2003, pp. 736–748,
- Poole, Mackworth & Goebel 1998, pp. 408–414,
- Luger & Stubblefield 2004, pp. 453–505,
- Nilsson 1998, chpt. 3

[128] kernel methods such as the support vector machine, Kernel methods:

- Russell & Norvig 2003, pp. 749–752

[129] K-nearest neighbor algorithm:

- Russell & Norvig 2003, pp. 733–736

[130] Gaussian mixture model:

- Russell & Norvig 2003, pp. 725–727

[131] Naive Bayes classifier:

- Russell & Norvig 2003, pp. 718

[132] Decision tree:

- Russell & Norvig 2003, pp. 653–664,
- Poole, Mackworth & Goebel 1998, pp. 403–408,
- Luger & Stubblefield 2004, pp. 408–417

[133] Classifier performance:

- van der Walt & Bernard 2006

[134] Backpropagation:

- Russell & Norvig 2003, pp. 744–748,
- Luger & Stubblefield 2004, pp. 467–474,
- Nilsson 1998, chpt. 3.3

[135] Feedforward neural networks, perceptrons and radial basis networks:

- Russell & Norvig 2003, pp. 739–748, 758
- Luger & Stubblefield 2004, pp. 458–467

[136] Recurrent neural networks, Hopfield nets:

- Russell & Norvig 2003, p. 758
- Luger & Stubblefield 2004, pp. 474–505

[137] Competitive learning, Hebbian coincidence learning, Hopfield networks and attractor networks:

- Luger & Stubblefield 2004, pp. 474–505

[138] Hierarchical temporal memory:

- Hawkins & Blakeslee 2005

[139] Control theory:

- ACM 1998, ~I.2.8,
- Russell & Norvig 2003, pp. 926–932

[140] Lisp:

- Luger & Stubblefield 2004, pp. 723–821
- Crevier 1993, pp. 59–62,
- Russell & Norvig 2003, p. 18

[141] Prolog:

- Poole, Mackworth & Goebel 1998, pp. 477–491,
- Luger & Stubblefield 2004, pp. 641–676, 575–581

[142] The Turing test:

Turing's original publication:

- Turing 1950

Historical influence and philosophical implications:

- Haugeland 1985, pp. 6–9
- Crevier 1993, p. 24
- McCorduck 2004, pp. 70–71
- Russell & Norvig 2003, pp. 2–3 and 948

[143] Subject matter expert Turing test:

- CITATION IN PROGRESS.

[144] Game AI:

- CITATION IN PROGRESS.

[145] Mathematical definitions of intelligence:

- Jose Hernandez-Orallo (2000). "Beyond the Turing Test" (http://citeseerx.ist.psu.edu/viewdoc/summary?doi=10.1.1.44.8943). *Journal of Logic, Language and Information* **9** (4): 447–466. doi:10.1023/A:1008367325700. . Retrieved 2009-07-21.
- D L Dowe and A R Hajek (1997). "A computational extension to the Turing Test" (http://www.csse.monash.edu.au/publications/1997/tr-cs97-322-abs.html). *Proceedings of the 4th Conference of the Australasian Cognitive Science jSociety*. . Retrieved 2009-07-21.
- J Hernandez-Orallo and D L Dowe (2010). "Measuring Universal Intelligence: Towards an Anytime Intelligence Test". *Artificial Intelligence Journal* **174** (18): 1508–1539. doi:10.1016/j.artint.2010.09.006.

[146] "AI set to exceed human brain power" (http://www.cnn.com/2006/TECH/science/07/24/ai.bostrom/) (web article). CNN.com. 2006-07-26. . Retrieved 2008-02-26.

[147] Brooks, R.A., "How to build complete creatures rather than isolated cognitive simulators," in K. VanLehn (ed.), Architectures for Intelligence, pp. 225–239, Lawrence Erlbaum Associates, Hillsdale, NJ, 1991.

[148] Hacking Roomba » Search Results » atmel (http://hackingroomba.com/?s=atmel)

[149] Philosophy of AI. All of these positions in this section are mentioned in standard discussions of the subject, such as:

- Russell & Norvig 2003, pp. 947–960
- Fearn 2007, pp. 38–55

[150] Dartmouth proposal:

- McCarthy et al. 1955 (the original proposal)
- Crevier 1993, p. 49 (historical significance)

[151] The physical symbol systems hypothesis:

- Newell & Simon 1976, p. 116
- McCorduck 2004, p. 153
- Russell & Norvig 2003, p. 18

[152] Dreyfus criticized the necessary condition of the physical symbol system hypothesis, which he called the "psychological assumption": "The mind can be viewed as a device operating on bits of information according to formal rules". (Dreyfus 1992, p. 156)

[153] Dreyfus' critique of artificial intelligence:

- Dreyfus 1972, Dreyfus & Dreyfus 1986
- Crevier 1993, pp. 120–132
- McCorduck 2004, pp. 211–239
- Russell & Norvig 2003, pp. 950–952,

[154] This is a paraphrase of the relevant implication of Gödel's theorems.

[155] The Mathematical Objection:

- Russell & Norvig 2003, p. 949
- McCorduck 2004, pp. 448–449

Making the Mathematical Objection:

- Lucas 1961
- Penrose 1989

Refuting Mathematical Objection:

- Turing 1950 under "(2) The Mathematical Objection"
- Hofstadter 1979

Background:

- Gödel 1931, Church 1936, Kleene 1935, Turing 1937

[156] This version is from Searle (1999), and is also quoted in Dennett 1991, p. 435. Searle's original formulation was "The appropriately programmed computer really is a mind, in the sense that computers given the right programs can be literally said to understand and have other cognitive states." (Searle 1980, p. 1). Strong AI is defined similarly by Russell & Norvig (2003, p. 947): "The assertion that machines could possibly act intelligently (or, perhaps better, act as if they were intelligent) is called the 'weak AI' hypothesis by philosophers, and the assertion that machines that do so are actually thinking (as opposed to simulating thinking) is called the 'strong AI' hypothesis."

[157] Searle's Chinese Room argument:

- Searle 1980. Searle's original presentation of the thought experiment.
- Searle 1999.

Discussion:

- Russell & Norvig 2003, pp. 958–960
- McCorduck 2004, pp. 443–445
- Crevier 1993, pp. 269–271

[158] Robot rights:

- Russell & Norvig 2003, p. 964
- "Robots could demand legal rights" (http://news.bbc.co.uk/2/hi/technology/6200005.stm). *BBC News*. 21 December 2006. . Retrieved 3 February 2011.

Prematurity of:

- Henderson, Mark (24 April 2007). "Human rights for robots? We're getting carried away" (http://www.timesonline.co.uk/tol/news/uk/science/article1695546.ece). *The Times Online* (London). .

In fiction:

- McCorduck (2004, p. 190-25) discusses *Frankenstein* and identifies the key ethical issues as scientific hubris and the suffering of the monster, i.e. robot rights.

[159] Independent documentary Plug & Pray, featuring Joseph Weizenbaum and Raymond Kurzweil (http://www.plugandpray-film.de/en/content.html)

[160] Ford, Martin R. (2009), *The Lights in the Tunnel: Automation, Accelerating Technology and the Economy of the Future* (http://www.thelightsinthetunnel.com), Acculant Publishing, ISBN 978-1448659814, . *(e-book available free online* (http://www.thelightsinthetunnel.com/).*)*

[161] "Machine Learning: A Job Killer?" (http://econfuture.wordpress.com/2011/04/14/machine-learning-a-job-killer/)

[162] AI could decrease the demand for human labor:

- Russell & Norvig 2003, pp. 960–961
- Ford, Martin (2009). *The Lights in the Tunnel: Automation, Accelerating Technology and the Economy of the Future* (http://www.thelightsinthetunnel.com). Acculant Publishing. ISBN 978-1-4486-5981-4. .

[163] In the early 70s, Kenneth Colby presented a version of Weizenbaum's ELIZA known as DOCTOR which he promoted as a serious therapeutic tool. (Crevier 1993, pp. 132–144)

[164] Joseph Weizenbaum's critique of AI:

- Weizenbaum 1976
- Crevier 1993, pp. 132–144
- McCorduck 2004, pp. 356–373
- Russell & Norvig 2003, p. 961

Weizenbaum (the AI researcher who developed the first chatterbot program, ELIZA) argued in 1976 that the misuse of artificial intelligence has the potential to devalue human life.

[165] Technological singularity:

- Vinge 1993
- Kurzweil 2005
- Russell & Norvig 2003, p. 963

[166] Transhumanism:

- Moravec 1988
- Kurzweil 2005
- Russell & Norvig 2003, p. 963

[167] AI as evolution:

- Edward Fredkin is quoted in McCorduck (2004, p. 401).
- Butler, Samuel (13 June 1863). *the Press* (Christchurch, New Zealand). http://www.nzetc.org/tm/scholarly/tei-ButFir-t1-g1-t1-g1-t4-body.html, Letter to the Editor.
- Dyson, George (1998). *Darwin among the Machiens*. Allan Lane Science. ISBN 0738200301.

References

AI textbooks

- Luger, George; Stubblefield, William (2004). *Artificial Intelligence: Structures and Strategies for Complex Problem Solving* (http://www.cs.unm.edu/~luger/ai-final/tocfull.html) (5th ed.). The Benjamin/Cummings Publishing Company, Inc.. ISBN 0-8053-4780-1.
- Nilsson, Nils (1998). *Artificial Intelligence: A New Synthesis*. Morgan Kaufmann Publishers. ISBN 978-1-55860-467-4.
- Russell, Stuart J.; Norvig, Peter (2003), *Artificial Intelligence: A Modern Approach* (http://aima.cs.berkeley.edu/) (2nd ed.), Upper Saddle River, New Jersey: Prentice Hall, ISBN 0-13-790395-2
- Poole, David; Mackworth, Alan; Goebel, Randy (1998). *Computational Intelligence: A Logical Approach* (http://www.cs.ubc.ca/spider/poole/ci.html). New York: Oxford University Press. ISBN 0195102703.
- Winston, Patrick Henry (1984). *Artificial Intelligence*. Reading, Massachusetts: Addison-Wesley. ISBN 0201082594.

History of AI

- Crevier, Daniel (1993), *AI: The Tumultuous Search for Artificial Intelligence*, New York, NY: BasicBooks, ISBN 0-465-02997-3
- McCorduck, Pamela (2004), *Machines Who Think* (http://www.pamelamc.com/html/machines_who_think.html) (2nd ed.), Natick, MA: A. K. Peters, Ltd., ISBN 1-56881-205-1

Other sources

- "ACM Computing Classification System: Artificial intelligence" (http://www.acm.org/class/1998/I.2.html). ACM. 1998. Retrieved 2007-08-30.
- Aleksander, Igor (1995). *Artificial Neuroconsciousness: An Update* (http://web.archive.org/web/19970302014628/http://www.ee.ic.ac.uk/research/neural/publications/iwann.html). IWANN. Archived from the original (http://www.ee.ic.ac.uk/research/neural/publications/iwann.html) on 1997-03-02. BibTex (http://dblp.uni-trier.de/rec/bibtex/conf/iwann/Aleksander95) Internet Archive (http://web.archive.org/web/19970302014628/http://www.ee.ic.ac.uk/research/neural/publications/iwann.html)
- Brooks, Rodney (1990). "Elephants Don't Play Chess" (http://people.csail.mit.edu/brooks/papers/elephants.pdf) (PDF). *Robotics and Autonomous Systems* **6**: 3–15. doi:10.1016/S0921-8890(05)80025-9. Retrieved 2007-08-30..
- Buchanan, Bruce G. (2005). "A (Very) Brief History of Artificial Intelligence" (http://www.aaai.org/AITopics/assets/PDF/AIMag26-04-016.pdf) (PDF). *AI Magazine*: 53–60. Retrieved 2007-08-30.
- Dennett, Daniel (1991). *Consciousness Explained*. The Penguin Press. ISBN 0-7139-9037-6.
- Dreyfus, Hubert (1972). *What Computers Can't Do*. New York: MIT Press. ISBN 0060110821.
- Dreyfus, Hubert (1979). *What Computers* Still *Can't Do*. New York: MIT Press. ISBN 0262041340.
- Dreyfus, Hubert; Dreyfus, Stuart (1986). *Mind over Machine: The Power of Human Intuition and Expertise in the Era of the Computer*. Oxford, UK: Blackwell. ISBN 0029080606.
- Dreyfus, Hubert (1992). *What Computers* Still *Can't Do*. New York: MIT Press. ISBN 0-262-54067-3.
- Edelman, Gerald (23 November 2007). "Gerald Edelman - Neural Darwinism and Brain-based Devices" (http://lis.epfl.ch/resources/podcast/2007/11/gerald-edelman-neural-darwinism-and.html). Talking Robots.
- Fearn, Nicholas (2007). *The Latest Answers to the Oldest Questions: A Philosophical Adventure with the World's Greatest Thinkers*. New York: Grove Press. ISBN 0802118399.
- Forster, Dion (2006). "Self validating consciousness in strong artificial intelligence: An African theological contribution" (http://www.spirituality.org.za/files/D Forster doctorate.pdf). Pretoria: University of South Africa.
- Gladwell, Malcolm (2005). *Blink*. New York: Little, Brown and Co.. ISBN 0-316-17232-4.
- Haugeland, John (1985). *Artificial Intelligence: The Very Idea*. Cambridge, Mass.: MIT Press. ISBN 0-262-08153-9.
- Hawkins, Jeff; Blakeslee, Sandra (2005). *On Intelligence*. New York, NY: Owl Books. ISBN 0-8050-7853-3.
- Hofstadter, Douglas (1979). *Gödel, Escher, Bach: an Eternal Golden Braid*. New York, NY: Vintage Books. ISBN 0394745027.
- Howe, J. (November 1994). "Artificial Intelligence at Edinburgh University: a Perspective" (http://www.inf.ed.ac.uk/about/AIhistory.html). Retrieved 30 August 2007..
- Kahneman, Daniel; Slovic, D.; Tversky, Amos (1982). *Judgment under uncertainty: Heuristics and biases*. New York: Cambridge University Press. ISBN 0521284147.
- Kolata, G. (1982). "How can computers get common sense?". *Science* **217** (4566): 1237–1238. doi:10.1126/science.217.4566.1237. PMID 17837639.
- Kurzweil, Ray (1999). *The Age of Spiritual Machines*. Penguin Books. ISBN 0-670-88217-8.
- Kurzweil, Ray (2005). *The Singularity is Near*. Penguin Books. ISBN 0-670-03384-7.

- Lakoff, George (1987). *Women, Fire, and Dangerous Things: What Categories Reveal About the Mind.* University of Chicago Press. ISBN 0-226-46804-6.
- Lakoff, George; Núñez, Rafael E. (2000). *Where Mathematics Comes From: How the Embodied Mind Brings Mathematics into Being*. Basic Books. ISBN 0-465-03771-2..
- Lenat, Douglas; Guha, R. V. (1989). *Building Large Knowledge-Based Systems.* Addison-Wesley. ISBN 0201517523.
- Lighthill, Professor Sir James (1973). "Artificial Intelligence: A General Survey". *Artificial Intelligence: a paper symposium.* Science Research Council.
- Lucas, John (1961). "Minds, Machines and Gödel" (http://users.ox.ac.uk/~jrlucas/Godel/mmg.html). In Anderson, A.R.. *Minds and Machines.* Retrieved 30 August 2007.
- Maker, Meg Houston (2006). "AI@50: AI Past, Present, Future" (http://www.engagingexperience.com/2006/07/ai50_ai_past_pr.html). Dartmouth College. Retrieved 16 October 2008.
- McCarthy, John; Minsky, Marvin; Rochester, Nathan; Shannon, Claude (1955). "A Proposal for the Dartmouth Summer Research Project on Artificial Intelligence" (http://www-formal.stanford.edu/jmc/history/dartmouth/dartmouth.html). Retrieved 30 August 2007..
- McCarthy, John; Hayes, P. J. (1969). "Some philosophical problems from the standpoint of artificial intelligence" (http://www-formal.stanford.edu/jmc/mcchay69.html). *Machine Intelligence* **4**: 463–502. Retrieved 30 August 2007.
- McCarthy, John (November 12, 2007). "What Is Artificial Intelligence?" (http://www-formal.stanford.edu/jmc/whatisai/whatisai.html).
- Minsky, Marvin (1967). *Computation: Finite and Infinite Machines.* Englewood Cliffs, N.J.: Prentice-Hall. ISBN 0131654497.
- Minsky, Marvin (2006). *The Emotion Machine*. New York, NY: Simon & Schusterl. ISBN 0-7432-7663-9.
- Moravec, Hans (1976). "The Role of Raw Power in Intelligence" (http://www.frc.ri.cmu.edu/users/hpm/project.archive/general.articles/1975/Raw.Power.html). Retrieved 30 August 2007.
- Moravec, Hans (1988). *Mind Children*. Harvard University Press. ISBN 0674576160.
- NRC, (United States National Research Council) (1999). "Developments in Artificial Intelligence". *Funding a Revolution: Government Support for Computing Research.* National Academy Press.
- Needham, Joseph (1986). *Science and Civilization in China: Volume 2*. Caves Books Ltd..
- Newell, Allen; Simon, H. A. (1963). "GPS: A Program that Simulates Human Thought". In Feigenbaum, E.A.; Feldman, J.. *Computers and Thought*. New York: McGraw-Hill.
- Newell, Allen; Simon, H. A. (1976). "Computer Science as Empirical Inquiry: Symbols and Search" (http://www.rci.rutgers.edu/~cfs/472_html/AI_SEARCH/PSS/PSSH4.html). *Communications of the ACM*. **19**..
- Nilsson, Nils (1983), "Artificial Intelligence Prepares for 2001" (http://ai.stanford.edu/~nilsson/OnlinePubs-Nils/General Essays/AIMag04-04-002.pdf), *AI Magazine* **1** (1), Presidential Address to the Association for the Advancement of Artificial Intelligence.
- Penrose, Roger (1989). *The Emperor's New Mind: Concerning Computer, Minds and The Laws of Physics*. Oxford University Press. ISBN 0-198-51973-7.
- Searle, John (1980). "Minds, Brains and Programs" (http://www.bbsonline.org/documents/a/00/00/04/84/bbs00000484-00/bbs.searle2.html). *Behavioral and Brain Sciences* **3** (3): 417–457. doi:10.1017/S0140525X00005756.
- Searle, John (1999). *Mind, language and society*. New York, NY: Basic Books. ISBN 0465045219. OCLC 231867665 43689264.
- Serenko, Alexander; Detlor, Brian (2004). "Intelligent agents as innovations" (http://foba.lakeheadu.ca/serenko/papers/Serenko_Detlor_AI_and_Society.pdf). *AI and Society* **18** (4): 364–381. doi:10.1007/s00146-004-0310-5.

- Serenko, Alexander; Ruhi, Umar; Cocosila, Mihail (2007). "Unplanned effects of intelligent agents on Internet use: Social Informatics approach" (http://foba.lakeheadu.ca/serenko/papers/AI_Society_Serenko_Social_Impacts_of_Agents.pdf). *AI and Society* **21** (1–2): 141–166. doi:10.1007/s00146-006-0051-8.
- Shapiro, Stuart C. (1992). "Artificial Intelligence" (http://www.cse.buffalo.edu/~shapiro/Papers/ai.pdf). In Shapiro, Stuart C.. *Encyclopedia of Artificial Intelligence* (2nd ed.). New York: John Wiley. pp. 54–57. ISBN 0471503061.
- Simon, H. A. (1965). *The Shape of Automation for Men and Management*. New York: Harper & Row.
- Skillings, Jonathan (3 July 2006). "Getting Machines to Think Like Us" (http://news.cnet.com/Getting-machines-to-think-like-us/2008-11394_3-6090207.html). *cnet*. Retrieved February 3, 2011.
- Turing, Alan (October 1950), "Computing Machinery and Intelligence" (http://loebner.net/Prizef/TuringArticle.html), *Mind* **LIX** (236): 433–460, doi:10.1093/mind/LIX.236.433, ISSN 0026-4423, retrieved 2008-08-18.
- van der Walt, Christiaan; Bernard, Etienne (2006<!—year is presumed based on acknowledgements at the end of the article—>). "Data characteristics that determine classifier performance" (http://www.patternrecognition.co.za/publications/cvdwalt_data_characteristics_classifiers.pdf) (PDF). Retrieved 5 August 2009.
- Vinge, Vernor (1993). "The Coming Technological Singularity: How to Survive in the Post-Human Era" (http://www-rohan.sdsu.edu/faculty/vinge/misc/singularity.html).
- Wason, P. C.; Shapiro, D. (1966). "Reasoning". In Foss, B. M.. *New horizons in psychology*. Harmondsworth: Penguin.
- Weizenbaum, Joseph (1976). *Computer Power and Human Reason*. San Francisco: W.H. Freeman & Company. ISBN 0716704641.

Further reading

- TechCast Article Series, John Sagi, Framing Consciousness (http://www.techcast.org/Upload/PDFs/634146249446122137_Consciousness-Sagifinalversion.pdf)
- Boden, Margaret, Mind As Machine, Oxford University Press, 2006
- Johnston, John (2008) "The Allure of Machinic Life: Cybernetics, Artificial Life, and the New AI", MIT Press
- Myers, Courtney Boyd ed. (2009). The AI Report (http://www.forbes.com/2009/06/22/singularity-robots-computers-opinions-contributors-artificial-intelligence-09_land.html). Forbes June 2009
- Serenko, Alexander (2010). "The development of an AI journal ranking based on the revealed preference approach" (http://foba.lakeheadu.ca/serenko/papers/JOI_Serenko_AI_Journal_Ranking_Published.pdf) (PDF). *Journal of Informetrics* **4** (4): 447–459. doi:10.1016/j.joi.2010.04.001.
- Sun, R. & Bookman, L. (eds.), *Computational Architectures: Integrating Neural and Symbolic Processes*. Kluwer Academic Publishers, Needham, MA. 1994.

External links

- What Is AI? (http://www-formal.stanford.edu/jmc/whatisai/whatisai.html) — An introduction to artificial intelligence by AI founder John McCarthy.
- Logic and Artificial Intelligence (http://plato.stanford.edu/entries/logic-ai) entry by Richmond Thomason in the *Stanford Encyclopedia of Philosophy*
- AI (http://www.dmoz.org/Computers/Artificial_Intelligence//) at the Open Directory Project
- AITopics (http://aaai.org/AITopics/) — A large directory of links and other resources maintained by the Association for the Advancement of Artificial Intelligence, the leading organization of academic AI researchers.
- Artificial Intelligence Discussion group (https://www.researchgate.net/group/Artificial_Intelligence)

Fighting_game

Fighting game is a video game genre where the player controls an on-screen character and engages in close combat with an opponent. These characters tend to be of equal power and fight matches consisting of several rounds, which take place in an arena. Players must master techniques such as blocking, counter-attacking, and chaining together sequences of attacks known as "combos". Since the early 1990s, most fighting games allow the player to execute special attacks by performing specific button combinations. The genre is related to but distinct from beat 'em ups, which involve large numbers of antagonists.

Although *Street Fighter II* was not the first fighting game, it popularized and established the gameplay conventions of the genre.

The first game to feature fist fighting was *Heavyweight Champ* in 1976, but it was *Karate Champ* and *The Way of the Exploding Fist* which popularized one-on-one martial arts games in 1984 and 1985 respectively. Also in 1985, *Yie Ar Kung-Fu* featured antagonists with differing fighting styles, while 1987's *Street Fighter* introduced hidden special attacks. In 1991, Capcom's highly successful *Street Fighter II* refined and popularized many of the conventions of the genre. The fighting game subsequently became the preeminent genre for competitive video gaming in the early to mid-1990s, especially in arcades. This period spawned numerous popular fighting games in addition to *Street Fighter*, including the successful and long running franchises *Mortal Kombat* and later *Virtua Fighter* and *Tekken*.

The genre's popularity stagnated as games became more complicated and as arcades began to lose their audience to increasingly powerful home consoles near the end of the 1990s, though new franchises such as *Dead or Alive*, and the *Soul* series achieved success. In the new millennium, the genre remains popular but retains a much smaller proportion of enthusiasts than it once did, due to the increasing popularity of other genres and internet multiplayer gaming.

Definition

Fighting games are a type of action game where on-screen characters fight each other.[1] [2] [3] These games typically feature special moves that are triggered using rapid sequences of carefully timed button presses and joystick movements. Games traditionally show fighters from a side-view, even as the genre has progressed from two-dimensional (2D) to three-dimensional (3D) graphics.[1] *Street Fighter II*, though not the first fighting game, popularized and standardized the conventions of the genre,[4] and similar games released prior to *Street Fighter II* have since been more explicitly classified as fighting games.[3] [4] Fighting games typically involve hand-to-hand combat, but may also feature melee weapons.[5]

This genre is distinct from beat 'em ups, another action genre involving combat, where the player character must fight many weaker enemies at the same time.[3] During the 1980s publications used the terms "fighting game" and "beat 'em up" interchangeably, along with other terms such as "martial arts simulation" (or more specific terms such as "judo simulator").[6] [7] [8] With hindsight, critics have argued that the two types of game gradually became dichotomous as they evolved, though the two terms may still be conflated.[3] [9] Fighting games are sometimes grouped with games that feature boxing or wrestling.[5] [9] Serious boxing games belong more to the sports game genre than the action game genre, as they aim for a more realistic model of boxing techniques, whereas moves in fighting games tend to be highly exaggerated models of Asian martial arts techniques.[1] As such, boxing games and wrestling games are often described as distinct genres, without comparison to fighting games.[10] [11]

Game design

Virtua Fighter is rendered in 3D, but is typical of most fighting games in that most action takes place in a 2D plane of motion. Here, one player ducks the other's attack.

Fighting games involve combat between pairs of fighters using highly exaggerated martial arts moves.[1] They typically revolve around primarily brawling or combat sport,[2] [5] though some variations feature weaponry.[5] Games usually display on-screen fighters from a side view, and even 3D fighting games play largely within a 2D plane of motion.[1] Games usually confine characters to moving left and right and jumping, although some games such as *Fatal Fury: King of Fighters* allow players to move between parallel planes of movement.[1] [12] Recent games tend to be rendered in three dimensions and allow side-stepping, but otherwise play like those rendered in two dimensions.[5]

Attacking and defending

Aside from moving around a restricted space, fighting games limit the player's actions to different offensive and defensive maneuvers. Players must learn which attacks and defenses are effective against each other, often by trial and error.[1] Blocking is a basic technique that allows a player to defend against attacks.[13] Some games feature more advanced blocking techniques: for example, Capcom's *Street Fighter III* features a move termed "parrying" which causes the attacker to become momentarily incapacitated (a similar state is termed "just defended" in SNK's *Garou: Mark of the Wolves*).[14] [15] In addition to blows such as punches and kicks, players can utilize throwing or "grappling" to circumvent "blocks". Predicting opponents' moves and counter-attacking, known as "countering", is a common element of gameplay.[5] Fighting games also emphasize the difference between the height of blows, ranging from low to jumping attacks.[16] [17] Thus, strategy becomes important as players attempt to predict each others' moves, similar to rock-paper-scissors.[1]

An integral feature of fighting games includes the use of "special attacks", also called "secret moves",[16] that employ complex combinations of button presses to perform a particular move beyond basic punching and kicking.[18] Combos, in which several attacks are chained together using basic punches and kicks, are another common feature in fighting games and have been fundamental to the genre since the release of *Street Fighter II*.[19] Some fighting games display a "combo meter" that displays the player's progress through a combo. The effectiveness of such moves often relate to the difficulty of execution and the degree of risk. These moves are often beyond the ability of a casual gamer and require a player to have both a strong memory and excellent timing.[1] Taunting is another feature of some fighting games and was originally introduced by Japanese company SNK in their game *Art of Fighting*.[20] [21] It is used to add humor to games, but can also have an effect on gameplay such as improving the strength of other attacks.[22] Sometimes, a character can even be noted especially for taunting (for example, Dan Hibiki from *Street Fighter Alpha*).[23] [24]

Goals

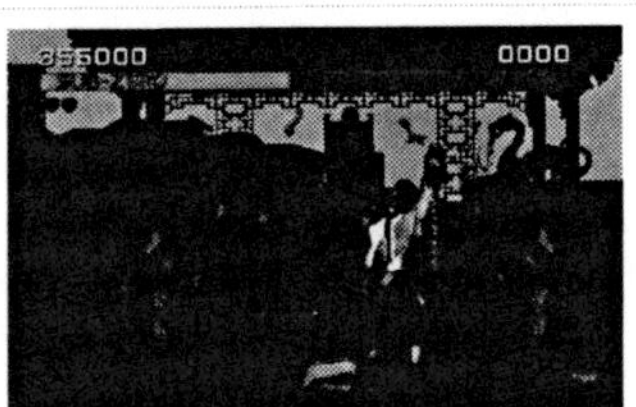
The player's objective in a fighting game is to win a match by depleting their rival's health. *Mortal Kombat* even allows the victor to perform a gruesome finishing maneuver called a "fatality".

Fighting game matches generally consist of several rounds; the player who wins the most rounds wins the match.[25] Fighting games widely feature life bars, which are depleted as characters sustain blows.[12] [26] Each successful attack will deplete a character's health, and the game continues until a fighter's energy reaches zero.[1] Hence, the main goal is to completely deplete the life bar of one's opponent, thus achieving a "knockout".[15] Beginning with Midway's *Mortal Kombat* released in 1992, the *Mortal Kombat* series introduced "fatalities" in which the victor kills a knocked-out opponent in a gruesome manner.[27] Games such as *Virtua Fighter* also allow a character to be defeated by forcing them outside of the fighting arena, awarding a "ring-out" to the victor.[13] Round decisions can also be determined by time over (if a timer is present), which judges players based on remaining vitality to declare a winner.

Fighting games often include a single player campaign or tournament, where the player must defeat a sequence of several computer controlled opponents. Winning the tournament often reveals a special story–ending cutscene, and some games also grant access to hidden characters or special features upon victory.[28]

Multiple characters and players

In most fighting games, players may select from a variety of characters who have unique fighting styles and special moves. This became a strong convention for the genre with the release of *Street Fighter II*, and these character choices have led to deeper game strategy and replay value.[29] Although fighting games offer female characters, their image tends to be hyperfeminized, and they have even been featured as pin-up girls in game magazines.[30] [31] Male characters in fighting games also tend to be hypersexualized, with extra-broad chests and shoulders, huge muscles, and prominent jaws.[1]

Custom creation, or "create–a–fighter", is a feature of some fighting games which allows a player to customize the appearance and move set of their own character. *Super Fire Pro Wrestling X Premium* was the first game to include such a feature,[32] and later fighting games such as *Fighter Maker*,[33] *Soulcalibur III*,[34] *Mortal Kombat: Armageddon*,[35] and *Dragon Ball Z: Budokai Tenkaichi 2* adopted the concept.[36] Some fighting games allow the player to create other types of game content: *Super Smash Bros. Brawl*, for example, features the ability to create custom stages.[37]

Fighting games may also offer a multiplayer mode in which players fight each other, sometimes by letting a second player challenge the first at any moment during a single player match.[2] A few titles allow up to four players to compete simultaneously.[38] Several games have also featured modes that involve teams of characters; players form "tag teams" to fight matches in which combat is one-on-one, but a character may leave the arena to be replaced by a team mate.[39] Some fighting games have also offered the challenge of fighting against multiple opponents in succession, testing the player's endurance.[28] Newer titles take advantage of online gaming services, although lag created by slow data transmission can disrupt the split-second timing involved in fighting games.[28] [40]

History

Early fighting games: 1970s to 1980s

Fighting games find their origin in boxing games but evolved towards battles between characters with fantastic abilities and complex special maneuvers.[41] The first two-player video game based on human-to-human combat was Taito's shooter-based *Gun Fight* in 1975.[42] Sega's black and white boxing game *Heavyweight Champ*, which was released in 1976, is considered the first video game to feature fist fighting.[43] It used boxing glove controls that could be moved up and down for high and low punches, and inwards for striking. Like later fighting games, it was viewed from a side-view perspective.[44] 1979's *Warrior* is another title sometimes credited as one of the first fighting games.[19] In contrast to *Heavyweight Champ* and most later titles, *Warrior* was based on sword fighting duels and used a bird's eye view.[3] In 1983, Sega released another boxing game *Champion Boxing*,[45] which was Yu Suzuki's debut title at Sega AM2.[46] However, Technōs Japan's *Karate Champ* from 1984 is credited with establishing and popularizing the one-on-one fighting game genre.[47] A variety of moves could be performed using the dual-joystick controls, it used a best-of-three matches format like later fighting games,[48] [49] and it featured training bonus stages. It went on to influence Konami's 1985 release *Yie Ar Kung Fu*,[47] which expanded on *Karate Champ* by pitting the player against a variety of opponents, each with a unique appearance and fighting style.[47] [50] The player could also perform up to sixteen different moves,[51] including projectile attacks,[52] using a combination of buttons and joystick movements while standing, crouching or jumping.[53] Also released in 1985, martial arts game *The Way of the Exploding Fist* achieved critical success and subsequently afforded the burgeoning genre further popularity.[8] [54] Numerous other game developers tried to imitate the financial successes of *Karate Champ*, *Yie Ar Kung-Fu* and *The Way of the Exploding Fist* with similar games; Data East took unsuccessful legal action against Epyx over the computer game *International Karate*.[55]

Both *Karate Champ* and *Yie Ar Kung Fu* provided a template for Capcom's *Street Fighter* in 1987.[4] *Street Fighter* found its own niche in the gaming world,[4] partially because many arcade game developers in the 1980s focused more on producing beat-em-ups and shoot 'em ups.[56] Part of the game's appeal was the use of special moves that could only be discovered by experimenting with the game controls, which created a sense of mystique and invited players to practice the game,[57] although similar controller motions used for grappling maneuvers in the earlier *Brian Jacks Uchi Mata* were deemed too difficult.[8] Following *Street Fighter's* lead, the use of command-based hidden moves began to pervade other games in the rising fighting game genre.[57] *Street Fighter* also introduced other staples of the genre, including the blocking technique as well as the ability for a challenger to jump in and initiate a match against a player at any time. The game also introduced pressure-sensitive controls that determine the strength of an attack, though due to causing damaged arcade cabinets, Capcom replaced it soon after with a six-button control scheme offering light, medium and hard punches and kicks, which became another staple of the genre.[44] Meanwhile, home game consoles largely ignored the genre. *Budokan: The Martial Spirit* was one of few releases for the Sega Genesis but was not as popular as games in other genres.[56] Technical challenges limited the popularity of early fighting games. Programmers had difficulty producing a game that could recognize the fast motions of a joystick, and so players had a hard time executing special moves with any accuracy.[4] [56]

Rise and peak: early 1990s

Enthusiasts playing fighting games at a video arcade

The release of *Street Fighter II* in 1991 is often considered a revolutionary moment in the fighting game genre. Yoshiki Okamoto's team developed the most accurate joystick and button scanning routine in the genre thus far. This allowed players to reliably execute multi-button special moves, which had previously required an element of luck. The game was also highly successful because its graphics took advantage of Capcom's CPS arcade chipset, with highly detailed characters and stages. Whereas previous games allowed players to combat a variety of computer-controlled fighters, *Street Fighter II* allowed players to play against each other. The popularity of *Street Fighter II* surprised the gaming industry, as arcade owners bought more machines to keep up with demand.[4]

SNK released *Fatal Fury* a few months before *Street Fighter II*.[12] It added a two-plane system where characters could step into the foreground or background. Meanwhile, Sega experimented with *Dark Edge*, an early attempt at a 3D fighting game where characters could move in all directions. Sega however, never released the game outside of Japan because it felt that unrestrained 3D fighting games were unenjoyable. Several fighting games achieved greater commercial success, including SNK's *Art of Fighting* and *Samurai Shodown* as well as Sega's *Eternal Champions* . Nevertheless, *Street Fighter II* remained the most popular,[56] spawning a special *Champion Edition* that improved game balance and allowed players to use additional characters.[4] The popularity of *Street Fighter II* led it to be released for home game consoles and allowed it to define the template for fighting games.[4] [56]

Many American developers tried to capitalize on the template established by *Street Fighter II*,[4] but it was Chicago's Midway Games who achieved unprecedented notoriety when they released *Mortal Kombat* in 1992. The game featured digital characters drawn from real actors, numerous secrets,[56] [58] and a "fatality" system of finishing maneuvers with which the player's character kills their opponent. The game earned a reputation for its gratuitous violence,[58] and was eventually adapted for home game consoles.[56] The home version of *Mortal Kombat* was released on September 13, 1993, a day that was promoted as "Mortal Monday". The advertising resulted in line-ups to purchase the game and a subsequent backlash from politicians concerned about the game's violence.[58] The *Mortal Kombat* franchise would ultimately achieve iconic status similar to that of *Street Fighter* with several sequels as well as movies, television series, and extensive merchandising.[27] [59] Numerous other game developers tried to imitate *Street Fighter II* and *Mortal Kombat*'s financial success with similar games; Capcom took unsuccessful legal action against Data East over the arcade game *Fighter's History*.[16] Data East's largest objection in court was that their 1984 arcade game *Karate Champ* was the true originator of the competitive fighting game genre, which predated the original *Street Fighter* by three years.[60] That same year saw the release of the Sega Activator, a Mega Drive (Genesis) peripheral that allowed motion detection of the player's physical movements,[61] based on the Light Harp invented by Assaf Gurner.[62] It was marketed as a new way to play fighting games,[63] such as *Street Fighter* and *Mortal Kombat*, but the device was a commercial failure due to its "unwieldiness and inaccuracy."[61]

Sega began to attract attention with the 1993 release of *Virtua Fighter* in arcades. It was the first fighting game with 3D polygon graphics and a viewpoint that zoomed and rotated with the action. Despite the graphics, players were confined to back and forth motion as seen in other fighting games. With only three buttons, it was easier to learn than *Street Fighter* and *Mortal Kombat*, having six and five buttons respectively. By the time the game was released for the Sega Saturn in Japan, the game and system were selling at almost a one-to-one ratio.[56] Meanwhile, the 1993 title *Mortal Kombat II* captivated American audiences, and was considered the best *Mortal Kombat* game in retrospect during a 2008 review.[64]

The 1994 PlayStation launch title *Battle Arena Toshinden* is credited for taking the genre into "true 3-D" due to its introduction of the sidestep maneuver, which IGN described as "one little move" that "changed the fighter forever."[65] The same year, SNK released *The King of Fighters '94* in arcades, where players choose from teams of three characters to eliminate each other one by one.[66] Eventually, Capcom released further updates to *Street Fighter II*, including *Super Street Fighter II* and *Super Street Fighter II Turbo*. These games featured more characters and new moves, some of which were a response to people who had hacked the original *Street Fighter II* game to add new features themselves. However, criticism of these updates grew as players demanded a true sequel. By 1995, the dominant franchises were the *Mortal Kombat* series in America and *Virtua Fighter* series in Japan, with *Street Fighter Alpha: Warriors' Dreams* unable to match the popularity of *Street Fighter II*.[4] Throughout this period, the fighting game was the dominant genre in competitive video gaming, with enthusiasts popularly attending arcades in order to find human opponents.[27]

Relative decline: late 1990s

In the latter part of the 1990s, the fighting game genre began to decline in popularity, with specific franchises falling into difficulty. *Electronic Gaming Monthly* awarded the excess of fighting games the "Most Appalling Trend" award of 1995.[67] Although the release of *Street Fighter EX* introduced 3D graphics to the series and continued the success of *Street Fighter II* and *Street Fighter Alpha*,[68] [69] [70] the *Street Fighter: The Movie* arcade game was regarded as a failure. *Street Fighter: The Movie* used digitized images from the *Street Fighter* film starring Jean-Claude Van Damme.[4] Capcom released *Street Fighter III* in 1997 which featured improved visuals and character depth, but was also unable to match the impact of *Street Fighter II*.[4] Despite excitement in Japan over *Virtua Fighter 3* in arcades, the limited hardware capabilities of the Sega Saturn led Sega to delay a console release.[56] Sega eventually released the game for its Dreamcast console,[71] but the company became unprofitable and was forced to discontinue the console.[72] Meanwhile, SNK released several fighting games on their Neo-Geo platform, including *Samurai Shodown II* in 1994, *Real Bout Fatal Fury* in 1995, *The Last Blade* in 1997, and annual updates to their *The King of Fighters* franchise.[73] *Fatal Fury: Mark of the Wolves* from 1999 was considered one of SNK's last great games,[74] and the company announced that it would close its doors in 2001.[75]

In retrospect, multiple developers attribute the decline of the fighting genre to its increasing complexity and specialization. This complexity shut out casual players, and the market for fighting games became smaller and more specialized.[76] [77] Furthermore, arcades gradually became less profitable throughout the 1990s due to the increased technical power and popularity of home consoles.[16] [73] Even as popularity dwindled, the fighting game genre continued to evolve; several strong 3D fighting games also emerged in the late 1990s. Namco's *Tekken* (released in arcades in 1994 and on the PlayStation in 1995) proved critical to the PlayStation's early success, with its sequels also becoming some of the console's most important titles.[78] The *Soul* series of weapon-based fighting games also achieved considerable critical success, beginning with 1997's *Soul Edge* (known as *Soul Blade* outside of Japan) to *Soulcalibur IV* in 2008.[79] [80] Tecmo's *Dead or Alive* (released in 1996 in Japanese arcades and 1998 on the PlayStation) spawned a long running franchise, known for its fast paced control system and innovative counter attacks. The series again included titles important to the success of their respective consoles, the latest installment being *Dead or Alive 4* for the Xbox 360.[28] [81] [82] In 1998, *Bushido Blade*, published by Square, introduced a realistic fighting engine that featured three-dimensional environments while abandoning time limits and health bars in favour of an innovative Body Damage System, where a sword strike to a certain body part can amputate a limb or decapitate the head.[83]

Video game enthusiasts took an interest in gaming crossovers which feature characters from multiple franchises in a particular game.[84] An early example of this type of fighting game was the 1998 arcade release *Marvel vs. Capcom: Clash of Super Heroes*, featuring comic book superheroes as well as *Street Fighter* characters.[4] In 1999, Nintendo released the first game in the *Super Smash Bros.* series, which allowed match-ups such as Pikachu versus Mario.[84]

Recent history: 2000s

The early part of the decade saw the rise of major international fighting game tournaments such as Tougeki – Super Battle Opera, and famous players such as Daigo Umehara.[85] [86] Several more fighting game crossovers were released in the new millennium. The two most prolific developers of 2D fighting games, Capcom and SNK, combined intellectual property to produce *SNK vs. Capcom* games. SNK released the first game of this type, *SNK vs. Capcom: The Match of the Millennium*, for its Neo Geo Pocket Color handheld at the end of 1999. GameSpot regarded the game as "perhaps the most highly anticipated fighter ever" and called it the best fighting game ever to be released for a handheld console.[87] [88] Capcom released *Capcom vs. SNK: Millennium Fight 2000* for arcades and the Dreamcast in 2000, followed by sequels in subsequent years. Though none matched the critical success of the handheld version, *Capcom vs. SNK 2 EO* was noted as the first game of the genre to successfully utilize internet competition.[88] [89] Other crossovers from 2008 included *Tatsunoko vs. Capcom: Ultimate All-Stars* and *Mortal Kombat vs. DC Universe*.[90] [91] The most successful crossover, however, was *Super Smash Bros. Brawl*, also released in 2008 for the Wii. Featuring characters from Nintendo's various franchises, the game was a runaway commercial success in addition to being lavished with critical praise.[37] [38] [92]

In the new millennium, fighting games are less popular and plentiful than in the mid 1990s, with multiplayer competition shifting towards other genres.[27] [93] However, SNK reappeared in 2003 as SNK Playmore and continues to release games.[73] Arc System Works received critical acclaim for releasing *Guilty Gear X* in 2001, as well as its sequel *Guilty Gear XX*, as both were 2D fighting games featuring striking anime inspired graphics.[94] The fighting game is currently a popular genre for amateur and doujin developers in Japan. The 2002 title *Melty Blood* was developed by then amateur developer French-Bread and achieved cult success on the PC. It became highly popular in arcades following its 2005 release, and a version was released on PlayStation 2 the following year.[95] While the genre is generally far less popular than it once was,[27] arcades and their attendant fighting games are still reasonably popular in Japan. *Virtua Fighter 5* lacked an online mode but still achieved success both on home consoles and in arcades; players practiced at home and went to arcades to compete face-to-face with opponents.[96] *Tekken 6* was positively received, selling more than 3 million copies worldwide as of August 6, 2010.[97] In addition to *Virtua Fighter* and *Tekken*, the *Soul* and *Dead or Alive* franchises continued to release installments.[28] [80] Classic *Street Fighter* and *Mortal Kombat* games have been re-released on PSN and Xbox Live Arcade, allowing internet play, and in some cases, HD graphics.[27] [98] [99] *Street Fighter IV*, which incorporated online multiplayer modes, was released in early 2009 to critical acclaim,[100] having garnered praise since its release at Japanese arcades in 2008.[101] *Street Fighter's* successful revival has sparked a renaissance for the genre,[102] [103] [104] followed by titles such as *Super Street Fighter IV*,[103] *Deadliest Warrior: The Game*,[105] *Marvel vs. Capcom 3: Fate of Two Worlds*,[103] [104] *Mortal Kombat*,[103] *Street Fighter X Tekken*, and *Tekken X Street Fighter*.[102]

See also

- List of fighting games
- List of fighting game companies

References

[1] Rollings, Andrew; Ernest Adams (2006). *Fundamentals of Game Design* (http://wps.prenhall.com/bp_gamedev_1/54/14053/3597646.cw/index.html). Prentice Hall. .

[2] Ashcraft, Brian (2008). *Arcade Mania! The Turbo-Charged World of Japan's Game Centers*. Kodansha International. p. 90.

[3] Spencer, Spanner. "The Tao of Beat-'em-ups" (http://www.eurogamer.net/articles/the-tao-of-beat-em-ups-article). EuroGamer. . Retrieved 2009-04-29.

[4] "The History of Street Fighter" (http://www.gamespot.com/features/vgs/universal/sfhistory/history.html). GameSpot. . Retrieved 2008-10-11.

[5] Treit, Ryan. "Novice Guides: Fighting" (http://web.archive.org/web/20090515013224/http://www.xbox.com/en-US/games/tips/noviceguides/fighting.htm). Xbox.com. Archived from the original (http://www.xbox.com/en-US/games/tips/noviceguides/fighting.htm) on 2009-05-15. . Retrieved 2009-01-15.

[6] *Way of the Tiger*. Crash. 1986-05-28.

[7] Bielby, Matt (1990-05). *Oriental Games*. Your Sinclair. p. 31.

[8] Candy, Robin & Eddy, Ricky (1987-10). *Run it Again!*. Crash.

[9] Staff (2006-05-03). "E3 Feature: Fighting Games Focus" (http://www.next-gen.biz/features/e3-feature-fighting-games-focus). Edge Online. . Retrieved 2009-02-11.

[10] Bramwell, Tom (2003-02-13). "Fighting in the Backyard" (http://www.eurogamer.net/articles/article_47405). EuroGamer. . Retrieved 2009-02-11.

[11] Walters, Stefan (2004-04-26). "Let's play: Mike Tyson Heavyweight Boxing" (http://news.bbc.co.uk/sport2/hi/funny_old_game/1855789.stm). BBC Sport. . Retrieved 2009-02-11.

[12] Provo, Frank (2007-10-11). "Fatal Fury: King of Fighters Review" (http://uk.gamespot.com/wii/action/fatalfury/review.html?om_act=convert&om_clk=gssummary&tag=summary;read-review). GameSpot. . Retrieved 2009-01-11.

[13] "The Essential 50: Virtua Fighter" (http://www.1up.com/features/essential-50-virtua-fighter). 1UP. . Retrieved 2009-01-30.

[14] Gerstmann, Jeff (1999-12-29). "Street Fighter III: Double Impact Review" (http://uk.gamespot.com/dreamcast/action/streetfighter3doubleimpact/review.html?om_act=convert&om_clk=gssummary&tag=summary;read-review). GameSpot. . Retrieved 2009-01-15.

[15] Chau, Anthony (2001-12-11). "Fatal Review: Mark of the Wolves" (http://uk.dreamcast.ign.com/articles/166/166258p1.html). IGN. . Retrieved 2009-01-15.

[16] "The Essential 50: 32. Street Fighter II" (http://www.1up.com/features/essential-50-street-fighter-ii). 1UP. . Retrieved 2009-01-15.

[17] Ekberg, Brian (2007-09-22). "TGS '07: K-1 World Grand Prix Hands-On" (http://uk.gamespot.com/ds/action/k1worldgp/news.html?sid=6179720&mode=previews&tag=result;title;0). GameSpot. . Retrieved 2009-01-15.

[18] Towell, Justin. "The Best Special Attacks Ever" (http://www.gamesradar.com/the-best-special-attacks-ever/). GamesRadar. . Retrieved 2009-01-29.

[19] "The making of *Street Fighter II*". *Edge presents Retro* ('The Making of...' Special). 2003. ""[Combos] became the base for future fighting titles""

[20] Ashcraft, pp. 100–101.

[21] Park, Andrew (2007-06-05). "Art of Fighting Anthology Review" (http://uk.gamespot.com/ps2/action/artoffightinganthology/review.html?om_act=convert&om_clk=gssummary&tag=summary;read-review). GameSpot. . Retrieved 2009-01-11.

[22] Rose, Martyn. "Designing Kung-Fu Chaos, Part 3" (http://web.archive.org/web/20081205175426/http://www.xbox.com/en-US/games/k/kungfuchaos/themakers5.htm). Xbox.com. Archived from the original (http://www.xbox.com/en-US/games/k/kungfuchaos/themakers5.htm) on 2008-12-05. . Retrieved 2009-01-11.

[23] "Top 20 Street Fighter Characters of All Time" (http://web.archive.org/web/20090301120425/http://www.gamedaily.com/articles/galleries/top-20-street-fighter-characters-of-all-time/?page=3). GameDaily. Archived from the original (http://www.gamedaily.com/articles/galleries/top-20-street-fighter-characters-of-all-time/?page=3) on 2009-03-01. . Retrieved 2009-01-11.

[24] "Top 25 Most Bizarre Fighting Characters" (http://web.archive.org/web/20090206220705/http://www.gamedaily.com/articles/galleries/top-25-most-bizarre-fighting-characters/?page=24). GameDaily. Archived from the original (http://www.gamedaily.com/articles/galleries/top-25-most-bizarre-fighting-characters/?page=24) on 2009-02-06. . Retrieved 2009-01-11.

[25] Kasavin, Greg (2004-11-16). "Capcom Fighting Jam Review" (http://uk.gamespot.com/ps2/action/capcomfightingjam/review.html?om_act=convert&om_clk=gssummary&tag=summary;read-review). GameSpot. . Retrieved 2009-02-05.

[26] Staff (2008-03-06). "The Making of... Japan's First RPG" (http://www.next-gen.biz/features/making-japans-first-rpg). Edge Online. . Retrieved 2009-01-15.

[27] Gertsmann, Jeff (2008-10-24). "Ultimate Mortal Kombat 3 Review" (http://uk.gamespot.com/xbox360/action/ultimatemortalkombat3/review.html?om_act=convert&om_clk=gssummary&tag=summary;read-review). GameSpot. . Retrieved 2009-01-11.

[28] Kasavin, Greg (2006-01-01). "Dead or Alive 4 Review" (http://uk.gamespot.com/xbox360/action/deadoralive4/review.html). GameSpot. . Retrieved 2009-02-02.

[29] "GameSpot: The History of Street Fighter - Street Fighter II: The World Warriors" (http://web.archive.org/web/20040902095224/http://www.gamespot.com/features/vgs/universal/sfhistory/games_02_02.html). GameSpot. Archived from the original (http://www.gamespot.com/features/vgs/universal/sfhistory/games_02_02.html) on 2004-09-02. . Retrieved 2009-04-29.

[30] Katie Salen, Eric Zimmerman (2003). *Rules of Play: Game Design Fundamentals*. MIT Press. p. 526. ISBN 0262240459, 9780262240451.

[31] "Most Attractive Female Fighters From Popular Games" (http://www.gamesreviews2010.com/most-attractive-female-fighters-from-popular-games/). .

[32] Craig Glenday, ed (2008-03-11). "Record-Breaking Games". *Guinness World Records Gamer's Edition 2008*. Guinness World Records. Guinness. p. 84. ISBN 978-1-904994-21-3.

[33] "Fighter Maker review" (http://uk.psx.ign.com/articles/153/153775p1.html). IGN. . Retrieved 2007-07-22.

[34] "Soulcalibur III Review" (http://palgn.com.au/article.php?id=3356). PALGN. . Retrieved 2007-07-22.

[35] "Mortal Kombat screens and Create-A-Fighter details" (http://www.gamingtarget.com/article.php?artid=5037). Gaming Target. . Retrieved 2007-07-22.

[36] "Dragon Ball Z Budokai 3(PS2) preview" (http://uk.ps2.gamespy.com/playstation-2/dragon-ball-z-tenkaichi-neo/725582p2.html). GameSpy. . Retrieved 2007-07-22.

[37] Casamassina, Matt (2008-03-04). "Super Smash Bros. Brawl Review" (http://uk.wii.ign.com/articles/856/856580p3.html). IGN. . Retrieved 2009-01-31.

[38] Anderson, Lark (2008-03-08). "Super Smash Bros. Brawl Review" (http://uk.gamespot.com/wii/action/supersmashbros/review.html?om_act=convert&om_clk=gssummary&tag=summary;read-review). GameSpot. . Retrieved 2009-02-02.

[39] Zdyrko, David (2000-10-23). "Tekken Tag Tournament" (http://uk.ps2.ign.com/articles/163/163946p1.html). IGN. . Retrieved 2009-02-02.

[40] Ashcraft, p. 108.

[41] Olli Leino, Hanna Wirman, Amyris Fernandez (2008). *Extending Experiences*. Lapland University Press. p. 53. ISBN 9789524841979 9524841975.

[42] Shirley R. Steinberg (2010), Shirley R. Steinberg, Michael Kehler, Lindsay Cornish, ed., *Boy Culture: An Encyclopedia* (http://books.google.co.uk/books?id=XRGEIqzv5rsC), **1**, ABC-CLIO, p. 451, ISBN 0313350809, , retrieved 2011-04-02

[43] Ashcraft, p. 94.

[44] Nadia Oxford, 20 Years of Street Fighter (http://www.1up.com/features/20-years-street-fighter), 1UP.com, 12/11/2007

[45] *Champion Boxing* (http://www.arcade-museum.com/game_detail.php?game_id=7308) at the Killer List of Videogames

[46] IGN Presents the History of SEGA: Coming Home (http://uk.retro.ign.com/articles/974/974695p2.html), IGN

[47] Ryan Geddes & Daemon Hatfield (2007-12-10). "IGN's Top 10 Most Influential Games" (http://games.ign.com/articles/840/840621p1.html). IGN. . Retrieved 2009-04-14.

[48] *Karate Champ* (http://www.arcade-museum.com/game_detail.php?game_id=8279) at the Killer List of Videogames

[49] *Karate Champ - Player Vs Player* (http://www.arcade-museum.com/game_detail.php?game_id=8280) at the Killer List of Videogames

[50] Hjul, Alison (1986-03). *Yie Ar Kung Fu*. Your Sinclair. p. 19.

[51] Game of The Week: Yie Ar Kung-Fu (http://classicgaming.gamespy.com/View.php?view=GameMuseum.Detail&id=82), *GameSpy*, accessed 2011-02-27

[52] Ste Curran (2004). *Game plan: great designs that changed the face of computer gaming* (http://books.google.co.uk/books?id=TXcWIWkIZ0AC&pg=PA40). Rotovision. p. 40. ISBN 2880466962. . Retrieved 2011-04-10

[53] *Yie Ar Kung-Fu* (http://www.arcade-museum.com/game_detail.php?game_id=10513) at the Killer List of Videogames

[54] Davies, Jonathan (1988-10). *Karate Ace*. Your Sinclair. p. 46.

[55] *Data East USA, Inc. v. Epyx, Inc.*, 862 F. 2d 204, 9 U.S.P.Q.2d (BNA) 1322 (http://bulk.resource.org/courts.gov/c/F2/862/862.F2d.204.87-2294.html) (9th Cir. 1988).

[56] "History of Sega Fighting Games" (http://www.gamespot.com/features/vgs/sat/segafight/page2.html). GameSpot. . Retrieved 2008-10-11.

[57] "Game Design Essentials: 20 Mysterious Games" (http://www.gamasutra.com/view/feature/3485/game_design_essentials_20_.php?page=16). Gamasutra. . Retrieved 2008-10-12.

[58] "Monday Bloody Monday" (http://www.1up.com/features/monday-bloody-monday). 1up. . Retrieved 2009-04-29.

[59] O'Neill, Cliff & Greeson, Jeff (1999-11-01). "History of Mortal Kombat" (http://uk.gamespot.com/gamespot/features/video/hist_mortalk/index.html). GameSpot. . Retrieved 2009-01-12.

[60] *GAMEST Magazine* **134**. December 30, 1994.

[61] Horowitz, Ken (2004-08-03). "Top 10 Tuesday: Worst Game Controllers" (http://web.archive.org/web/20100206204922/http://www.sega-16.com/feature_page.php?id=85&title=Genesis Accessory &Peripheral Guide). *Sega-16* (http://www.sega-16.com/). Archived from the original (http://www.sega-16.com/feature_page.php?id=85&title=Genesis Accessory &Peripheral Guide) on 2010-02-06. . Retrieved 2010-12-04.

[62] *Light Harp at CES 1993* (http://www.youtube.com/watch?v=YoxsnCiX05k). YouTube. . Retrieved 2010-07-06.

[63] Harris, Craig (2006-02-21). "Top 10 Tuesday: Worst Game Controllers" (http://xbox360.ign.com/articles/690/690449p1.html). *IGN*. . Retrieved 2009-08-07.

[64] "Mortal Kombat II Review" (http://ps3.ign.com/articles/780/780492p1.html). IGN. . Retrieved 2008-10-11.

[65] "Battle Arena Toshinden takes the fighter into true 3-D, but is it enough?" (http://uk.psx.ign.com/articles/150/150716p1.html). IGN. November 21 1996. . Retrieved 31 August 2011.

[66] "IGN: King of Fighters '94" (http://wii.ign.com/articles/852/852523p1.html). IGN. . Retrieved 2008-10-17.

[67] *Electronic Gaming Monthly's Buyer's Guide*. 1995.

[68] "Street Fighter EX Plus Alpha for PlayStation" (http://www.gamerankings.com/ps/198825-street-fighter-ex-plus-alpha/index.html). GameRankings. 1997-09-30. . Retrieved 2011-06-01.

[69] "Street Fighter EX Plus Alpha - PlayStation Review at IGN" (http://uk.psx.ign.com/articles/152/152120p1.html). Uk.psx.ign.com. 1997-10-26. . Retrieved 2011-06-01.

[70] "Imagine Media's PSM Names Top 25 PlayStation Games of All Time" (http://www.thefreelibrary.com/Imagine+Media's+PSM+Names+Top+25+PlayStation+Games+of+All+Time.-a050215867). Imagine Media. .

[71] "Dreamcast Virtua Fighter 3 Ships" (http://www.gamespot.com/dreamcast/action/virtuafighter3tb/news.html?sid=2448719). GameSpot. . Retrieved 2008-10-12.

[72] "Sega Scraps the Dreamcast" (http://news.bbc.co.uk/2/hi/business/1145936.stm). BBC. 2001-01-31. . Retrieved 2008-08-22.

[73] "The History of SNK" (http://uk.gamespot.com/features/6089278/index.html). GameSpot. . Retrieved 2009-01-12.

[74] Andrew Seyoon Park (11-05-2001). "Fatal Fury: Mark of the Wolves Review" (http://www.gamespot.com/dreamcast/action/fatalfurymarkofthewolves/review.html). GameSpot. . Retrieved 2008-11-12.

[75] "The Best and Worst of 2001 - BEST FIGHTING GAME" (http://www.gamespot.com/gamespot/features/video/bestof_2001/p3_16.html). GameSpot. 2001. . Retrieved 2008-11-12.

[76] Johnny Minkley (2008-11-26). "Rare "may do" new Killer Instinct" (http://www.eurogamer.net/articles/rare-may-do-new-killer-instinct). EuroGamer. . Retrieved 2008-11-28.

[77] "Saving Street Fighter: Yoshi Ono on Building Street Fighter IV" (http://www.gamasutra.com/view/feature/3791/saving_street_fighter_yoshi_ono_.php?page=2). GamaSutra. . Retrieved 2008-10-12.

[78] Gerstmann, Jeff (1998-03-30). "Tekken 3 Review" (http://uk.gamespot.com/ps/action/tekken3/review.html?om_act=convert&om_clk=gssummary&tag=summary;read-review). GameSpot. . Retrieved 2009-01-11.

[79] Robertson, Ed (1997-04-03). "Soul Blade Review" (http://uk.gamespot.com/ps/action/soulblade/review.html?om_act=convert&om_clk=gssummary&tag=summary;read-review). GameSpot. . Retrieved 2009-01-11.

[80] Calvert, Justin (2008-07-31). "Soulcalibur IV Review" (http://uk.gamespot.com/xbox360/action/soulcaliburiv/review.html?om_act=convert&om_clk=gssummary&tag=summary;read-review&page=2). GameSpot. . Retrieved 2009-01-11.

[81] Staff (1998-03-27). "Dead or Alive (PS)" (http://uk.psx.ign.com/articles/152/152306p1.html). IGN. . Retrieved 2009-01-12.

[82] Rorie, Matthew (2006-01-09). "Two Men Enter, One Man Leaves..." (http://uk.gamespot.com/features/freeplay/index.html?story=6142102&tag=result;title;2). GameSpot. . Retrieved 2009-01-12.

[83] "Top 25 Beat-'Em-Ups: Part 1" (http://web.archive.org/web/20110714203743/http://www.nowgamer.com/features/434/top-25-beat-em-ups-pt-1). *Retro Gamer*. 2 October 2009. Archived from the original (http://www.nowgamer.com/features/434/top-25-beat-em-ups-pt-1) on 2011-07-14. . Retrieved 2011-03-17.

[84] "Nintendo designs fighting game for its Wii console" (http://www.financialexpress.com/news/Nintendo-designs-fighting-game-for-its-Wii-console/284337/). The Financial Express. . Retrieved 2008-10-12.

[85] Kevin Gifford (2010-06-23). "Being The Very Best at Fighting Games" (http://www.1up.com/do/newsStory?cId=3180053). 1UP. . Retrieved 2010-06-24.

[86] "Daigo Umehara: The King of Fighters" (http://www.eurogamer.net/articles/daigo-umehara-the-king-of-fighters-interview). eurogamer.net. . Retrieved 2010-05-18.

[87] Mielke, James (2000-01-28). "SNK vs. Capcom: Match of the Millennium Review" (http://uk.gamespot.com/ngpc/action/snkvscapcommatchoftm/review.html?om_act=convert&om_clk=gssummary&tag=summary;read-review). GameSpot. . Retrieved 2009-02-05.

[88] Lopez, Miguel (2000-09-14). "Capcom vs. SNK Review" (http://uk.gamespot.com/dreamcast/action/capcomvssnkmf2000/review.html). GameSpot. . Retrieved 2009-02-05.

[89] Kasavin, Greg (2003-02-14). "Capcom vs. SNK 2 EO Review" (http://uk.gamespot.com/xbox/action/capcomvssnk2eo/review.html). GameSpot. . Retrieved 2009-02-05.

[90] Miller, Greg (2008-11-15). "Mortal Kombat vs. DC Universe Review" (http://uk.xbox360.ign.com/articles/930/930344p1.html). IGN. . Retrieved 2009-04-29.

[91] Tanaka, John (2008-12-11). "Tatsunoko VS Capcom Playtest" (http://uk.wii.ign.com/articles/937/937796p1.html). IGN. . Retrieved 2009-02-05.

[92] "Super Smash Bros. Brawl Smashes Nintendo Sales Records" (http://www.nintendo.com/whatsnew/detail/AU8xLess7wISKbSMpYCj_HThii8UiBzG). Nintendo.com. 2008-03-17. . Retrieved 2009-02-06.

[93] "GameSpot's Best of 2007: Best Fighting Game Genre Awards" (http://web.archive.org/web/20090213194247/http://www.gamespot.com/best-of/genreawards/index.html?page=5). GameSpot. Archived from the original (http://www.gamespot.com/best-of/genreawards/index.html?page=5) on 2009-02-13. . Retrieved 2008-10-12.

[94] Kasavin, Greg (2003-02-01). "Guilty Gear X2 Review" (http://uk.gamespot.com/ps2/action/guiltygearx2/review.html?om_act=convert&om_clk=gssummary&tag=summary;read-review). GameSpot. . Retrieved 2009-02-05.

[95] Ashcraft, pp. 109–112.

[96] Ashcraft, pp. 108–109.

[97] "Tekken 6 breaks 3 million sales" (http://www.eurogamer.net/articles/2010-08-06-tekken-6-breaks-million-marker). Eurogamer. 2010-08-06. . Retrieved 2010-08-08.

[98] "Xbox Live: Street Fighter II Hyper Fighting" (http://web.archive.org/web/20090430183304/http://www.xbox.com/en-US/games/s/streetfighter2livearcadexbox360/default.htm). Xbox.com. Archived from the original (http://www.xbox.com/en-US/games/s/streetfighter2livearcadexbox360/default.htm) on 2009-04-30. . Retrieved 2009-02-06.

[99] "Super Street Fighter II Turbo HD Remix, Super Street Fighter II Turbo HD Remix PS3 - GameSpot.com" (http://uk.gamespot.com/ps3/action/superstreetfighteriiturbohdremix/index.html). Uk.gamespot.com. . Retrieved 2011-06-01.

[100] Chiappini, Dan (2009-02-18). "Street Fighter IV Review" (http://uk.gamespot.com/ps3/action/streetfighteriv/review.html?om_act=convert&om_clk=gssummary&tag=summary;read-review&page=2). GameSpot. . Retrieved 2009-02-26.

[101] Rogers, Tim (2008-10-12). "The 20 Best Games at TGS" (http://web.archive.org/web/20081015034257/http://www.edge-online.com/features/the-20-best-games-tgs?page=0,3). Edge Online. Archived from the original (http://www.edge-online.com/features/the-20-best-games-tgs?page=0,3) on 2008-10-15. . Retrieved 2009-02-09.

[102] "Top 10 Franchises Capcom Should Bring Back" (http://www.411mania.com/games/columns/175683/The-10th-Hour-02.25.2011:-Top-10-Franchises-Capcom-Should-Bring-Back.htm). 411mania.com. 2011-02-25. . Retrieved 2011-02-26.
[103] "Marvel vs Capcom 3" (http://www.computerandvideogames.com/287197/previews/marvel-vs-capcom-3-the-most-bonkers-beat-em-up-of-the-year/). *Computer and Video Games*. 5 February 2011. . Retrieved 2011-02-26.
[104] "Marvel vs. Capcom 3: Fate of Two Worlds Review" (http://www.shacknews.com/article/67542/marvel-vs-capcom-3-fate). Shacknews. . Retrieved 2011-02-26.
[105] "Deadliest Warrior: The Game Review" (http://uk.ps3.ign.com/articles/113/1132247p1.html). IGN. November 3, 2010. . Retrieved 2011-03-24.

Single-player_video_game

A **Single-player video game** is a video game where input from only one player is expected throughout the course of the gaming session. "Single-player game" usually implies a game that can only be played by one person, while "single-player mode" usually refers to a game mode for a single player, where several players may play in a different mode.[1] The earliest video games, such as *Tennis for Two*, *Spacewar!*, and *Pong*, were two-player, with single-player games gaining popularity soon after with titles such as *Speed Race* and *Space Invaders*. In 1978, the first multiplayer role-playing games, known as MUDs, were created. The early 1990s introduced many games which utilized local area networks and null modems for multiplayer mode. *Doom* is a notable example of one such game.

The major selling points of larger single-player games are interesting storylines, impressive graphics, and realistic non-player characters and opponents. Notable examples include action-adventure games such as *The Legend of Zelda*, platform games such as *Mario* and *Sonic*, stealth games such as *Metal Gear*, survival horror such as *Resident Evil* and *Silent Hill*, and first-person shooters such as *Doom*, *Half-Life* and *Deus Ex*. Selling points of the smaller games are low learning curve and availability (many are free to play on various sites).

Certain game genres are inherently oriented towards single-player in their design. Such genres include puzzle games, such as *Tetris*, and plot-based role-playing games (RPGs), such as *Dragon Quest* and *Final Fantasy*. Recent additions to these genres, such as the massively multiplayer online games (MMOGs) or online multiplayer versions of *Tetris*, are serving to undo this trend.

The vast majority of modern console games and arcade games are designed so that they can be played by a single player; although many of these games have modes that allow two or more players to play (not necessarily simultaneously), very few actually require more than one player for the game to be played. The *Unreal Tournament* series is an example of such.

See also

- Multiplayer video game
- List of video gaming topics

References

[1] Oosterhuis, Kas; Feireiss, Lukas (March, 2006). *The Architecture Co-laboratory: Game Set and Match II : on Computer Games, Advanced Geometries, and Digital Technologies* (http://books.google.co.uk/books?id=tXBdOoZ-faYC&pg=RA2-PA108&dq=single+player+simulations#PRA2-PA108,M1). Delft: Delft University of Technology. p. 180. ISBN 9059730364. .

Multiplayer_video_game

A **multiplayer video game** is one which more than one person can play in the same game environment at the same time. Unlike most other games, computer and video games are often single-player activities that put the player against preprogrammed challenges and/or AI-controlled opponents, which often lack the flexibility and ingenuity of regular human thinking. Fantge has members and non mebers Multiplayer components allow players to enjoy interaction with other individuals, be it in the form of partnership, competition or rivalry, and provide them with a form of social communication that is almost always missing in single-player oriented games. In a variety of different multiplayer game types, players may individually compete against two or more human contestants, work cooperatively with a human partner(s) in order to achieve a common goal, supervise activities of other players, or engage in a game type that incorporates any possible combination of the above. Examples of better-known multiplayer gametypes include deathmatch and team deathmatch, MMORPG-associated forms of PvP and Team PvE, capture the flag, **domination** (competition over control of resources), co-op, and various objective-based modes, often expressed in terms of "assault/defend a control point". Multiplayer games typically require the players to share resources of a single game system or use networking technologies that allow players to play together over greater distances.

History

Some of the earliest known video games were two-player games, including early sports games like *Tennis For Two* (1958) and *Pong* (1972), early shooter games like *Spacewar!* (1962), and early racing video games like *Astro Race* (1973).[1] First known examples of massively multi-player real time games based around real time networking were developed on the PLATO system starting around 1973. Important multi-user games developed on this system included Empire from 1973 and Spasim from 1974. The latter was a pioneering first-person shooter.

The first large scale serial sessions based around a single computer were STAR (based on the series Star Trek), OCEAN (a battle of ships, submarines and helicopters with multiple players divided up between the two combating cities) and CAVE (based on Dungeons and Dragons), created by Christopher Caldwell (with art work and suggestions by Roger Long and some assembly coding by Robert Kenney) in 1975 on the University of New Hampshire's DECsystem-1090. The University's computer system had hundreds of terminals connected via serial lines through cluster PDP-11s for student, teacher and staff access. The games worked by having one instance of the program running on each terminal (for each player), sharing a segment of shared memory (known as the "High segment" in the OS TOPS-10). Due to their popularity, the games were frequently banned by the University's Computer Services since they could easily take up all available RAM and cycles. STAR was based on the original single-user turn oriented BASIC program STAR written by Michael O'Shaughnessy at UNH in 1974.

Digital Equipment Corporation soon distributed another multi-user version of Star Trek called Decwars though not featuring real-time screen updating. Decwars was widely distributed to universities with DECsystem-10s. In 1981, Cliff Zimmerman wrote a more detailed homage to Star Trek in Macro-10 for DECsystem-10s and -20s using VT100 series graphics. "VTtrek" pitted 4 Federation players against 4 Klingons in a three dimensional universe.

MIDI Maze was an early first-person shooter released in 1987 for the Atari ST. It was unique in featuring network multiplayer through the MIDI interface long before mainstream Ethernet and Internet play became commonplace. It is considered the first multiplayer 3D shooter on a mainstream system and the first major network multiplayer action game, with support for as many as 16 players. It was followed up by ports to various platforms in 1991 under the title *Faceball 2000*, including the Game Boy and Super NES, making it possibly the first handheld and multiplatform first-person shooter and an early console example of the genre.[2]

Called "netplay" to refine the meaning. The first popular videogaming title to release a LAN version was Spectre in 1991 for the Apple Macintosh, featuring AppleTalk support for up to eight game players. Spectre's multiplayer popularity was attributed in part by the display of a player's name above their cyber tank vehicle. This was followed by Doom in 1993, when the first network version of the game allowed a total of four simultaneous gamers.[3]

Playing networked multiplayer games via LAN often eliminates problems common in Internet play, such as lag and anonymity of players. As a result, multiplayer games usually are the focus of LAN parties. Play-by-email games are multiplayer games that use email as the method of communication between computers. Other turn-based variations which do not require players to be online at the same time are Play-by-post gaming and Play-by-Internet. Some online games are "massively multiplayer" games, which means that a large number of players participate simultaneously. The two major genres are MMORPG (massively multiplayer online role-playing game) such as *World of Warcraft* or *EverQuest* and MMORTS (massively multiplayer online real-time strategy).

Some networked multiplayer games do not even feature a single-player mode. For example, MUDs and massively multiplayer online games, such as RuneScape are multiplayer games by definition. First-person shooters have become very popular multiplayer games and games like *Battlefield 1942* and *Counter-Strike* gained their fame despite not featuring extensive (or any) single-player plot or gameplay. Casual gaming has also seen an emergence in networked multiplayer games as shown by developer and gaming site OMGPOP, whose library of offerings only include multiplayer flash games. The biggest MMOG in the world is Lineage out of South Korea with 14 million registered gamers which is played in several mostly Asian countries. The biggest Western MMOG in 2008 is World of Warcraft with over 10 million registered gamers worldwide. This category of games currently requires multiple machines to connect to each other over the Internet, but before the Internet became popular, MUDs were played on time-sharing computer systems, and games such as *Doom* were played on a LAN.

Gamers often refer to latency by the term ping, which measures round-trip network communication delays (by the use of ICMP packets). For example, a player on a DSL connection with a 50 ms "ping" will be able to react faster to game events than a modem user with 350 ms average latency. Another popular complaint is packet loss and choke, which can render a player unable to "register" their actions with the server. In first-person shooters, this problem usually manifests itself in the problem of bullets appearing to hit the enemy, but the enemy taking no damage. Note that the player's connection is not the only factor; the entire network path to the server is relevant, and some servers are slower than others. While latency is frequently complained about, many players believe a lack of finesse and decent tactics is more damaging than a slow connection in most games. Major and frequent variations in latency, however, can be another story; these can make it very difficult to properly play the game.

Starting with the Sega NetLink in 1996, Game.com in 1997 and Dreamcast in 2000, game consoles have also begun to support network gaming, over both the internet and LANs. Many mobile phones and handheld consoles also offer wireless gaming through Bluetooth or similar technologies.

Turn-based multiplayer

Other early video games sometimes included turn-based multiplayer modes, popular in tabletop arcade machines. In such games play is alternated at convenient points, often after the loss of a life. Scores of all players are often retained onscreen in order that players can gauge their relative standing.

Single-system

In modern console games, arcade games, and personal computer games, *multiplayer* usually implies that the players play together by using several controllers plugged into the same game system. Home console games often use split-screen so that each player has an individual view of the action (important for genres such as first person shooter and racing video games), although most arcade games and some console games (ranging from *Pong*) do not.

See also

- Cooperative gameplay
- Dedicated Game Server
- Multiplayer online game
- Online game
- Spawn installation
- Video game center

References

[1] *Astro Race* (http://www.arcade-museum.com/game_detail.php?game_id=6949) at the Killer List of Videogames
[2] Parish, Jeremy, The Essential 50: Faceball 2000 (http://www.1up.com/do/feature?cId=3133741), *1UP*, Accessed April 24, 2009
[3] Doom (computer game) on Britannica (http://www.britannica.com/EBchecked/topic/932958/Doom)

Multiplayer_online_game

A **Multiplayer Online Game** is a multiplayer video game which can be played via a game server over the internet, with other players around the world. Some prominent examples of this are *Battlefield 2*, *Halo 3*, *Counter-Strike*, *Quake 3*, *Warcraft 3*, *StarCraft*, *Call of Duty Black Ops*, *NBA 2K11*, and the Sega Meganet.

These games differ from MMORPGs in that they do not create a persistent world, but create a playing arena for the purpose of a single game or round. In other words, they rely on a game listen server used only for that round, and there can be numerous servers all around the world. MMORPGs on the other hand, rely on dedicated servers, as these games must be running continuously.

Server structure and gameplay

The existence of a wide variety and number of servers has made possible several variations on gameplay.

For example, in *Battlefield 2*, various servers have their own names, websites and gaming groups known as "clans." Often a list of rules will display when a player first logs on.

Browser-based MOG

A browser-based multiplayer online game (BMOG) is a special case of multiplayer online game (MOG) in the form of a browser game. The term could also be applied to many other browser-based competitions.

In order to run in a web browser, the client-side implementation must be a client side solution such as HTML, JavaScript, Adobe Flash, Java or a browser plug-in. Unlike a stand-alone client or video game, being confined to a browser limits to some degree the extent to which 3-D rendering can be supported. BMOG's can be seen as an evolutionary development of browser-based implementations of board games or forum games.

Many types of MOGs can potentially be browser-based. Popular examples include simple web sites such as a simple website for a prediction market or games that involve some 3D rendering such as *Quake Live* or MMORPGs such as *RuneScape*.

See also

- Comparison of MMORPGs
- History of MMORPGs
- History of online games
- List of free MMOGs
- List of free multiplayer online games
- List of freeware video games
- List of MMORPGs
- List of multiplayer browser games
- List of text-based MMORPGs
- MMOFPS
- MMOG
- MMORPG
- MMORTS
- MUD
- Multiplayer video game
- Online game
- Sega Meganet

External links

- MMORPG.com [1] Overview site on MMORPGs.

Article Sources and Contributors

Dragon Ball Z: Ultimate Tenkaichi *Source*: http://en.wikipedia.org/w/index.php?title=Dragon_Ball_Z:_Ultimate_Tenkaichi *Contributors*: Banjotron2000, Blackgaia02, Chiefmartinez, Coolboy3452, Crustyshampoo, Darksimmsy, Djashford1, Grafen, GroovySandwich, HJ Mitchell, Jeangabin, JohnCD, Kingrizvi23, Mahewa, MechaTeddy417, MelbourneStar, MigrantP, Munem939, Sarujo, ShadowRatchet92, Sjones23, Sole Soul, Spinrad, StarScream18, SupremeEvan, Taelus, Tbhotch, The Earwig, Thedarkknight57, Veniks, Vitor Summer, Wknight94, X201, ZamorakO o, 192 anonymous edits

Dragon Ball *Source*: http://en.wikipedia.org/w/index.php?title=Dragon_Ball *Contributors*: 11vegeta11, 123Hedgehog456, 1337h4h4h4, 1oddbins1, 317sokr, 333baddog123, 949paintball, 95j, Abdi 909, Addihockey10, Adidas159, Aeonx, Afro239, Agent007ravi, Aitias, Ajs111, AkankshaG, Allen4names, AllyUnion, Andrewcrawford, Andros Amatakis, Andyjsmith, AnmaFinotera, Anna Lincoln, Answerwriter21, Arman Cagle, Arthena, AtlaOS, Avenged Eightfold, Ayrton Prost, Azips, B7T, Banime, Bardan03, Barts1a, BatteryIncluded, Beamathan, Beeswaxcandle, Bigman5667, Billy Liakopoulos, Blake-, Blast your rage, Blehfu, Blood sliver, Bobblehead, Boflexan, Bomac, Bonadea, Bongwarrior, BornonJune8, Boudinot b95, Brandonbee, Broly555, BryanBR01, Bull-Doser, Buspar, CBLee1978, CFLeon, CTZMSC3, CageBakur501, CanadianLinuxUser, Canarias, Causa sui, Chamal N, Chelo61, Chiefmartinez, Chimpflyer, ChipmanX, Chowbok, Clarince63, Coasterlover1994, Coentor, Cold Skin, ConcernedVancouverite, Coolboyjohn, Coralmizu, Cruzis619, Curtdude10, D4c3nt3n0, DARTH SIDIOUS 2, DBZ MANGA, DBZ power levels, DBZROCKS, DBZfan29, DMacks, DWP102589, Da Vynci, Daftpunkboy93, Damnedfan1234, Damonstration, Daniel 993, Daniel Carrero, Daniel Case, Danielgonz, Danlip1, Danman8760, Dantman, Dark Rain, Dark dragon474, Darktower 12345, Darkwind, Darth Goku, Darth Knight, David31584, Deerstop, Demonkoryu, DestradoTensai, Dfsghjkgfhdg, Diannaa, Dinoguy1000, Dion8114, Dirjarnocksorz, Discospinster, DisillusionedBitterAndKnackered, Disturbed92893, Dlohcierekim, Don4of4, Dougweller, Douken, DrVonDre, Dragonballaf32, Drmies, Dwanyewest, ERAGON, ETLamborghini, EWikist, EddieMcfluff, Edokter, Efficient vegetariannnwf607, Elchoaro, Elm-39, Empty Feeling, EoGuy, Epbr123, Erachima, Erebus Morgaine, Eurokazi, Excirial, Falcon8765, Fanatix, Fanpoptvshows, Fantastic fred, Fastily, Favonian, Ferrior, Fieldday-sunday, Figureskatingfan, FreshCorp619, Friginator, Fuhghettaboutit, Funkamatic, Fæ, G.A.S, GB fan, GTBacchus, Gaius Cornelius, Galilsnap, Gashius, GeoffCapp, Geoffman13, Geopgeop, Ghaleonh41, Giants27, Gnbfd0, Gogo Dodo, Goku1st, Gokuthesaiyan127, GoldPro, Gonnafly1, Gravitywheel, Greeney, Griffinofwales, GroovySandwich, Gscshoyru, GuKeltke, Guoguo12, HalfShadow, Halmstad, Ham236, Han218, Happysailor, Hellp201, Hu12, Imanimanad2, Infinitysend, Instagator, Interfear2, Intheeventofstructuralfailure, Inxsfan, Iridescent, Issyl0, Italik, Ivansanchez, J'onn J'onzz, J. M., J.delanoy, JHunterJ, JNW, JV Smithy, Jab843, Jackol, James26, Jamesontai, Jay32183, Jericho190, Jienum, JimVC3, Jj98, Jlaking, Jlochoap, Jmlk17, John Smith (test), Joku43, Jolelle, Jon21488, Jonathan Hall, Jonathancoles11, Jonny2x4, JordashTalon, Joshschr, Jrp, JuangM, Judgesurreal777, Jump Guru, Justme1altF4, Jwhyte1, Kahsiv, Kaijucole, Kanabekobaton, Kangaroo186, Kayau, Kdodds22, Kflester, Khairul hazim, Kimchi.sg, Kinaro, King of knives, Kingofthering1994, Kiwichris, Kogsquinge, KojiDude, KrebMarkt, Kurtle, Kyaa the Catlord, Kylie jensen, L Kensington, LHCD1, Lacrymocéphale, Landon1980, Laurinavicius, Laws dr, LeaveSleaves, Legion fi, Leonard^Bloom, Leonopteryx, LilHelpa, Linkdude20002001, Livebrick, Logan, LoganTheGeshrat, Lord Crayak, Lordrockman, Louis the Rogue, Loyalist Cannons, Ltiler, Lucia Black, Luk, MC10, MJ94, Magegg, Majinbuu54321, Malkinann, Mandady12, MangaRJ, Mariansavu, Marshmello123, Martarius, MastaFighta, Materialscientist, Maxim, Mazca, Mclay1, Mebtora, Melaen, Menswear, Metricopolus, Migrations, Mike Rosoft, MikeWazowski, Miles Blues, Minna Sora no Shita, Mister Universe, Mjames4040, Mk32, Moloch09, Moshechoc, Mr Vinx, Mr.BoBo Boob, Mr.Z-man.sock, Mrscnheebly 09, Mrtvshow, Mtirkmane, Muchi, MuzRat, My76Strat, NHRHS2010, Nandunr, NarutoD444, Nathan, NawlinWiki, Ned Scott, NeoSuperBlissey, Neustradamus, Newone, Nightscream, Nihonjoe, Nintendoman01, Nohansen, Noliver, Nsaa, Nymf, Ocaasi, Odie5533, Ojoe117, Oldirtygrza, OliverDF, Ollice, Only)(uman, Onore Baka Sama, Ophois, Orange Suede Sofa, Orangesodakid, Oroso, Oxymoron83, PTJoshua, PWWII, Pachang, Panicpack121, Paralympic, PeRshGo, Peepnklown, Pennyadodumuss, Peterdude4, Philip Trueman, Piano non troppo, Pinkadelica, Pkisme, Plau, Poop2000, Porporal1, Propaniac, Puffin, Quaeler, Qwyrxian, R000t, Raging goddess, RaidonMakoto, RandomAct, Ratchetcomand, RattleMan, Reach Out to the Truth, Reaper Eternal, Redbird 41, Relly Komaruzaman, RexLoner, Rickinasia, Rjwilmsi, Robathruz, Robwar, Rodeosmurf, Rrburke, Rror, Rukhairy, Ruy Pugliesi, Ryu Ematsu, Ryulong, SSB Fan, STFX1046190, Sakmode, Samiibnjamil, Santryl, Sarujo, Scapler, SchfiftyThree, Scottyhoohow, Sean The Spartan93, Secret (renamed), Sesshomaru, Sevilledade, Sgt.pepper54321, ShadowRatchet92, Shadowjams, Shake320, Shawnjohnson3, ShelfSkewed, Shicoco, Sirius85, Sjones23, Sk Nator, Skrapz2xw, Sktlz123, SkyWalker, Smijes08, So God created Manchester, SoleX sy, Someguyatyou, Son waleed, SonGoku786, Sperril, SpikeToronto, SpookYouGhost, StaticGull, Stevenrasnick, Stigler, StrangerAtaru, Stroppolo, Suavelyphe, Subhaan123, Subhkanish, Sumit sony145, SuperCurseSealSauske, SuperReaper13, SuperTiencha, Supermann, Supermanx93, Syst3mfailur3, TJ Spyke, Takuyaono, Tapion99, Techdawg667, Tekinszilla, Teles, Tezzdawggz, The Master of Mayhem, The Rogue Penguin, The Thing That Should Not Be, TheBlackPaladin, TheEdgartv, TheFarix, Thecheesykid, Thegamemuster, Theo10011, Theoneintraining, Therexbanner, Thingg, Tide rolls, Tigerclaws98, Tintor2, Tiptoety, Tombomp, Tony1, Topher385, Toshio Yamaguchi, Toxicroak, Tpbradbury, TreyC11, Trollololol, Tslocum, Ttc817, Turtleheart, TvGenious, Tylerob7, Uhai, Ultix, UnicornTapestry, UnqstnableTruth, UrbanEpitaph, UzEE, Valoem, VampireBaru, Vanitas100, VegetaSaiyan, Vicarious, Victory93, Vinyldestination, Vivify, Voku0427, Vrblitz, Vrenator, Wasbeer, Washburnmav, Wayne Slam, We hope, Whisky drinker, WhisperToMe, Wikifordbz, Wikipedian2, Wikipelli, Wikiwiki1234567891011, Will Beback Auto, William Avery, Winrarover9000, Winston365, Wipqozn113, Woohookitty, Worldwidewaffle, WtW-Suzaku, XRK, Xanzzibar, Xezbeth, Xoger, Xxnightbarkxx, Xxshadow9303xx, Xxzeromercyxx, Yahia.barie, Yamamoto Ichiro, Yerp19, Yottamol, Zaharous, Zero no Kamen, Zuzuu, 1269 anonymous edits

Manga *Source*: http://en.wikipedia.org/w/index.php?title=Manga *Contributors*: 000peter, 07michaelg, 100110100, 12ab3, 17tracks, 24.252.8.xxx, 2Receive, 4hodmt, 63.192.137.xxx, 65.82.160.xxx, ABF, ADS190, ARUNKUMAR P.R, Ace2209, Active Banana, Adashiel, Addshore, Ahoerstemeier, Aiko Loveli, Aimabe, AirBa, Aitias, Ajl772, Akanemoto, Akira-otomo, Aknorals, Akuyume, Alackobama, Alai, Alansohn, Aldnonymous, Ale jrb, AlexTiefling, Alexthe5th, Allen4names, Allstarecho, Alpha Quadrant, Altiego, Alton.arts, Amazonfire, Amoi, AnOddName, Anaxial, AndrewTJ31, Andycjp, AngelSanguinius, Angelzfromhell0364, Anger22, Anime2828, Animechristian, AnmaFinotera, Anna Lincoln, Anonymous Dissident, Ans, Antaeus Feldspar, Aponar Kestrel, Applejaxs, Arman Cagle, Arthur96green, Ashmoo, Ashwai95, Ask123, Astral, Atlant, Aubrey4650, Audiosmurf, Avoided, Awien, Awigo50, AxelBoldt, Axl, Aznhobo11, Azukimonaka, BCKILLa, Bact, BakaElf, Baloonkaz4life!, Batz-Maru, Bbenzon, Bdesham, Beckmcq, Beeblebrox, Ben-Zin, Bendono, Benjwong, Bensin, Bestlyriccollection, Beta m, Beyond silence, Bhadani, BiT, Bibleprophecytruth, Bibliomaniac15, Binary TSO, Bishonen, Black779, Bobabobabo, Bobbinally, Bobblehead, Bobdoe, Bobo192, Boi217, Bongwarrior, Boogieboy, Booshakla, Branddobbe, Brianga, Bryan Derksen, Bte99, Bubwater, Bugo30, Burzmali, Buster7, CIreland, CJLL Wright, CR85747, Calaschysm, Calton, Can't sleep, clown will eat me, Canadian-Bacon, CanadianLinuxUser, Canderson7, Canonmaster24, CapitainMoose, Captain Disdain, Carefulcollie, Carey Evans, Cartoonmaster, Caspian blue, Cast, Catesys, Catgut, Catsownface, Cattus, Cbm, Cenarium, CesarB, Cetacean21, Chedderchelsea, Chensiyuan, Chmod007, Chris83, ChuChu, Chuunen Baka, Cirt, Ck lostsword, Ckorff, Claytonian, Clevomon, Closedmouth, Cluth, Collectorjake12, ColourBurst, Conversion script, Corington, Corvus cornix, Courcelles, Crazycomputers, Crossedwires, Ct lance, Ctjf83, Curly Turkey, CutieNakky, Cxz111, CyberSkull, Cybercobra, Cyktsui, Cyrus Sixguns, DCAnderson, DHN, DVD R W, DVdm, DaL33T, DabMachine, Dabomb87, Dalgspleh, DanB DanD, Danceswithzerglings, Dandy Sephy, Danger, Danielrocks15, DannyWilde, Dante Alighieri, DarkAdonis255, DarkTripe, Darkstar949, Darkwind, Darth Panda, Darth Shogun XX, Daveryan, DeadEyeArrow, Deadcandy, Debresser, Deerstop, Deflective, Dekimasu, Denelson83, DenisMoskowitz, DerHexer, DerekKearney, Design, Dey Out In Da Cuh, Diderot, Difu Wu, Dijxtra, Dinobrava, Dinoguy1000, DionysosProteus, Dippee, Discospinster, Djbryce3, DocWatson42, DopefishJustin, Dr31, Dreamafter, Droid b33, Dromioofephesus, Dtortot, DustFormsWords, EEMIV, EamonnPKeane, Easternknight, Ebricca, Echoray, Edward, Egan Loo, Ehccheehcche, Ekotkie, Elijya, Elvenscout742, Emc2, Emin3md0gg, Emo tears, Emperor, Emperorbma, Endroit, Enigmaman, Epbr123, Erachima, Erikku, Erisie, Escape Orbit, Esn, Eternal Pink, Euryalus, EvanProdromou, Everyking, Evice, Evil Monkey, Evil saltine, Evilembz, Evoludo, Exploding Boy, Ezhiki, FF2010, Fadeblue, Fangfufu, Fantasy, Faradayplank, Feinoha, Fennec, Fg2, Finite, Fire Kamer, Fireflow77, FisherQueen, Flatterworld, Fnfd, Francois Trazzi, Frank101, Fredrik, Fritz Saalfeld, From-cary, Fsamkieth, Fubar Obfusco, Fujitofu, Fullstop, FurubaGoddess, Gaara kun Naruto, Gail, Gaius Cornelius, GaryNigel, Geekdiva, Gene.arboit, Georgiacatcrimson, Gerbrant, GiantSloth, Gideon2, Gigemag76, Gilliam, Glane23, Gogo Dodo, Google908, Graham87, GregorB, Greyweather, Gross, Gtrmp, GunnarRene, Gurch, Gwalla, Gwern, Gwernol, HMP22, Ha ha ha whitey, Hadal, Haitike, Halfasgood, Hamstermage, Harry, Haseywho, Hatenar, Haukurth, Havoc8844, Hayabusa future, Headbomb, HedgeHog, HellecticMojo, Herbythyme, Herostratus, Hersfold, Hertz1888, HexaChord, Hiding, Hinoku, Hmccasla, Hoary, Holothurion, HonestTom, Hoodboy Shadzar, Hoshi.sasori, Hottentot, HoundsOfSpring, Howcheng, Huddley, Hut 8.5, Hydrogen Iodide, Hyju, I dream of horses, IRP, Ian Pitchford, IanOsgood, Iceberg3k, Igoldste, Ike9898, Ikiroid, Imontopoftheworld, Impala2009, InShaneee, Infosocialist, Inka 888, InnocuousPseudonym, Instantnood, Inter, Internetinternet, InverseHypercube, Io Katai, Ipleadthe5th, Iritscen, Iron Ghost, Isadoraam, Iuggernaut, Iwishiwashere, Ixfd64, Izalithium, J.delanoy, JForget, Jabelicious, Jackaranga, Jacker 22, Jackmacmillan, Jackrs, Jaeger5432, James.S, Jangsta Mmkay, Jasonbrice, Jaybob1222, Jazzy252, Jc37, Jeff G., Jeffq, Jerzy, Jim Douglas, Jj137, Jjjjjjjjjj-10, Jmlk17, Jniemenmaa, Joanofark7, Jockesama, John Ryan, Jojit fb, Jorge Stolfi, Jpark3909, Jrp, Jtpackersrule, JuJube, JudeOedipus, Juhachi, Julian Grybowski, Jump Guru, Jup184, JustAGal, Kainightroad, Kaiwanxiao, Kariteh, Kasaalan, Kasuga, Kazu-kun, KeNsHiN, Kemiv, Ken Arromdee, Kewp, Khirysly, Khym Chanur, Kiba, Killer of whites, Kilo-Lima, Kimberry352, Kingpin13, KirbyMeister, Kitty1983, Kneegrowlong, KnighTeen87, KnowledgeOfSelf, Konstable, Koolabsol, Korath, Kotjze, Kukini, Kusunose, Kwamikagami, KyuuA4, L Kensington, La goutte de pluie, LaggedOnUser, LapoLuchini, LeLoupeNoir, Leandrod, LeaveSleaves, Lenny-au, Leszek Jańczuk, Leucius, Leuko, Leuqarte, LibLord, Lightmouse, Lightsage, LilHelpa, Llamangry, Llywrch, Loginnow, Londonfi, Longhair, Looxix, Lord Air, LordHoborgXVII, Lordshmeckie, Lovemyshoes, Lowellian, Lošmi, Lucid, Luna Santin, Lyhana8, Lysdexia, MER-C, Mackeriv, Macy, Madrat, Maestro25, Mailer diablo, Mako098765, Malkinann, MangaCreator, Mangartist, ManiF, Manop, Maple Leaf, Marek69, Margin1522, Mariko, MarkGallagher, Markkawika, MarsRover, Masssiveego, Master of the Orichalcos, Mathsgeek, Matt Thorn, MattTM, Matteh, Maugreas, Maximus Rex, MayaParty, MaybeJJonzalso, Mboverload, McGeddon, Mdgeist747, Meeve, MegX, Mel Etitis, Melkiades, Melodia, Mentifisto, Mgcsinc, Miaow Miaow, Michael Devore, Michael Hardy, Michaelmas1957, Mike Klaassen, Minghong, Mintguy, Mistyhut, Mitsukai, Modulatum, Moonriddengirl, Morganfitzp, Movementarian, MrWashaboo, Mranostay, Mrh30, Ms2ger, Msleeman, Murlock, Mursaatmesmer, Mushroom, Mx3, Mythologia, N5iln, NAHID, Nacho beast, NarSakSasLee, Naraht, Natkeeran, Nccarter, Ned Scott, NeilN, Neilworms, NellieBly, Neurolysis, Nezu manga, Nihiltres, Nihonjoe, Nimur, Nirakka, Nizarck, Nn123645, Nnoitra, NoSeptember, Norm mit, NotAnonymous0, Notinasnaid, NuclearWarfare, Nufy8, Nuttycoconut, Nv8200p, OChiBiAyAo, OJSlaughter, OOODDD, Oatmeal batman, Octane, Oda Mari, Ohnoitsjamie, Omokage, Oneiros, Opelio, Orchis29, OsacA-Kanzai, Otaku-Sempai, Oxymoron83, PAntoni, Palica, Paracel63, Pasuke, Paul Soth, Paul o'grady, Pb30, PeaceNT, Peanut4, Pepsisux, Pepso, Pepso2, Peregrine Fisher, Perfectblue97, Peruvianllama, Petman3000, Pfahlstrom, Phoenix7777, Phoenixrod, PhotoBox, Phthoggos, Pi nerd, Piano Bench Boy, Picapica, Pilotguy, Pinethicket, Pisomojado, Platypusdara, Pm215, Pmsyyz, Poetic Decay, Poochy, Poohea, Porty, Pourboy, Ppk01, Pschemp, Psycho Kirby, Pyrop, Qtoktok, Quantpole, Quasirandom, Qwertyuiop908, R, R'n'B, R.E. Freak, RICARDOX, RadicalBender, Radiomaru, Ragesoss, RandomP, Raven23, Rawling, Ray Radlein, Razorflame, Rdislandred, Rdsmith4, Reach Out to the Truth, Red Alien, RedWolf, RedWordSmith, Reinis, Reinosuke, Remember the dot, Remurmur, Repku, Reverend Loki, Revolutionary, Revth, RexNL, Rgoodermote, Rhobite, Rich Farmbrough, Rich257, Richard Barrett, Richard cocks, Rippakrator0.7, Rivemont, Rjwilmsi, Robby.is.on, Robert Skyhawk, Robocoder, Rockonppl, RodC, Ronbo76, Ronhjones, Roo72, RoseParks, Rst20xx, Rtty, Running, Ryoho, Ryulong, Rémih, SEWilco, SJP, SKopp, Sagaciousuk, Samtheboy, Samuelsen, Sango123, Scarian, SchfiftyThree, Scohoust, Sconway596, Seaphoto, Sephiroth BCR, Sesshomaru, Sferrier,

Shadow1, Shaina lee, Shalom Yechiel, Shay Guy, Shibboleth, Shikeishu, Shinlin3000, Shiromatt666, Shishishin, Shnorb, Shogo Kawada, Shoshonna, Siawase, SidP, Sietse Snel, Sintonak.X, Skeetch, Sketch-The-Fox, Skyefoxfire, Sloties888, Smack1011, Snarfies, Snowolf, Snoyes, SoWhy, Some jerk on the Internet, Spacecat2, Sparkit, Sparky the Seventh Chaos, Spencer.vdm, Splash, Splashkid2, Squilibob, Srikeit, Ssipseki2, Steel, SteinbDJ, Stephen B Streater, Stephenb, Stormwatch, Stuartajt1234, Stupid2, Stylecrime, Sue, Summoning, SuperHamster, Superbeecat, Superjustinbros., Suruena, Susvolans, Suva, Svanslyck, Swind, Symane, Sytrus, Szczerbiec01, THEN WHO WAS PHONE?, THF, Taichi, Taw, Tbhotch, Tdo1992, Teemeah, Tempodivalse, Tentacle Monster, The Minister of War, The MoUsY spell-checker, The Thing That Should Not Be, TheFarix, TheOtherJesse, Therealpw2, Theresa knott, Think outside the box, Thor 7, Thue, Tide rolls, Tikki hit, Timo Honkasalo, Timothy Perper, Tiptoety, Tlotoxl, Tohru Honda13, Toiims, Tokyo pocky, Toma merta bonado, Tomchiukc, Tommy2010, Tongpoo, Tony1, Tothetruth, Totoro35, Trusilver, Turnip, TvfanO1nine, Tylerob7, Uchihasharingan, Ulquiorra, Ulric1313, Uncle Scrooge, Unyoyega, Utcursch, VQuakr, Valereee, Vapour, Vbs, Vdog14, Ventusa, Vibiesh, Vincivinci, Vindicata, ViperSnake151, Vishahu, VityUvieu, Volpeculus sagacis, Vsmith, WHeimbigner, WVRMad, Wakuran, Wareware, Wayland, Wayne Slam, Werdan7, WereSpielChequers, Weregerbil, Wereon, Where next Columbus?, WhisperToMe, WhiteArcticWolf, WikHead, Wiki alf, Wikieditor06, Wikiklrsc, Wikiloop, Wikipeditor, Wikitumnus, Wilker, Willofwii, Wingchi, WinterRose, Wintran, Wirbelwind, Wknight94, Wldms, WolvenOne, WoodElf, Work permit, Worromp Warg, Woseph, Writtenright, Wwiifanatic, Xaliqen, Xedaf, Xiglofre, Xinchu, Xino, Xiong, Xnuala, Y0u43v3a, Yamamoto Ichiro, Yechan.K.B., Yellowpika, Yelyos, Yoji Hajime, Youssefsan, Yuanchosaan, Yueni, Yueti, Yug, Z E U S, ZAPEROR, Zack Holly Venturi, Zephipps, Zereshk, Ziiv, Zooweee, Zzuuzz, Σ, Александър, , 1844 anonymous edits

Anime *Source*: http://en.wikipedia.org/w/index.php?title=Anime *Contributors*: -OOPSIE-, 000peter, 041744, 10987, 165.123.179.xxx, 17tracks, 1qaz-pl, 203.12.60.xxx, 213.145.178.xxx, 2Receive, 404notfound, 98E, A Softer Answer, A-giau, ABCD, ACSE, Aaron Brenneman, Aaron Einstein, Abdullais4u, Abtract, Abyss of enchantment, Academic Challenger, Acewolf359, Acidburn24m, Adam Bishop, Adambiswanger1, Adashiel, Aeconley, Aertis, Aeusoes1, After Midnight, Ahoerstemeier, AirKingdom102, Aiuhtiauhret, Akamad, Akemi2.0, Akira-otomo, Aknorals, Akradecki, Al-malik-myth, Aldnonymous, Aldy, AlexChurchill, Alexstraz, Alison, Alison Rowe, AlistairMcMillan, AllyUnion, Alpalfour, Amcaja, Amibite, Amire80, Amulet Heart, AnOddName, Anaraug, Anders Torlind, Andonic, Andre Engels, Andres, Andrew Filipe, AndrewHows, AndrewTJ31, Andrewjackson09, Andrewlp1991, Andryono, Andy M. Wang, Andy5, Andypandy.UK, AngelFire3423, AngelSanguinius, Angelofdeath275, Angie Y., Angusmclellan, Animedow, Animegirl265, Animehater1, Animelover12, Animetreehouse, Animevandal, Animez3, AnmaFinotera, Anonymous editor, Ans, Antaeus Feldspar, Antandrus, Apocalyptic Destroyer, Apostrophe, Applecore91, Applejaxs, Appraiser, Aranel, Arb976, Archer7, Area88, Aresmo, Arjun01, Arsenal666, Article editor, Ashmoo, Ask123, Asrghasrhiojadrhr, Assmafia, AtaruMoroboshi, Augustus Sabius, Axley, Axlq, Ayls, Ayyu, Azaghal, Aznguypimp, B&W Anime Fan, Babajobu, Babak 91, Bantosh, Barneobanjo, Batz-Maru, Belphegor, Belscb, Ben D., Ben Tibbetts, Ben-Zin, BenAveling, Benjaminx, Benjwong, BesigedB, Beska, Bestlyriccollection, Bevo, Beyond silence, Bfbbrown, BiT, BigElvis, BigHouse08, Bigblah, BiggKwell, Bignose, Billdorr, Bishonen, Bitonchen, Blackkid1, Blackmamba000, Blean, Bleubeatle, BlueMario1016, Bobblehead, Bobblewik, Bobbo, Bobdoe, Bobet, Bobo192, Boleslav Bobcik, Bongwarrior, BorgHunter, Borisbaran, Bpnpx, Bradeos Graphon, Brain40, Brian Kendig, Brianabxy, Brion VIBBER, Brona, Brtkrbzhnv, Bryan Derksen, Bubba hotep, Bubbachuck, Bull-Doser, Bumblezellio, Bunny-chan, Buspar, Byron D., C12H22O11, C777, CALR, CBDunkerson, CJGB, CJMiller, CLW, CWH, Calliopejen1, Caltas, CambridgeBayWeather, Can't sleep, clown will eat me, Canadian-Bacon, Cantras, Capt Jim, Carsonauto, Cattus, CelticJobber, Centrx, Ceyockey, Chairlunchdinner, Chal7ds, Chappy-Kun, Charklecharkle, Charlieage, CharlotteWebb, Chase me ladies, I'm the Cavalry, Cheessey, Chico rico, Chmod007, Chocolateboy, Chris 73, Chris Roy, Chrism, Chugger1992, Citicat, Ckiddicus, Clarp, Closedmouth, Cmprince, Coffee4binky, Cokesmoke, Collectorjake12, ColourBurst, CommonsDelinker, Conversion script, Coolarooni, Coredesat, CorvusCorax, Count3D, Countdown to Extinction, Covington, Crann777, Crazlunatic, Crazycomputers, Crazyeddie, Creol, Crooshi, Cryptic, Csba22, Curps, Cyanoa Crylate, CyberSkull, Cynix, D-Rock, DBZGirl, DBailey635, DHN, DKqwerty, DMG413, DRaGZ, Dabomb87, Daekharel, Daftpunkboy93, Dali, Dalziel 86, Damocles 123, Damse, Dancter, Daniel, Daniel J. Leivick, DanielCD, DannyWilde, Dannychoo, Danorux, Dantes Servant, DarKnEs5 WaRr0r, Dark Shikari, Darkeffekt, DarklyCute, Darkstar949, DarshaAssant, Dave-ros, Davepealing, David Gerard, Dawg80, Deljr, Dcoetzee, Deadcandy, Deadkid dk, Deadworm222, Death blaze, Deathno, Debresser, Defender of the Law, Dekimasu, Delia19, Deltaneos, Denelson83, DenisMoskowitz, DerHexer, Destroy-all-nerds, Desulover, Dforest, Dfrg.msc, Dhp1080, DiamondDragon, Dieboybun, Dimpleswho, Dina, Dinoguy1000, Dippee, Discospinster, Disko, Dispenser, Djfinn, Dmlandfair, Doberman Pharaoh, DocWatson42, Dozols, Dposse, DrKiernan, DragonflySixtyseven, Drcwright, Dreadstar, Dream Focus, Dreamyshade, DroEsperanto, DropDeadGorgias, Drumnbass, DtEW, Dustice, Dwanyewest, Dwoodwoo, E Wing, EEMIV, ESkog, EarthboundOnett, Echidnaboy, Ed Poor, Edgar181, Edisonwhite, Editor at Large, Edward Z. Yang, Edward321, Egan Loo, Egrim, Eltargrim, Eluchil, Elvenscout742, Elvis, Elysdir, Emin3md0gg, Emperorbma, Enigmaman, Entangle, EnthusiastFR, Enviroboy, Ephert, Erachima, Ericloca, Etaph, Eva 01, Everyking, Evice, EvilCouch, ExTincTioNz, Excirial, Exert, Facts707, Fadeblue, Faerychild, Fafas, Familyman69, Fang Aili, Fantasy, Farken, Favonian, Fckjapanophiles, Feezo, Feinoha, Felve, Fennec, Ffaker, Fiction Alchemist, Filby, Filthybutter, Finite, Fireaza, Firsfron, FisherQueen, Flearosca, Flintmecha, Flockmeal, Flyer22, Flyguy649, Fnfd, Foant, Foggymindedhero, Folken de Fanel, Foobaz, For3v3rzero, Forgotten gentleman, Fourohfour, Fox, FrankCostanza, Frungi, Fryed-peach, Fschoenm, Fubar Obfusco, Furrykef, FurubaGoddess, Future Perfect at Sunrise, Fuzheado, Fvw, G.W., GBYork, Gaius Cornelius, Galactor213, Galenus, Gamaliel, Gamerzlimited, Gandalf88, Ganryuu, Gdo01, Gene.arboit, Gentgeen, GeoGreg, Georgette2, Gerbrant, Giancarloguerrero, Gilgamesh, Gilliam, Ginsengbomb, Glen, Glenn Magus Harvey, Gmaxwell, Gmcfoley, Goatbear, Godfoster, Godsmack7991, Gogo Dodo, Goldom, Goncalopp, Gooderham1, GorillaWarfare, Graham87, Gregseko, GreyCat, Greyweather, GroovySandwich, Grunt, Gtrmp, Guanaco, Gundamforce, Gunmetal Angel, GunnarRene, Guppy, Gurubrahma, GuyZani, Gwern, Gwernol, Gwonam the Wizard, Hadal, Hai There, Halfasgood, Hall Monitor, Hallpriest9, Halofan101, Halogod666, Hanasato, Hanchi, Hansag, Harrybean96, Haseo9999, Hateless, Haukurth, Hayabusa future, Headbomb, Healthwise, HedgeHog, Hiberniantears, Hikoto, Hikui87, Himalstorch, Hitsuji Kinno, Holizz, Holonboy, Homerjay, Homomofo, HoundsOfSpring, Hucz, Hut 8.5, Hutch517, Hvn0413, Hwalee76, I didn't push her, Iceberg3k, Icedog, IcemanGrrrr, Idyls, Ikarus14, Immunize, In Defense of the Artist, InShaneee, InfallibleRoxi, Inter, Inwake, Ipatrol, Ipridian, Irishguy, IsaacAA, Isida1028, Itachi guy, Iuggernaut, Ixfd64, Izukio, J-DOGG 94, J.delanoy, JHunterJ, JQF, JSH-alive, JYOuyang, Jaadowgg, Jackhopalong, Jacoplane, Jagged 85, James26, Janechra, JasonAQuest, Jayblivion, Jayden54, Jaydubayubee, Jdcooper, Jdstroy, Jdude222, Jeffrey O. Gustafson, Jeffthejiff, Jerlbaum, JetLover, Jibbley, Jinnai, Jj98, Jjjsixsix, Jlaking, Jobsearch, Joe blo76, John, John254, JohnWittle, Johnbrownsbody, JohnnyMrNinja, Jojit fb, Jonih, Jordanhass, Jorge Stolfi, JoshG, Jossi, Jpatokal, Jpgordon, Jsmith, Jsteph, JuJube, Jube, Juhachi, JulianDalloway, Juliegilbert, Jumbuck, Jump Guru, Jusdinos, Jv821, Jzcool, K.A.David, KEJ, KODAMAYSHR, KPH2293, Kaare, Kainraziel, Kanon6996, Kappa, Karn, Kasaalan, Katana4455, KathrynLybarger, KawaiiDragoness, Kazu-kun, Kchishol1970, KeNsHiN, Keeper Garrett, Keeper76, Keero, Kelly Martin, Ken Arromdee, KenoSarawa, Kernow, Kesac, Kess, Ketiltrout, Khukri, Khym Chanur, Kieff, Kihunter, Kilhan, Killt, Kim Bruning, Kimberry352, KinHikari, King Arthur6687, Kingghost, Kintetsubuffalo, Kitt2006, Kitteh2007, Kitten-Kati, Kitty1983, Kjoonlee, Kkailas, KnighTeen87, KnowledgeOfSelf, Koavf, Koolabsol, Kopf1988, Koukag, Koyaanis Qatsi, KramarDanIkabu, Krnboy1009, Ktsquare, Kufei-san, Kukini, Kungfuadam, Kunzite, KurtKotzur, Kuru, Kusunose, Kwamikagami, Kwasiamoako, Kweh!, Kyaa the Catlord, Kyokujitsu no Tenshi, Kyorosuke, KyuuA4, Kzaral, LOTRrules, LQ, LaMenta3, Laacaa, Ladyofspain, Ladysybilla, Laibcoms, Lao commando, Laudaka, Lazybook, Leafnin789, Lectonar, Lee Daniel Crocker, Leiko, Leisuresuit, LeoNomis, Lightazurik, Lightmouse, Limesuit, Lindosland, Link25, Linkdude20002001, Linsauke, Litefantastic, Lithoderm, Little Mountain 5, LittleChib, LittleOldMe, Llort, Lolurface, Londonexpo, Lord Of Demise, Lord Voldemort, LrdAzrael, Lskkat, Lucien the Librarian, Luk, Lulzed, Lumines, Luna Santin, Lupinoid, Lurkingfate, Luxor, Lysis rationale, M4bark, MER-C, MZMcBride, MacGyver07, Mackeriv, Mad Hatter, MagicJuggler, Mahanga, Mahdimus, Majorly, Makomk, Malber, Malkinann, Maln1995, Malo, Malomeat, Mamolu, Mandragorae, Marceki111, Marixist101, Martarius, Martin Osterman, MartinDK, Marudubshinki, Mary Read, Masamage, Masaruemoto, Masssiveego, Master Deusoma, Master of the Orichalcos, Mastereditoratlarge, MatLu, Math321, Matsurika, Mattbr, Matteh, Matthew0028, Mav, Maverick Leonhart, Maxinuss, MayaParty, Mayooranathan, MayumiTsuji, Mazca, McAusten, McGeddon, Mcmillin24, Mdgeist747, Meelar, Melsaran, Mentis, Mentisock, Merope, Metal man69, Metalhead0505, Metamagician3000, Mib, Michael Hardy, Michig, Michitaro, MightyOtaKing, Mightywayne, Mike Rosoft, Mike18xx, Mikespedia, Mikieminnow, Mild Bill Hiccup, Mimiko desu, MindStorM, Mingan, Minghong, Ministry of truth 02, Mirv, MisoAnimeClub, Misortie, Misterx2000, Mithrilmoon, Mitsukai, Miwasatoshi, Mjordan, Mjpieters, Mlk, Mocchi, Modemac, Monobi, Monocrat, Montrealais, MoogleFan, MoonShadow, Morris7, Mr. ATOZ, Mr.Majikjr, MrNexx, Mrgarrisson13, Ms2ger, MuZemike, Muchi, Mulukhiyya, Mushroom, Mvent2, Myscrnnm, N-Denizen, NSLE, NachoCheese, Naddy, Nakon, Naniwako, NatalieS, Nate Speed, Nateji77, NatsukiGirl, NawlinWiki, Ned Scotland, Ned Scott, NeilN, Neilworms, NeoDeGenero, Neofcon, Neojessica, Nerd65536, Netkinetic, Netsnipe, Neuromancien, Neutral Milk Hotel, Neverquick, New Age Retro Hippie, Nick, Nick L., Nick90210, Nickzakharyan, Nightmarelh, Nihonjoe, Nikk0, Nikki chan, Ninja Delia, Ninjawolf, NinyaEvu, Nivix, Nmohan 101, Nneonneo, Node ue, None Error, Nono64, Norborb, Norm, Num1dgen, Numbo3, Nyletak, OChiBiAyAo, Obdo, Oliver Pereira, Oliver202, Omegatron, Omnicog, Oneiros, Onesimos, Oni Lukos, Optichan, Optimistic Dude, Orangemike, Orz, Ost316, Otaku1994, Ottava Rima, Otto heinemann vundafarks, OwenX, Owl320, Oxymoron83, P4k, PGSONIC, PP Jewel, PYBS, Pagrashtak, Palmiped, Panchitaville, Panicpack121, Papacha, Park-j, Patrick, Patstuart, Paul Christian B. Yang-ed, Paul the greatest, Paulnwatts, Paurullan, Pavel Vozenilek, Pax:Vobiscum, Pedant17, Pegasus1138, Pheonix1974, Phil Boswell, Philip Trueman, Philosophistry, Phorque, PiglnTheMud, Pigman, Piotrus, Pipian, Pitman6787, Pixelperf, Plasticup, Plau, PlueXD, Poetic Decay, Pokemononfyarr, Polluxg, Pooooooooooooooooooop, Popious, Poszest16, Power level (Dragon Ball), Ppk01, Prede, Premeditated Chaos, Princess Moonray, Prof Wrong, Prosfilaes, Prosperosity, PseudoSudo, Pshent, Punarbhava, Pxma, Pyromania153, Pyrop, Qirien, Quadell, Quasirandom, QubitOtaku, Quiddity99, Quintote, Quuxplusone, Qwe, R'n'B, R.G., RJN, RP9, RaGnaRoK SepHirOtH, Radical Dreamer, RadicalBender, Radioactive afikomen, Radiomovie, RainbowCrane, RainbowOfLight, Raiyuu, Random (usurped), Ranma9617, Rasmuffin, Ratteler50, RattleMan, Raul654, Ravila, Razimantv, Red Director, RedCoat10, RedRollerskate, RedWolf, RedWordSmith, Reinis, Relly Komaruzaman, Remurmur, Renegadeviking, Repku, Retardsrox, RevolverOcelotX, Revth, RexNL, Riana, Richard W.M. Jones, Richardi, Rizzleboffin, Rjanag, Rje, Rjwilmsi, Rm dash r, Rmhermen, Roastytoast, Robertvan1, Robneild, RockMFR, Romanm, Ronin Aquila, Rory096, RossPatterson, Rowaasr13, Royalguard11, Rvalles, Ryan, Ryan Postlethwaite, Ryanasaurus007, Rydar, Ryo, Ryulong, Ryuuseipro, Rzrscm, SERSeanCrane, SEWilco, SKaze, SODA JUNKY, Saderlius, Saikano II, SailorAlphaCentauri, SaiyanEmperor2008, Sam, Samir, Samuel Blanning, Samurai 004, Samurai Drifter, Sanatherandom, SangeYasha, Sango123, Sankalpdravid, Saosei, Sarke, Sarpicaste, Satanael, Saturnfan32x, Savidan, Savius, Sazabirules, Sbluen, ScifiterX, Scineram, Scm83x, Scott Paeth, Sectorzz9pluralzalpha, Seibzehn, Seishirou Sakurazuka, Selmo, Senkei242, Sensei chan, Senti mental, Sephiroth kills, Serenaxiyade, Serinde, Servant Saber, Seryass, Sesshomaru, Shadovar, Shadow Angle, Shaina lee, ShakataGaNai, ShakingSpirit, Shanes, Shamak, Shatner, Shawnc, ShelfSkewed, Shibboleth, Shieber, Shii, Shimei, Shinjiman, Shinorei1, Shiori Aburame, Shiroi Hane, Shizni, Shobu244, Shoessss, Shokatsu-sensei, ShortShadow, Shrine Maiden, Siddharth Mehrotra, Silentl337, Silver Edge, SiobhanHansa, Skier Dude, Skierpage, SkyWalker, Skyraider, Slady, Slp1, Smash, Snarfies, Snowmanradio, Sobolewski, Sobreira, Soccer boi87, Solitude, Someguy1221, Sonic Mew, Sonicsuns, Sonjaaa, Soosed, Soraciel, Soul Slayer, Sp, Spacecat2, SparrowsWing, Spazz Star, SpookyMulder, Squilibob, SquishyBuddy123, Srikali, Srnec, Srpnor, Ss112, Starwind Amada, Stealthound, Steel, SterlingNorth, Stifle, StuffOfInterest, Subtlecoolness, Suede, Sukmydik, Superrodent, Supers, Supersaiyanplough, Supertigerman, Surindrag, Susan Davis, SusanLarson, Susvolans, Sverdrup, Swatjester, Synchrite, System Sniper, Szczerbiec, Szczerbiec01, T-borg, TAS, Tablizer, Tacubus, Tainted Angel, TakuyaMurata, Tamariki, TangentCube, Tangotango, Taras, Tartarus, Taw, Teflon Don, Teh Ch8t, Teiler Köregäten, Telescope, Tenka Muteki, Tenmei, Tentacle Monster, Teo64x, Tesi1700, Th1rt3en, The Haunted Angel, The house that jack built, The misha, TheAmericanizator, TheBlueFox, TheFarix, TheKoG, TheStripèdOne, Thepulse2007, Theresa knott, Thewikipedian, Thiseye, Thisismypage, Thisisnotthejew, Thue, Thumperward, TigressofIndia, Tikki hit, Tim Ivorson, Time for action, Timotab, Tintero, Titi, Tjbk tjb, Tobias Wolter, Tom X. Tobin, TomStar81, Tomcat16789, Tondog, Tony Fox, Tony1, Too Cool, Tooclose, Torx, Toya, Tpbradbury, Trafford09, Trainra, Trisar, Trivialist, Trunksoul, Tsange, Ttc345, Tubular, Turtleheart, TuxedoInferno, Twipie, Twitchamoo, Tyciol, Tzaquiel, Ubardak, Uberdiot713, Ugh12345, Ugur Basak, Un12341, Unibini, Unimath, Unyoyega, Urutapu, Username314, Username84, Vampirekisses13, VegaDark, Verdamondo, Versageek, VeryVerily, Vik-Thor, Vindictive Warrior, VioAdmin, Viperpray, Vitz-RS, Vivvo, Vknightbd, Vmk, Vodak, Voltzz, Volvaiga, Vonn, W, WOSlinker, Waffenbaum, Wakuran, Walter

Nwaokolo, Wapcaplet, WarChild, Wareware, Warmaster, Warreed, Warreneaston123, Wavelength, Wayward, Wcd, Webdinger, Weisskreuzgluhenx, Wereon, Wetman, WhisperToMe, Whitepaw, Wik, WikHead, Wiki alf, Wikipedia Team, Wildstar2501, Willkommen Benutzer, Wilson(cc), Wimt, Winsia, Wmahan, Wmorrow, Wolfs2122, Wonkothesane42, Woohookitty, Worldatlas1989, WurdBendur, XBloodyXNobodyX, Xaliqen, Xander89, Xanzzibar, Xaosflux, Xenofan 29A, Xeonxl, Xezbeth, Xiahou, Xino, Xiong Chiamiov, Xjaymanx, Xxlalavialxx, Xxpor, Xym, Xyzyxx, Xyzzyplugh, Yamamoto Ichiro, Yami Takashi, Yancyfry jr, Yellow1996, Yellowshat, Yelyos, Yintan, Yiplop stick stop, Yippiekiyeh, Ymirfrostgiant, Youspasm69, Yuckfoo, Yuuki Mayuki, Zakosai, Zaqxsw76063, Zazou, ZeWrestler, Zereshk, ZhangKhaiEn, Ziiv, Zikar, Zippedmartin, Zsinj, Zuanzuanfuwa, Zundark, Zzuuzz, Δ, 2031 anonymous edits

Spike_(company) *Source*: http://en.wikipedia.org/w/index.php?title=Spike_%28company%29 *Contributors*: Aknorals, Alexander Vince, Aspects, Banej, Blindman shady, Breno, CyberSkull, DBZROCKS, Dieubussy, Elephant Talk, Fg2, Hibana, Hydao, Jeangabin, Kerotan, KokoroTechnix, Lvivske, Martarius, Munem939, Nydas, Parrothead1983, Salavat, SamXX1X, SoulSlayer, TJ Spyke, Vhoskins, Waacstats, Zachorious, 25 anonymous edits

Namco_Bandai_Games *Source*: http://en.wikipedia.org/w/index.php?title=Namco_Bandai_Games *Contributors*: Alphathon, Amakase, Archimëa, Asher196, Axaul Conners, Axem Titanium, Beem2, Beetstra, Bongomatic, Canderson7, Chan Yin Keen, Coyets, CyberSkull, Daftpunkboy93, Ecs-Tactic, Elm-39, Enterryan, Evice, Faramir1138, Farine, FromtheWordsofBR, GamerPro64, Halfd, Ham Pastrami, Hero of legend, ICEYOU88, Inuboy1000, IronGargoyle, JDC808, Jamesontai, Jeff Silvers, Kariteh, Kenshinflyer, KiasuKiasiMan, Kipuk, Martarius, Mastercaster, Mattb90, Michaeltobbylee, Moocowsrule, Mps, Msh.spike, Nintendogeek05, Ominae, Pdiddyjr, Raito89, Rangoon11, Rich Farmbrough, Salavat, Silver Edge, Silver Sonic Shadow, Smash, Soredewa, Spock2266, TJ Spyke, TaerkastUA, Tobz1000, Tombomp, Toshio Yamaguchi, WhisperToMe, Wiki yann, Woohookitty, Yurayurateikoku, Zys saab, 109 anonymous edits

PlayStation_3 *Source*: http://en.wikipedia.org/w/index.php?title=PlayStation_3 *Contributors*: (aeropagitica), (jarbarf), -5-, -Anthony-, -Majestic-, -calum-r-, 00iddy, 0612, 1(), 10plakon, 11loopholes11, 123dylan456, 123shadow123, 123wiki123, 1337 XXIII, 1800Allen, 1NeverKnows, 233, 31Gabe, 4RM0, 4twenty42o, 64773R, 77Seven77, 95j, A Clown in the Dark, A Man In Black, A Nobody, A man alone, A.T LOVER, A0me, AA7, AB1992, ABarnett, ACSE, AIR DJ, AJCham, AJR, AMFilmsInc, AMHR285, AMIGA dUDE, ARC Gritt, ASHTONZANECKI, ASOTMKX, AZNJOKER, Aaaaaabbb, Aaron Bowen, Aaron Schulz, Aaron117, AaronSOLDIER, Aaronliendo, Aarozone, Aashish950, Aazeez, Abdy batman, Abolabo, Abolabo1, Abscissa, Academic Challenger, Acdx, Ace of Sevens, AceCard Jones, Acolyte of Discord, Acowsik, ActivatedWaffle, Adam1213, Adamy Wadamy, Adashiel, Addict 2006, Ade2005, Ademola Adebowale, AderakConsteen, Admeister200x, Admiral Maxtreme, Adrado, Advent swords, Aeluwas, Aericanwizard, Afghana, Afghanballer, Aformalevent, After Midnight, Agar762, Agbwiki, Ahmad90, Ahmadyk23, Ahoerstemeier, Aintaer, Air2data, Airconswitch, Ajay154, Ajcadoo, Ajgrovery2, Akamad, Akbarc, Akendall, Akhristov, Aknorals, Aksi great, Ale jrb, Alec92, Aleksael, Alex Stanek 9999, Alex.k k, Alex43223, AlexOvShaolin, Alexdeangelis86, Alexei-ALXM, Alexnottke, Alexville, Alexwcovington, Alfakim, Alfredosoto95, Ali@gwc.org.uk, Alias Flood, Alice Mudgarden, Alien1280, Alimn, AllTeam, Allenc28, Allmightyduck, Allstarecho, AlphaTwo, Alphaman, Alphathon, Alpta, Alra111, Altenmann, Alyssa3467, Aman27deep, Amanaplanacanalpanama, Ambuj.Saxena, Amer-aln7l, AmericanSights2008, Amish Gramish, Amoal, Amorker, Ampersandfrabjous(mark), Amren, Amunoz tico, Amux, An Sealgair, Anand Karia, Anantha92, Anaraug, Anclation, Anded991, Andersonma, Andre1985, Andrew.in.snow, Andrewjnyc, Andrewlaneyboi93, Andrewmatta, Andrewpmk, Andrewwski, Andy Marchbanks, Anetode, Angelic Wraith, Angie Y., AngryDespot, Angus Lepper, Anibius, Animal91X, Anish7, AnneBoleyn1536, Anomie, Anonymous Bassist, Anonymous anonymous, Anonymous editor, Anphanax, Anstice, Antandrus, Anthony Leppert, Anthony416, Anthonydazet, Antidemon, Antidrugue, Antisora, Antispy, Antoniosus, Antuan10, Apadilla80, Apmastersite, Apostrophe, Apple123, Apyule, AquaStreak, Arcaynn, Archanamiya, Arcticchill56, Arethia, ArielGold, Arite, Arknascar44, Arlika150792, Armaced, ArmadniGeneral, Armando, Armbrust, Arnold Go, ArnoldReinhold, Arnon Chaffin, Arranbhav, Arsonal, Art LaPella, Arthena, Arthur Smart, Arthur2045, Arunsagar1, Arwel Parry, Asgard Elite, AshSert, Ashdurbat, Ashenai, Ashmellor, Ashmon340, Ashwinr, AssassinXIII, Assassinist54, Asterion, AstroHurricane001, Asymmetric, Atomic Cosmos, AuDioFreaK39, AuburnPilot, Aufaber, Auger Martel, AustDan91, AvicAWB, Avillia, Awakefield, Awesomeness95, Awhit003, AxG, Axelace123, AySz88, Ayla, Ayrton Prost, AzaPutra, AzaToth, Azaan95, Azer Red, Azimuth1, Azn boi9832, Azrael6519, B-poc, BB660, BD2412, BKmetic, BMetts, BNichols1707, BTNH Fan, BUF4Life, BaRiMzI, Baa, Baardhimself, Babajazz!, Babakmomen, Bachrach44, BagHutch, Bahahs, Bahtoe27, BajillionGremlin, BakaPie, Bakarocket, Bal3d, Balling89, Ballsout, Banksbr2, BanyanTree, Barachel001, Barnolde, Barronlroth, Barrylb, Bart133, Battered and so misled, Baz34, Bbatsell, Bd8494, Bdhoff, Bearbear, Beardy10, Beebee49, Ben Di Luca, Ben5ive, BenDee, Benandorsqueaks, Bender235, Benhead 07, Benjaminhill, Benji64, Benjovi, Benman333, Benstown, Benw2112, Beoba, Bergles, Berkut, Berserkerz Crit, Beta, Bevo74, Bew, Bewareofdog, Bflora, Bgold4, Bhavinshah, Bhcecil, Bhobo, BiT, Big.P, BigBear431, BigBoned87, BigJoker, Bigbosspatch, Bigdottawa, Bigger Boss, Bigheadedboy, Bigman606, Biker14, Biker141, Bill2510, Billy574, Billywhack, Binary TSO, BinaryTed, Bingo88380, BizarreLoveTriangle, Bkawcazn, Bkd, Bkkbrad, Bkkeim2000, Black-Velvet, Blackkrash, Blackvelvetband, Bladestorm, Blahblahblahblah1y56, Blanchardb, Blaze Ya Dead Homie, BlightedEye, Blindman shady, Blinkstale, BlkStarr, Blobzorz, BloodAdo, Bloodyporn, Blound, Blu Aardvark, BlueAzure, BlueMint, Bluedude3011, BluesFan985, Bmicomp, Bob EggHead, BobFredJonesTwelve, Bobblehead, Bobblewik, Bobcheese411, Bobguy89, Bobo192, Bobthebobthetimothy, BocaJ, Boimaanderson, Bondolo, Bongwarrior, Bonno51, Bookoflove, Booksworm, Boram, Bordello, BorgQueen, Bornhj, Borszczuk, Bovineone, Bowmanjj, Boydoesknow, BrOnXbOmBr21, Bradwart, Bratant, Brazil4Linux, Breakdown2218, Breno, Brian Kendig, Brian0918, Brieeyeball, Brighat, Brighterorange, Britannic124, BrokenSegue, Brokendowntoyotacamry, Brolgeta7000, Broncostar, Broooce, BrotherFlounder, BrowardPlaya, BrownHairedGirl, Brownnoser15, BrownsRock10, Brusegadi, Bryan4everkh, BryanG, Bryces, Bryjwal13, Btboy500, Btilm, Buckeyes1186, Bueller 007, Bulletproof50, Bunbunss, Bunchofgrapes, Bungalowbill, Bunny-chan, Burgercat, Burner0718, BurningZeppelin, Burnte, Burpelson AFB, Burto88, Burtonsk8r125, Burwellian, Bwill6, C-town dude, C0nanPayne, CBDunkerson, CCORYY, CJ, CPO Pieman, CSA956, Cablebfg, Cactus.man, Cadsuane Melaidhrin, Cakster, Calderra, Caleb5617, Caleson, CaligoPravvus, CalumH93, CalusReyma, CalvinHobbes, Calvins48, Camann, Camoxide, Camster342, Can't sleep, clown will eat me, CanDo, Canadian-Bacon, CanadianLinuxUser, Cananian, Canderson7, CanesOL79, Canido11, CanisRufus, Canterbury Tail, CanucksRule, CapPixel, Capecodeph, CapitalLetterBeginning, Caprinoe, Capsulegim, Captain Kevles, Captain Wikify, CaptainHaZ, CarVac, Carbine, Carbon-16, Carbonite, Card Destruction Duelist, Cardsplayer4life, CarelessHair, Carface, CaribDigita, Carl.bunderson, Carl0s, Caseshy, CaseyPenk, Causa sui, Cbmaster, Cbourff31, Cdabrah, Cdr-13, Ceecookie, Celestianpower, Cenarium, Ceptizzy, Ceros, Cerowyn, Cg880054c, Ch0rx, Ch110593, ChadBC, Chalklit73, Chan Yin Keen, Chanjim, Chanlord, Chanman121, Chanting Fox, Chaoaretasty, Chaos Stein, ChaosAngelZero, ChaosControl, Chappy84, Charizard777, Charles Nguyen, Charlesr, CharlieA, CharlotteWebb, Charphil, Chavagnatze, Cheeseman Muncher, Cheesor, Chensiyuan, Chevinki, Chicken2456, ChicosBailBonds, Chill Pill Bill, Chimaera2005, ChimpanzeeUK, Chimuel, Chinesekid05, Chingang2006, Chipster2009, Chmod007, Chocobogamer, Chopper3, Chowbok, Chrcharcor, Chris 73, Chris S, Chris huh, Chris.hosuabi, Chris1219, ChrisErbach, ChrisW, Chrisandtaund, Chrisd0687, Christina Silverman, Christopher Parham, ChristopherROARTY360, Christus11, Chronocontrol, Chubes, Chuglur, Chzz, Ciao 90, CinchBug, Circeus, CityFeedback, Cj395, Cjoyce1980, Clam0p, Clashfrankcastle, Classicalgas, Clayheaddreslough, CleanEdit101, Clefomatic, Clerks, Cliché Online, Clintafer, Cloachland, ClosedEyesSeeing, Closetoeuphoria, Cloud668, Clr75, Clutzy jamie, Cm&bw2007, Cmc0, Cmcfarland, Cmsjustin, Cncplyr, Cncxbox, Cnota, CoJaBo, CobraBK, Cobracool, Cochonfou, Cocoloco, Coconutfred73, Codemonkey, Cody 2.0, Cody 88, Cody.schatzle, Coffeenotwar2, Col. Hauler, Colin Watson, Colincbn, College Watch, Colon3lCrunch, Colonies Chris, CombatBeaver, Combination, Combopizza, Combopk, Comerfordcathal, Cometstyles, CommonsDelinker, Compact105, Computer97, Computerdefinitions, Computerjoe, Concretechocobo, Condone, Connor Nintendawon, Connor oc, ConnorJack, Consumed Crustacean, Conversion script, Coolbho3000, Coolman06, CopperSquare, Coredesat, Corpsedust, Corpx, Corti, Coryisgay, Count Chocula, Countdown42, Cowman109, Cowwarrior, Cpc464, Cpl Syx, Cracker017, Craigsjones, Craptree, Crash3462, Crazyfrengy, Crazyguy86, Crazymaner2003, Crazytales, Create g77, Cresync, Crisss1104, Crocker22, Crofteh, Crowstar, CryptoDerk, Css, Ctbolt, Cubetriangle, Cur2702, Curps, Cute 1 4 u, Cutter, CyberFreq, CyberSkull, Cyberpsi88, Cyrus777, Czechnmymail, D Tunisia, D-Terminal, D.Kurdistani, D24, D3v4st4t0r, DANfan, DARQ MX, DARTH SIDIOUS 2, DBPhil, DCBen, DCHStrumpet, DChiuch, DDima, DEW 7, DGJM, DJ Mike TJG, DOOD1002, DOOMEDtoDOOM, DOSGuy, DRonalds, Dagibit, Dak is back, Damian Yerrick, Damien Russell, Damnnn itzz john, Dan Harkless, Dan100, Dan64, DanB91, Dancingweasals, Dancter, Dane Gustafun, Dang08, Danhash, Daniel J. Leivick, Daniel Olsen, Daniel.levine, Daniel.Cardenas, DanielGuirguis, DanielHawking, Danielb8047, Danj205, Danno uk, Danny15buzzin, DannyB!!, Dannybriggs93, Dannybu2001, Dannycas, Danoasis, Dante, Danyankees123, Danzio2020, Darin-0, DariuszT, Dark Duelist 5000, Dark Knight6, Dark jedi requiem, Dark spartan082, Dark verdant, Dark-Devil, DarkFalls, DarkMech, DarkWarrior, Darkcat1, DarkestBlade, Darkhunger, Darklock, Darkman007e, Darkmanx806, Darkpower, Darksniper404, Darkwind, Darlene4, Darranc, Darth Mike, Darth Newdar, DarthSmee, Darthkenobi, Darz Mol, Dascott, Dastardly 5656, Dave Coleman, Dave00000, DaveJB, Davedx, Daveswagon, Davewild, Davhorn, David Cat, David Fuchs, David Gerard, David Levy, David Underdown, David Woodward, David136a, DavidBailey, DavidGX, DavidHOzAu, Davidblouin, Davidjk, Davidsnoodles, Davinche, Davo88, Davud363000, Daxkelson, Dberliner, Dboyz-x.etown, Dbzkid, Dcandeto, Dck7777, Dcooper, Dctcool, DeAceShooter, DeVante', DeadPixels, Deadlypoe, Deadman1848, DeathSeeker, Deathtrap3000, Deathwiki, Debackerl, Debresser, Dec12, Decat, Declanthebullbull, Decora, DeeKay64, Deetdeet, Def1pony, Defenestrate, Definitionmesserupper626, Defufna, Dekabreak101, Deku Scrubby, Deleski, Dell9300, DellTG5, Delldot, Delta759, Deltabeignet, Denalin, Denbigh, Denier-of-Soup, DennyColt, Dentine, Denzelio, Depictionimage, DerHexer, Derek Cormier, Descendall, DesiMD, Destroy101, DevAnubis, Deville, Devoindahouse, DevonTheDude, Df747jet, Dfgsfgsdf, Dfrg.msc, Dg0896, Dgies, Dhiwin, DiGi-Masters, DiamondDragon, Dieboybun, Diefor, Diegofrgc, Digitalme, Dina, Dingledow, Dinohunter999, Dionyseus, Dios Aron, Directorstratton, Discoh8er, Discospinster, Dismas, Dispenser, DividedByNegativeZero, Dj b01us, Djbuhhda, Djgunn16, Djlollyb, Djungelurban, Dmdallas, Dmitry Malayev, DmitryKo, Dmsar, Dngnta, DoNo876, Dobisskill, Doc glasgow, DocWatson42, Dockersfan, Doctorinthahouse, Dodgyc, Dolda2000, Domino2097, Don t, DonEd, Dookie101, Doom127, DoomBringer, Doomed, DoomsDay349, Doradus, Dorvaq, DoubleCross, Douglasr007, Down10, Dp462090, Dpmarshall, Dposse, Dpotter, Dr doris, Dr. Blofeld, DrSmacky, DragonflySixtyseven, Drake2u, Drakgamer, Drat, Dreaded Walrus, Dream Rurouni, Dream-2291, DreamsReign, Dreedo, Drew08867, Drewboy64, Drmies, Dsd510, Dsnow77, Dtaw2001, Dtcdthingy, Duagloth, Dusty506, Dwarfpower, Dwayne, Dxco, Dycedarg, Dylanmoore44, Dylanspronck, Dynamo Matt, Dyold Knip, E to the b, E-Magination, EGOvoruhk, ENeville, ESkog, EarthPerson, East718, Ecksemmess, Eclipsed aurora, Ecwcwwe, Ed g2s, Ed411, Eddied123, Eddy Editor, Edit20, Eduardo Gutiérrez, Edurant, Edwardbains, Edwin1124, Edwy, Effer, Eggdude9, Einstein the afrodude, Eiskis, Ejg930, El C, El sand bag57, El3m3ntsk8r24, El3mentry, ElTyrant, Elagatis, Eldberg, Elecbullet, Electricmoose, EliasAlucard, Eliashc, Eliminatex, ElinorD, Elitesoilder, Elkano92, Elliotgoodrich, Elliottp216, Elvaangohst, Elven6, Elwell, EmGen76, Emc2, Emil baseleres, EmmSeeMusic, Emo goth, Empty2005, Enax99, Endofskull, Energon22, Enviroboy, Eo1spy, EoGuy, EonOmega, Epbr123, Epitaf, Epolk, Eraserhead1, Erencexor, Ergbert, Eric Heckman, Eric Wester, Ericjraack, Ericlaw02, ErikHaugen, ErrantX, Errickfoxy, Eruption1978VHt2, Esanchez7587, Escobar4Life, Esin0420, EsperHunter, Estel, Estemshorn, Eternal Pink, Eugene a, Evan-Amos, Evangelion883, Evanh, Evansam, EveryDayJoe45, Everyking, Everyone Dies In the End, Evice, Evil Merlin, EvilGoldfish, EvilReborn, EvilZak, Evilgohan2, EvocativeIntrigue, Ex-Nintendo Employee, Excirial, Exert, Exeunt, Exiledmalice, Exobyte, Exodite, ExplicitImplicity, Exukvera, Ezhiki, F9o0oly, FCYTravis, FF2010, FF810, FPAtl, Fagot39, FairlyEvident, FaithLehaneTheVampireSlayer, FaithRaven, Falcon866, Falcon9x5, Fallaway6554, Famstar8888, Farrerdave, Faster2010, FatFox, Fatone411, Favonian, Fawcett5, Fb767, Fcouk, Feduciary, Feebtlas, Felix the Hurricane, Fenrir-of-the-Shadows, Ferkelparade, Ferose2, Fewfeet, Ffgamera, Filipdr, Finalius, Finny95, Fintan123, Fire 55, FireballDWF2, Fireemblem, Firionel, Firsfron, Fish and karate, Fisher-s, Fizan, Fizzo0, Fjbfour, Flapjackboy, Flatman666, Flaviohmg, Fletch rox, Flewis, Flib0, FlieGerFaUstMe262, Flyoffacliff, Fnagaton, Fnerchei, Fnfd, FoH, Fobia, FocalPoint, Forteblast, Fosscape, Fotero, Fourthords, Fox, FoxAndFalco, Foxxygrandpa, FrYGuY, Frakistan, Fram, Frap, Freak42, FreakmanJ, Freakofnurture, Freaky Face Films, Frecklefoot, Freddie McPhyll, Fredrik, Freeps, Frencheigh, Frenrir1, FreplySpang, Fresheneesz, Freyr, Fribbler, Friday, Frvernchanezzz, FullMetal Falcon, FunPika, FungasUK, Funkypants, FunnyGuyAlex, Funnybunny, Funper, Furrycushion, Furrykef, Fuzheado, FvdP, Fvw, G.W., GCFreak2, GDInc., GDibyendu, GGreeneVa, GHe, GKJ, GLNelson, GLdK, GMScribe, GSK, GTAsoldier, GTubio, GVnayR, Gaara the Fifth Kazekage, Gabbe, Gabriellundmark, Gaius Cornelius, Galaxas0, Galoubet, GameMasterCor, Gamer416, Gamerforever, Gamerguy337, Gamerman, Gamiar, Gaming&Computing,

Gamingexpert, Ganryuu, Garaa447, Garethon, Gargaj, GargoyleMT, Garland Strife, Gary9corrigan, GaryNigel, Gatoatigrado, Gavinmusic, Gaystation hater, Gbadx, Gbeeker, Gdo01, Geardawg, GeeIWonderWhatGanonIsUpTo, Geneb1955, General Jakass, GeneralDuke, GenericGabriel, Genuinemartian, Geologik, George Ho, George Leung, George shrink, Georgejames5168, Gerk, Gerrish, Gershwinrb, Gfrewq12, Ggunit 911, Gherrington, Ghoseb, GhostPirate, Ghostreveries, Giantcalledgrawp, Giantgrawp, GigaPsych21, Gigalocus, Giggy, Gilly0824, Gilo1969, Gjd001, Glen, Globe-trotter, Globelovers, Glocks Out, Gmacbeth, Gmcfoley, Gmudge, Gnevin, Gninja92, Godheval, Gogo Dodo, Gohiking, GoingBatty, Gokoon, Golbez, Gold Stur, Golfman, Gonads3, Goodone121, GorbaLor, GorillaWarfare, Graaay, Gracefool, Gracenotes, GraemeL, Grafen, Grahn, Grand key, GrandDrake, Grandgrawper, Grandmasterka, Graniterock, Grapes911, GrassyFields296, Grawp, Grawp the Giant, GreatSnake666, Greenway0208, GregorB, Gregory Heffley, GregoryPSmith, Grim Reaper, GrimRepr39, Grm wnr, Gromreaper, GroovySandwich, Ground Zero, Grundler1, Gspawn, Gtaarslan, Gtosmw, Guanaco, Guardian007, Gudac, Guest9999, Guinea pig warrior, Gunmetal Angel, Gurch, Guticb, Guyinblack25, Guyspy, Gwernol, Gws898, Gzimm21575, H6056, HD123, HDCase, HHS.student, HJ Mitchell, HP465, HQ, Haakon, Hadal, HaeB, Hagerman, Haidbro, Haipa Doragon, Hairy Dude, HalfShadow, Hall Monitor, Halsteadk, Halzred, Ham Pastrami, Hamada333, Hamedog, Hamiltha, Hamiltondaniel, Handscombmp, Hannibal222, Happy-melon, HappyVR, Hardiment, Hariharan91, Harryboyles, Harsh h3, Hasek is the best, Haseo9999, Hatmonkey, Haunted360, Havok, Hawkypunk, Hayden120, Haymaker, Hazard-SJ, Hbdragon88, Hda3ku, Hdt83, Helhowskir, Hellcat fighter, Hellfire1014, HelmsC1978, Hempeater, Henchman 2000, HenkvD, Henrik, Henriok, Henry Flower, HenvY, Hep MYGEN, HereToHelp, Herr Gruber, Hersfold, Hervegirod, Hetar, Heymid, Hgld, Hi878, HideyoshiJP, Hijak, Himanks, Hindahouse, Hisashi 0080, Hnandrew, Hoaqbui, Hobadee, Hojimachong, HolyCowProductions, Honette, Hongooi, Hongshi, Hoodednood, Hornymanatee, Horserice, Horv.petya, Hossenfeffer, Hotmann, Hotwiki, HowardSF, Hpcanswers, Ht1848, Hu12, Hugh000270, Humphreys7, Hungrypinksquid, Hurgymcgurgygurg, Hurricane111, Hurricanefloyd, Husker10, HweeWieng, Hyad, HybridFlare, Hydrogen Iodide, Hylian pirate, Hyperhullo06, Hyperlink, Hyperscope, Hyperspacey, Hypo, I will rule, I'm armed with my point of view and I'm not afraid to use it!, I'm-Back-1001, IAMTHEEGGMAN, IAlta, IDuDaChaCha, IG-2000, IMABBALLPLAYER, INaNimAtE, IRWeta, Iahead, Iain k, Iamfscked, Iamjim999, IamtheFrohman, Ian Pitchford, IanManka, Ianbeatson, Ianbell330, Ianhowlett, Ianjones1900, Ianjones50, Icairns, Ice12, IceDrake523, Iceberg3k, Icecube52, Iceduck, Icemotoboy, Icestorm815, Ico92, IcyRealmster, Idamlaj, Idont Havaname, Idontgottaname, Idreamincode, IdreamofJeanie, IgorJansen, Igordebraga, Igorwindsor, Iiltoxic, Ikarus14, Ikigorogoro, Ilion2, Illcendiary, Ilovebacon337, Ilyanep, Imaninjapirate, Imdaking, Imfxxz, Imgaril, Imobius, ImperfectlyInformed, Imroy, InGearX, Inchfizz, Indiearmy, Indon, IndyLawSteve, Int21h, Intelligentsock, Inter, Interchange88, Interiot, InternetMeme, Introbulus, Invincible Ninja, Ipodandmp3s, Iraqidude, Irguywhogont33chul3sson, Iridescent, Irishguy, Ishikawa Minoru, Islander, Israel09, It is happy, It's-is-not-a-genitive, Italianboy10, Iuhkjhk87y678, Iupolisci, Iwalke, Ixfd64, Ixnayonthetimmay, Izehar, J 1982, J Di, J Grant Fox, J-masta, J-smith.17, J. Nguyen, J.delanoy, J.reed, J450NH3, J4lambert, JDC808, JDCAce, JDRomano2, JEN9841, JForget, JGerretse, JHoltzman, JKinney, JLaTondre, JLogan, JNW, JNeal, JONJONAUG, JPX7, JParis, JSM, JSpudeman, JSunrai, JTBX, JaGa, JaMeZiZbAcK, Jacek Kendysz, Jacj, Jack007, JackSparrow Ninja, Jackalt, Jackthomasiscool, Jacoplane, Jacques.viviers, Jacquesl, JadeGarnet, Jagged 85, Jagzthebest, Jake900, JakePrime, Jam1993, James henry Lee, James pic, James-short, James086, JamesHenstridge, Jamesalaska, JamieSl, Jamiepippin, Jan otto, Jannev, Japan\'s flag, Japugs2k2, Jaranda, Jaredroberts, Jarry1250, Jason One, Jason the Delicious, Jason.cinema, Jaspreetjohal, Jastein, Jauerback, Java13690, Jaxl, Jaxster66, Jay Pizz, Jay95, JayKeaton, Jbav1278, Jbluez27, Jccl5000, Jderadou, Jdisme, Jdm64, JeLuF, Jebus989, Jecates, Jedi767, Jeff G., Jeffler, Jennatar, Jennica, Jeremy Butler, Jeremy Visser, JeremyA, Jeremykemp, Jericho1337, Jermama, Jess Flaps, JesseGarrett, Jesusclownfox, Jety111, Jewbag72590, Jheinz, Jhenderson777, Jheppner, Jhhudso, Jhsounds, Jiang, Jigahurtz, Jijjin, Jiminicriquet, Jimmy278, Jimmy4001, Jimmy93211, JimmyBlackwing, Jingleboy, Jivecat, Jizzbug, Jjjsixsix, Jl1995, Jmaster14656, Jmccallum1401, Jmoney19, JoWal, JoeDaniel4, JoePulliam, JoeSmack, Joelawson14, Joelon, Joerite, Joey19930303, Joezamboni, Johann Wolfgang, John Kjos, John Mash, John Quincy Adding Machine, John of Reading, John-joe123, John.n-irl, John254, John2kx, John71, Johndude1321, Johnleemk, Johnm4, Johnmalc, Johnny1269, JohnnySquires, Jojoschlueter, Jok2000, Jol123, Jolly bolton, JollyCrisp, JonHarder, Jonabbey, Jonohill, Jonphamta, Joose ra, Jor, Jordinho, JosJuice, Joseph Solis in Australia, Joshbuddy, Joshjames1234, Joshkeat, Joshua Scott, JoshuaWalker, Josiph, Josquius, Jossi, José Gnudista, Jpgarcia6, Jpgordon, Jquindlen, Jrg dnn, Jrg dnn@yahoo.com, Jrisbara, Jrphayes, Jtalledo, Jtkiefer, Juckum, Judacris, Judesalmon, Juiceboy, Jusdafax, JustToClarify, JustinRossi, Jwadie, Jwagnerhki, Jwissick, JzG, Jzcool, Jzhang, K-UNIT, K1Bond007, KAMiKAZOW, KAPITALIST88, KD5TVI, KJBracey, KJS77, KNRasengan, KPalicz, KahranRamsus, Kain77502, Kairos, Kaisanxara2580, Kalamata6666, Kamasutra, Kanyewes 05, Karada, Karam.Anthony.K, Karn, Kashman, Katalaveno, Katana3, KathrynLybarger, Katieh5584, Katrianya, KazamaBoy, Kbolino, Kbonediggity, Kc2545, Kcgreatfox, Kcluri, Kcome, Kcx2233, Kdashsohel, Kdivad, Ke5crz, KeLe Mc, Kediaj, Keegan, Keithkesslerexp, Kelden, Kelvin Samuel, Kelvinc, Kemyou, Ken l lee, KenBest, Kencaesi, Kendolamanoderecha, Kenimaru, Kenny92190, Kenyon, Keristrasza, Kesac, Kesla, Kev19, KevXYZ, Kevin B12, Kevinkor2, Keyblade Wielder, Keyser Söze, Khalfani khaldun, Khfan93, KiasuKiasiMan, Kicking222, Kiddothekid, Kieranace, Killamajig, KillerCommz, KillerbunnyM, Killerfox, Killik Twilight, Killz0ner, Kilo-Lima, Kiltman67, Kimchi.sg, King Lopez, King of Hearts, Kingofgames2, KingpinE7, Kirikomimoon, Kirinsilva, Kirkinout, KitAlexHarrison, Kitch, KittenKiller, Kiyusoma, Klaus Kratchet, Kman618, Kmweber, Knight45, Knixx, KnowledgGar, Knowledge lover1123, KnowledgeOfSelf, Knowledgeum, Knoxer101, KoRngear, Kobol, KodakYarr, Konangrit, Konman72, Koolkid007, Koprider, Koratnia, Korolev61, Korosu, Kotjze, Koweja, Kozuch, Kramer3214, Krappie, Krauser415, Krazidude95, Krazy Soviet, Krazyson13, Krinkles, Krispyiiz, Kronza21, Kross, Krtek2125, Krummy2, Krypticmind, Krystyn Dominik, Ksandler3, KsprayDad, Kudosx, Kungfuadam, Kungming2, Kupos, Kurt000, Kurtz69, Kuru, Kurykh, Kwasiamoako, Kwekubo, Kyidro, Kyle, Kyle1278, KylePIB, Kylu, KyraVixen, Kyrios320, L Kensington, L1A1 FAL, L33tfella H, LAMEDED, LChrisUK, LN2, LPscottyDawg, LT83, Ladlergo, Laja25, Lalo22, Lamename3000, Lancerex6, Lanhoj, Lanky, Lanoitarus, Larsinio, Lavenderbunny, Le Déchaîné, Leapfrog314, LebanonChild, Lectonar, LedRush, LeeSimmons, Leemorrison, Legend78, LegendsDX7, Legrolls, Lehx, Lengis, LeoDV, LeopoldG, Lerdthenerd, Lesliephung, Leszek Jańczuk, Lexington50, Lezek, Lf2planet, Lianne777, Liasos, Liberty Miller, License2Kill, Lightdarkness, LighthouseJ, Lightmouse, Ligulem, LilPro22, Lile009, Lilphil007, Limelight022, Link 486, LinkinPark9297, Linkinlovprk, Linkinman199, Lipis, Liquidsnakejr, LittleDan, LiuSun, Livetodie122, Lkinkade, Lloydpick, Lmz00, Local hero, Locke Yggdrasill, LockeShocke, Locostes, Logan GBA, Logical2u, Lololololololol, Loof1, Lord Alda, Lord Crayak, Lord Hawk, Lord Trixster, Lordmetroid, Lordofchaosiori, Lordofthesheep, Lovi23, Lowkeysteve, Lowriekid, Lt. penguin, Luckrider7, Lucktle, Luckyherb, Lugiadoom, Luigi30, Luigi6138, Luigifan, Luigii249, Luismbs, Luk, Lukemadness, Lukep59, Luna Santin, Luph25, LurkingInChicago, Lutherx, Lydia Pryon, M cappeluti, M337ING, M412k, MAINEiac4434, MCKINLEYROX2, MER-C, MKD Majstor, MKoltnow, MPD01605, MR. CORNCOB, MSGJ, MSTCrow, MTC, MVP5, MZMcBride, Maaajid, Mac Davis, Mac Starkiller, Macara, Macarion, Mackan, Mactin, Mad Munky838, Madcowmnky, Madman, Maduin, Magasin, MagicMoose, Magicsgram, Magister Mathematicae, Magnum2037, Magog the Ogre, Mahanga, Maheeki, Mailer diablo, Majin Izlude, Makeemlighter, Malcolm, Malik Shabazz, Malo, Malomeat, Maltese Dragon, Manav 95, Mance, Mandarax, Manganeez, ManiacalMonkey, Maniadis, Manmonk, Manofiorn, MantleX, Mantracker25, Manyrameriz, Marblespire, Marblewonder, Marcika, Mareino, Marginoferror, Marina T., Mario1021, Mark0528, Mark87, MarkGallagher, MarkThreesixty, Markintellect, Markthemac, Markx, Marqueed, MarshalN20, Marskuzz, Mart98, Martarius, Martin1993, MartinPackerIBM, Martpol, Masaki7777, Masamunecyrus, Masran Silvaris, MassKnowledgeLearner, Master Jay, Master Of Ninja, Master Shake2, Master Thief Garrett, MasterEditor99, MasterRegal, Masterjedi645, Mastershake2.0, Mastiland, Mat Parker 117, MatGreenfield, MatLu, Matgn03, Mathmo, Matjet, Matmoo4, Matt Britt, Matt Heard, Matt.ledger, MattParker116, Mattclarkie, Mattgirling, Matticus78, Mattopia, Mauls, MaxEnt, Maxamegalon2000, Maximumlobster, Mazin07, Mazumbado, Mbg6231, Mboverload, Mbslrm, Mc razza, McDonaldsGuy, McGeddon, Mckaysalisbury, Mclarenaustralia, Mclean007, Mdd4696, Mdenham, Me676, Meaghan, Mech wolf23@hotmail.com, Mee Merone, Meeples, Megabassxz, MegastarLV, Megata Sanshiro, Melman101, Memodude, Mence Master, Menmau, Meow07, Merretto, Messer Derek, Metalindustrien, Metroidfan23, Mets501, Meweight, Mewtu, Mfm44, Mgd3434, Mgiganteus, Mgiganteus1, Mgillespie, Mgs4isamovie, Mhking, Micahblitz, Michael Daly, Michael Devore, Michael Greiner, Michael Mad, Michael Scofield 94941, Michaelas10, Michas pi, MickWest, Microdoted, Middleschoolisbad, MidnightScott, Mike Rosoft, Mike Yaloski, Mike mgoblue, Mike.lifeguard, Mike92591, MikeMeoff, Mikedot, Mikel Ward, Mild Bill Hiccup, Militant07, Millahnna, Millionairex, Miltopia, MindWraith, Minestrone Soup, Minesweeper.007, Minghong, Mini-Montana, Miniclipuk, Minna Sora no Shita, Minots59, Mipadi, Misocurdo, MisterHand, MisterMarioMan, Mitaphane, Mjohnnyg, Mkdw, Mmernex, Mmmdonuts210, Mnemnoch, Mnia786, Mnts, Mo0, Moe Epsilon, Mofomojo, Moggie mn, Mohaas05, Mohammad7410, Mohammed2696, Mohare1125, Mohmaaytah, Moiz.khan.123, Momo san, Momus, Moneyisgod, Monkeeman3, Monkey316, Moocha, MoogleDan, Mooshimanx, Mooshmooshman, MorbidKrabs, Mordien, Moreschi, Morte, Morten, Mortense, Mottylufc, Mousti, Movellon, Mowwow, Mozzerati, Mr Adequate, Mr Mo, Mr Pumblechook, Mr toasty, Mr. Good, Mr. Lefty, Mr.Blonde, Mr.Do!, Mr.bonus, MrBubbles, MrDaBucket, MrDenny, MrHumperdink, MrMan, MrVoluntarist, Mramo, Mrbean667, Mrbradford, Mrbrown, Mrcongojack, Mrcornpants, Mrhappyboy, Mrjiggaman, Mrmercy, Mrxg4, Mrzaius, Mschumacher69, Mstempin, MtB, MuZemike, Muchosucko, Mufka, Mugka, Muhand, Muhyemin Tarin, Mujklob, Mulder416, Munchlax, Mushroom, Mushroomer, Musical Linguist, Muslimer12, Mutableye, Mvas, Mvent2, Mvp15, Mvsrhollywood, Mwboyer, Mweflen, Mwn3d, Mxn, My76Strat, MyShine, Myanw, Myscrnnm, N i grunze, N. Harmonik, N00body, NASCARforlife, NBC, NFAN3, NJM, NT1.0, NYyankees51, Nado, Naja Haje, Naji Yousef, Nakon, NameIsRon, Nandesuka, Napoleon4444, Nari50, Natalie Erin, Nate Speed, Nate54o, NauticalNonsense, Navstar, NawlinWiki, Nazim450, Nczempin, NeOak, Ned Scott, Neelchauhan, Nehrams2020, Neilc, Neilka, NeoChaosX, Neodarksaver, Neofelis Nebulosa, Neogamer2610, Neomagus26, Neorock, New Age Retro Hippie, New guy, New*allusion, NewEnglandYankee, NewShinyCD, NewbieDoo, Newbrak, Newmhost, Newskooltrooper, Newsnight Watcher, Newton2, Nezu Chiza, Niall42, Nicblais, Nicholas spurway, Nicholasyzh, Nick, Nick thelot, Nick7213, NickBush24, Nickelbackrules1518, Nickjd331, NicoNet, Nidis, Nifboy, Nigeria01, Night Gyr, Nightmare SE, Nightmeare, Nikhilpaladugu, Nikkimaria, Nikknighten, Niklas a, Nil Einne, Nils, Nima1024, Nineinchsin, NinjaRooster, Nintendo1202, NintendoTheatre, Nintendog master 54, Nintendonly, Nintendude, Nishkid64, Nitropower, Njh, Nlu, No Guru, Nofate456, Noj9, Nomalam, Noops, Norbi, Notagoodname, Notdedyet, Notwist, Nousemercenary, Nrlund, Nrpf22pr, Nrvn93, Nu89, Nudimmud (usurped), Nuevo003, Nuggetboy, Nw77, Nyttend, Nzk, OAP boba, OSX, Oakster, Obli, Oblivious, Ocarinaoftimez64, Ocee, Octeron, Octopus-Hands, Oda Mari, Odii, Of, Ogioh, Ognir, Ohai111111111, Ohconfucius, Ohnoitsjamie, Oklonia, Olberding, Olborne, Old Guard, Oli5893, Oliver Cromwell, Oliverdl, Olorin28, Omarismail, Omicronpersei8, One-dimensional Tangent, OnePt618, Onef9day, Onesimos, Onetwo1, Oni Ookami Alfador, Onikage725, Only777, Onlytofind, Onorem, Onsly, Oogles, Oosh, Openbrain, Operation fluffy bunny, Optichan, Orangpelupa, Oriol003, Orion Minor, Oscarthecat, Oscinis, Osfn8, Osh33m, Oskar Mayer Nguyener, Ost316, Otashiro, Our Tiny Servant, OuterHeaven, Outlamed, Outlyer, Outoftunecornet, OverMatt, Overload7, Overlordneo, OwenX, Oxymoron83, Ozguroot, P dump, P924s88, PCPP, PJ Pete, PJM, PKFC, PS2pcGAMER, PS3 Addict, PS3 Lover, PS3 wins, PSP4LIFE, PSPMario, PSXer, PabloG, Pablothegreat85, PadawanMD, Padddy5, Paddyman1989, Padillah, Panda.1001, Pandamor, Pandapower.1, Papi900, Para82, Paragon, Paranomia, Pardthemonster, Paris By Night, Parl1987, PartTimeGnome, Passport90, Pat Yost, Patchey1000, Patesta, Patman, Patriarch, Patrice58, Patstuart, Paul1991, Paulcmnt, Paulinho28, Pauljr231, Paulm27, Paxsimius, Pcpp, PdDemeter, Pdc005, Pdias 92, PeaceNT, Peanut4, Pedro, Pegaoleg, PenCoinBoxLamp, Pennyroyaltea, Pepperkid, Pepsidrinka, Percy Snoodle, Persian Poet Gal, PeterSymonds, Peterdjones, Peterstannard, PetosGreatos, Pgan002, Pgk, Phaedriel, Phaldo, PhantomInsider, Phatom87, Phaze2010, Phefner, Phil024, PhilOsirus, Philip Trueman, Philip200291, Philly619, Philwelch, Phoenix-forgotten, PhoenixMourning, Phoenixrod, Phonzonappy, Phort99, Phuntsok2000, Pie pie pie piepp ie, Pieguy48, Pielover87, Pierco, Pikablu5, Pikawil, Pilif12p, Pillow2011, Pilot850, Pilotguy, PinchasC, Pirate Thom, Pirinolo, PizzaMan, Pjcard, Pker45, Pkwildy69, Plagana, Plasticup, Platinumfawkes, Playstationdude, Plonk420, Plugin1, Plugwash, Poetic Decay, Poetman22, Pointer1, Poirier273, Poke mudkip, PokeMaster001, Pokemaster Justin, Polnewsa, Pooface13, PopUpPirate, Portillo, Poulsen, Powerslide, Pp1029, Ppntori, Prankboy973, Prem555, PresN, Pretender2j, PrimeHunter, Profcallum, Professional Gamer, Prolifix - Zaretser, Prolog, Promethean, Pronoun, Prosecreator, Protector of Wiki, Prozzaknation, Ps2fanboy, Ps3gamer, Psz, Puddin28, Puffin, Pugimee, PulpDood, Pumapayam, Punjabi101, Punk Rocka 07, PureLegend, Purplees, Pyroven, Qtheking, Quackshot, QuaestorXVII, QuannyG, QuantumSingularity, Quantumchaos, Quaque, QuasiAbstract, Quindraco, Quintin3265, Quuxplusone, Qwerasdfzxcvvcxz, Qwikjus69, R.E. Freak, R3i6nm8n, RABicle, RC Master, RG104, RG2, RHaworth, RJASE1, RJaguar3, RK, ROLO, RP459, RadioKirk, Radolph, Rafichamp, Ragnarok Addict, Rahlgd, Rahman2094, Rahzel, Railer 760, RainbowOfLight, Rainwarrior, Rairaunak, Rajrajmarley, Ramis84, Ramurf, Ran, RandomBSOD, RandomXYZb, Randoman412, Randwicked, Randy Jaiyan, Rangek, RangerOfFire, Ranmoth, Rarutorocks, Ratchetcomand, Ratiocinate, RattleMan, Rattlesnake, RattusMaximus, Raucanum, Ray andrew, Raza514, Rbarreira, Rcandelori, Rddflag, Rdsmith4, ReCkLeSs X, Realberserker, Realkyhick, Reapermage1990, Rebroad, Red dwarf, RedEyesMetal, RedWolfX,

Redpatcher, Redsparta, Redvers, Rekenner, Relow, Remdabest, Remeberthisaccount, Remurmur, Renegadeviking, Replaystay, Retardedmonkey, Revancher, RevolverOcelotX, RexImperium, RexNL, ReyBrujo, Reyn116, Rfl, Rgoodermote, Rgsgww, Rhallanger, Rhe br, Rhobite, Ribeirowaldo, Rich Farmbrough, RichAromas, Richard Matthews, Richard beer, Richardcavell, Richiekim, Richmeister, Rick Browser, Rick Sidwell, Rickington, Ricky81682, Rik201, Riksweeney, Rillian, Rincebrain, Rish.aggarwal158, Risk one, Riurik, Rjd0060, Rje, Rjwilmsi, RkOrton, Rkoed, Rlee0001, Rmccue, Roadstaa, RobJ1981, Robbiekarlson, RobbyCohen, Robchurch, Robert Merkel, Robert Nagle, RobertG, RobertsKong, Robivy64, Robost, Robth, RockMFR, Rocker afi91, Rodahue, Rodhullandemu, RodneyLove, Rogerd, Rogiciserk, Rojugu, Rolltide3738, Romario Maxwell, Ronan1888, Ronnie2, Roonaldo10, Rory096, Rorymanzaroli, RossPatterson, Roux, RoyBoy, RoyalFool, Royalbroil, Royboycrashfan, Rray, Rreagan007, Rsdugan, Rst20xx, Rtvoll, Rubiscous, Rubso, Rumblpak, RunedChozo, Running, Russborj, RuthlessOne, Ruyn, Rwallen, Rwiggum, Ryan Paddy, RyanCross, Ryanasaurus007, Ryanrab1, Ryanrossiter, Ryu Ematsu, Ryu Kaze, Ryulong, S, SCB '92, SCEhardt, SDC, SF007, SJP, SNS, ST47, SURIV, SVG, Sachinwalia, Sacredepitaph, Saddhiyama, Sagaciousuk, Saimcheeda, Sakuraz, Sam Burke, Sam Korn, Sam Li, Sam-Tyson92, Sam8, SamBishop, SamJ, SamLee0519, Samboro, Sammal11, Samuel Curtis, Samuel Sol, Samuraid, Sandstig, SangeYasha, Sango123, Sankarj12, Sanke88, Sanyasi, Sarenne, Sasquatch, Satanael, Satula, Sawblade5, Saxbryn, Scammy2010, Scarfus, Scarian, Scatman839, Scepia, Sceptre, Schmancy47, Schmeater, SchumiChamp, Schwnj, Schzmo, Scientizzle, Scott Nash, ScottSteiner, Scotty.n, Scubafinch, Sdinstrell, Sdonato2@twcny.rr.com, Sdornan, Sdt, Sean The Spartan93, Seanor3, Secret Saturdays, Seegoon, Seena1993, Seifer100, SeparateEntity, SeraphimXI, Seresin, Sesu Prime, Setti, Sfgiants320, Sgkay, ShK, Shaak ti jedi master, Shaddack, Shade28, Shades Of Grey51, Shadoom1, Shadow El, Shadow Hog, Shadow Siren, Shadow demon, ShadowInferno, Shadowfax1993, Shadowlightn, Shadowlink1014, Shadowofb33, Shadowriku, Shadowvirus17, Shagywashere, Shahiboomboom, ShakataGaNai, Shallandhaar, Shamusl, ShanE159, Shanel, Shanes, Shannonr, Shark96z, Sharkrocket, ShaunES, Shawnc, Shayde1014, Shehzad17, Shentino, Shiconia, Shifty36, Shiggy, Shimerdron, Shinnawy, Shinobizu, ShinraiTS4, Shizane, Shodown19, Shoessss, Shogunu, Shookvitals, Shoopdawoop, Shortyred12hd, ShotokanTuning, Shreshth91, Shtinky, Sibpin, Sickthiele, Sidasta, Siddanator, Siddharth Prabhu, Sidewinder468, Sietse Snel, Sig0, SigmaEpsilon, Silenthill addict, Silly rabbit, Sillyantman, Silver Edge, Silver Sonic Shadow, SilverKiento, Siman1990, Simdal, Simone, Sin-man, Singularity, Sinkillerj, Sion8, Sir Crimson, Sir Nicholas de Mimsy-Porpington, SirMetal, Sixteen Left, Sj, Sjakkalle, Sjdb5, Sjones23, Skate boarded ex7, Skater0913, Skew-t, SkidzKid, Skitut Master, Skorp, Skorpio-88, Skullcap, SkyCaptain, SkyWalker, SkylineEvo, Skylinegtrgsl, Skysmith, Slainbaby, Slapshot24, Sleep Question, Sligocki, Slivester, Slo-mo, Sly Curado, Slyang, Slydevil, Slyguy, Smallville123, Sman789, Smatts, SmeagoI.i.am, Smeggysmeg, Smilesau, Smoker68x, Smokeyfire, Smokizzy, Smoth 007, Smugface the untrustworthy dwarf, Snake121, Snakes12, SniperCS, Snoopdop, Sobi3NY, Socby19, Soccerisawesome, Sockatume, Socstudent, Soetermans, Someone42, SonicBenji, Sonnevillej, Soosed, Sora45, Sorcre, SorryGuy, Soulhack, Soumyasch, Souphanousinphone, Sp33dyphil, Spaceboy492, Spaced1999 AKA, Spacexplosion, Spangineer, Sparksalot, Sparky2002b, Spartanhelmet, Sparton 10, SpawnOfEvil, Spiderkeg, Spidey104, Spigot91, Spike105, Spitfire434, Splash, Spliffy, Spoonboy42, Spurfan15, SpuriousQ, Spyder810, Squamden, Squanderdalfast, Squibbles, Squideshi, SquirrelPhister, Srikeit, Srkhan2, Srose, Ssbbrules, Ssolbergj, StG1990, Starinvaders, Starionwolf, Starwarsrulez, Stat89, Steel, SteinbDJ, StellarFury, Stephenb, Steve Hart, SteveSims, Stevegallery, Steveketchup, Steven J. Anderson, Steven Zhang, Stevencho, Stevenup7002, Stevo1000, Stick it to the man, Stickeylabel, Stickyfox, Stifle, Stikhead, Stonedonkey, Stor stark7, Stormwatch, Strap, Stratadrake, StreetFighter, Strolls, Stryder7, Strykie-boy, Stuhacking, Stupidstupidstupid, Stx, Sub-zero-du, SubStandardDeviation, SubeyDes, Subhek, Submerge, SuicideDuchie, Sukiari, Sully1311, Sumeetg, Sumsum2010, Sunlitsix, Sunnmagic, Sunray, Sup22, Supaplex, Super hoops1967, Super waluigi, Super-Magician, Superiority, Supermaner, Superpiccolo, Supersaiyanplough, Supertyper, Suppakid, Supra2JZGTE, Suriel1981, Surtr, Suruena, Suso, SwK, Swarm, Sworder, Swotboy2000, SynergyBlades, Syracon30, Syrthiss, Syvanen, Szdg, Szimm, TAshkar18, TGilmour, TJ Spyke, TJD2, TJF588, TJRC, TMC1982, TRega123, TSO1D, Tachi 43, Tachyon01, Tactherapper, Taelus, TaerkastUA, Tagasaki, Tahirjon, Tails4, Tails623, Takuthehedgehog, Taladar, Talkie tim, Tamariki, Tamarzan, Tanaats, Tangotango, Tani unit, Taopman, Tapo, Targetter, Tariqabjotu, Taro James, Tashivana, Tastemyhouse, TastyCakes, Tawker, Tayger, Taylortbb, Tbhotch, Tbone762, Tchlouis, Tdnetworks, TeamOverload, Techman224, Techni, Technorj, Ted87, TedE, Tedcurly, Tedius Zanarukando, TedizKiller36, Teetzmagee, Teffen, Teggles, Teh man, Teh one111, Teh-Nathat, Tehw1k1, TeknoKid, Telcourbanio, TenPoundHammer, Tenpin477, Terence, Terminator50, Terraleone, Terrapin, Terroardon, Terror G, TexasAndroid, Tezero, Tfuji513, Th1rt3en, Th3heretic, ThE MoNk, Tha Web, Thardas, That Guy, From That Show!, Thax, The Anome, The Assimilator, The Berzerk Dragon, The Corruption of Lol, The Goat, The MP, The Marmalade, The Rambling Man, The Requiem, The Rogue Penguin, The Seventh Taylor, The Thing That Should Not Be, The Unstoppable Wikipedian 9, The Volcano, The Youth Counselor, The ed17, The evil one, The luigi kart assasions, The velociraptor, The wub, The1physicist, The359, TheCoffee, TheDeathCard, TheDevilYouKnow, TheEnlightened, TheJames, TheKittenEater, TheKoG, TheLoverofLove, TheMadRussian, TheN1Armyguy, TheUltimateHistorian, TheWakeUpCall, TheWarlock, Thecheaters, Theclassof007, Thecooldude101, Theda, Thedjt, Thedka, Thegreyanomaly, Thekoyaanisqatsi, Thelastneo, Theman00, Themasterofwiki, Themindset, Thend, Theonlyhuman, Theseven7, Thesmarterone, Theunknown42, Theweirdfan, Thewikipedian, Thewolrab, Thewyzewun, Thibbs, Thierry474, Thingg, Thingy28123, ThinkBlue, This user has left wikipedia, Thiseye, Thores, Thorpe, Thrallhascome, Thrashmetal3, Thrones, Thue, Thunderbrand, Thunderfox866, Thunderfox933, Thunderhawk89, Thylacine222, Tide rolls, Tig3rfang, Tigen, TigerShark, Tik, TimTim, Timberframe, Timdarbre, Time is running out, Timmeh, Timmytje, Timrem, Tineo03, Tiptoety, Titoxd, Tkessler, Tktktk, Tlj23, Tmcw, Tntnnbltn, Tnxman307, Tobylle, Tobz1000, Tocino, Tokyoship, Tom386, TomM374, TomXP411, Tomas4191, Tombias, Tombrabbin, Tomisaac, Tommo h, TommyBoi, Tomwchow, TonicBH, Tony Fox, Tony Myers, Tony1, Torturies, Tossinit, Tourusen12, Tozilla, Tpbradbury, Tr00st, Transce080, Travelbird, TravisBatos, Tree Biting Conspiracy, Treeout, Tregoweth, Treima, Trek222, Trevie, Trevor MacInnis, Trevor Saline, Trevorelsey, Trey56, TreyGeek, Treygdor, Tripacer99, Troll423, Trombe29, Troyoda1990, Trusilver, Tubby, Turbo1998, TurboBoom, Tuspm, Tutmosis, Tuxcantfly, Tv145033, Twexler, Twist05, Twistedchild1066, TwixDawg, TwoThumbsDown, Twobells, Twocooldudes, Txrat, Tyguy69, Tykell, Tyman 101, Typenolies, U.S.A.U.S.A.U.S.A., UBERFRY, UCS123, UKHack, URORIN, USEDfan, UberScienceNerd, Uch, Ugg600, Uilly2807, UkPaolo, Ukexpat, Ultimatejester, Ultimus, Ultrafez, Ummit, Umrguy42, UnDeRsCoRe, UnKnown X, UnKnownViper, Unchibaka, Uncle Dick, Uncle Milty, UncleDouggie, Unforgettableid, Unknownwarrior33, Unplugged68, Unreasonable, UnrivaledShogun, Unsterblich, Untraceable2U, Uopbound, UrbenLegend, Urhixidur, Usedbook, Useight, User aaqib12, Username21f, Usien6, Usman.shaheen, Utfreek185, Uttar, Uusitunnus, VJS, VTNC, Vader99, Vague Rant, Valdens, Valoem, Van helsing, Vapier, VargasUSMC, Varxx, Vash67, Vdub49, Veculous, VederJuda, Veemonkamiya, Vegardw, Veinor, Veldin963, Velkr0, Vertius, Vespine, Vf7993, Vghistorian, Vgy7ujm, Viamp, Vidbv, Video game fan11, VideoCWK, Viiv, Vikrant42, Vin789, Vindictive Warrior, Vinmartino, Violaguy5, Violetriga, Virchu, Virek, Viridae, Viriditas, Visor, Visual planet, Vitz-RS, VivaLaBen, Vladmirfish, Vliam, Vodka Martini, Shaken Not Stirred, Voidxor, Volchcha44, Volkner467, Volt M, Vorpheed, Votequimby, Vrdecimo, Vrray9000@yahoo, Vulcanstar6, Vulian, Vylen, W Tanoto, WAvegetarian, WMarsh, WODUP, WWEFreak666, WacoJacko, Wadezombie, Wageslave, WakiMiko, Walkingonthesun, Walshjd, Wangtjai, Wapbrit46, Warmpuppy2, Washod, Wavelength, Waycool27, Wayne Hardman, Wayward, Wbalmung, Wdflake, WebJunkie, Webhat, Webzpinner, Weezy504, Weihao.chiu, WelshMatt, Weltall445, Werdan7, Werdguy, Werdna, WereSpielChequers, West Brom 4ever, WesternActor, WhartoX, WhereAmI, While08, WhiplashInferno, Whisky drinker, Whitesanjuro, Who What Where Nguyen Why, Wholikeswood, Whomp, WiZZLa, Widkid85, Wierdo135, Wiesel 99, WiiLord, Wiifinder1, Wiimo, Wikada, Wikeel, Wiki alf, Wiki.user, WikiBone, WikiCats, WikiLeon, Wikiacc, Wikiborg, Wikidudeman, Wikiediter538, WikipediaGforce, WikipedianYknOK, Wikiposter0123, Wikirocks666, Wikischmedia, Wikitom97, Wikiuser387, Wilchett, Wilee, Wilester, Willhsmit, Willie512, Willieiscoolerthanchris, Willsmith342, Willy105, Willy888, Wimt, Windsok, WindyCityRockr, Winhunter, Winston365, Wisq, Without Due, Wkpd, WI219, Wmahan, Wnt, Wog7777, Wolf2p, Wolfscarr, WolvenOne, Wonchop, Wonphi, Woodsb02, Woofamania, Woohookitty, Wormbird1, Woseph, Wrcmills, Wrhsouth, Wtatour, WulfTheSaxon, Wuzupmyhomiz, Wwoods, X SMERDEL1Z3R x, X-Fi6, X201, XCore21, XFD, XMillar, XP1, XX55XX, XXFULLYLOADEDXx, Xabian40409, Xander756, Xangbar, Xaosflux, Xbox rules!, Xbox360isbetterbiches, Xbox6, Xchrisblackx, Xcom888, Xellxero, XenoL-Type, Xenoslayer, Xenoveritas, Xenus, Xeta991, Xevious, Xezbeth, Xiaoyang4673, Xino, Xinoph, Xiong Chiamiov, Xizer, Xkxdxmx, Xplosneer, Xs12345, Xtraxxl, Xtreme racer, XtremeGamer99, XvirusX, Xwiz, Xxaiolos, Xxpor, XxxFilipks, Xymor, Y.krzepicki, Y1k0, Y2j420, YUL89YYZ, Yachyx, Yafai jordan, Yamamoto Ichiro, Yamla, Yamster, Yankee112792, Yansa, Yamalgo, Yaughtmaster, Yeahbutnobut92, YellowMonkey, Yelyos, Yiwen017, Ylem, Yoelsilver, YolanCh, Yolexander, Yonatan, Yono, Yooo67, Yoosq, Yoslims, Youngjav, Yourboychris69, Yousou, Ysangkok, YumOooze, Yuubinbako, Yyyuuu, Z.E.R.O., ZBrannigan, ZIELDIABLOIZ, Zabrak, ZachPruckowski, Zachary, Zachary8222, Zahid Abdassabur, ZaidAhmed, Zain., Zain25, Zaita, Zakgrant, ZaknafienSpecial, ZakuSage, Zanibas, Zapdosxd, Zarrinnis, Zazaban, Zctyp18, Zeazzz, Zellin, Zeno McDohl, Zenpher, Zephead999, ZephyrAnycon, Zeppelin462, Zero1328, ZeroOne, ZeroRaider, ZeroUm, Zerokoolin, Zeta26, Zforsyth, Zhente, Zidane4028, Zilog Jones, Zim210, Zimbabweed, Zinck, Zoda-Z, Zodac, Zoe, Zoffdino, Zoken4, Zomno, ZookPS3, Zophras, Zoras, Zpb52, Zscout370, Zsinj, Zwee, Zxcvbasdfg, Zzuuzz, . , 4967 anonymous edits

Xbox_360 *Source*: http://en.wikipedia.org/w/index.php?title=Xbox_360 *Contributors*: -Majestic-, .:Alex:., 007craft, 041744, 0Mostel, 1(), 12345t, 123wiki123, 12hctawkcalB, 1mujin22, 23mike95, 2850935rey, 2mcm, 4finger, 8bitJake, A Man In Black, A Nobody, A s williams, A.A.M., ACupOfCoffee, AIR DJ, AMHR285, AP Shinobi, APAD, ARC Gritt, Aaaaaa, Aaron Brenneman, Aaron Schulz, Aaron117, Aaron20000, Aashish950, Aazeez, Abb615, Abby53, Abhishikt, Abm1996, Abomasnow, Abours, Academic Challenger, Ace 1000, Ace of Sevens, Aceramicbunny, AcesUpMaSleeve, Ackander, Acq3, ActivExpression, Adam Krellenstein, Adam24367, AdamJacobMuller, AdamSebWolf, Adambiswanger1, Adashiel, Addict 2006, Adilch, Adrianfogge, Adw2000, Aehrenreich, AerobicFox, Afftonz, Afghanballer, Agathoclea, Agent007ravi, Ah1298, Ahockey92, Ahoerstemeier, Aid1223, Ainsej, AintRight, Airconswitch, Airfoemoe, Airplaneman, Ajaxspray, Ajpralston1, Akadewboy, Akamad, Akessle1, Akshayvarma89, Alan Au, Alan Isherwood, Alansheaper, Alansohn, AlasdairJohnstone, Aldoman, Aleksael, Alerante, Alereon, Alex Stanek 9999, AlexJP, Alexander Vince, Alexandercode, Alexballiet, Alexbuster500, Alextrevelian 006, Alfa989, Alfiboy, Alias Flood, Alimadari, AlistairMcMillan, Allemannster, AlphaTwo, Alphachimp, Alphaman, Alphathon, Alpheta, Alsocal, Amakuru, AmericanSights2008, Amit g v, Amitch, Amren, AnOddName, Anaemic, AncientGreek, Anclation, Anded991, Andersonma, Andi Saleh, Andorhal, Andreas -horn- Hornig, Andrevan, AndrewSM, Andrewlaneyboi93, Andrewpmk, Andrewwg94, Andros 1337, Andy Marchbanks, Andy120, Andy13121990, AndyKing1, Andyahaziz, Andybogardng, Andyman33, Andypandy.UK, AnelZukic, Anetode, AngelOfSadness, Angela, Anilocra, Animal91X, Animekid3515, Anirvan, Annoyance134, Anomie, Anomymous0001, Anonymous Cow, Anonymous editor, Answerthis, AntL, AntiVanMan, Antimatterswarm, Antonioc, Apadilla80, Apau98, Apiniewski, Apmastersite, Apofisu, Appleboy, Apyule, Aquajester, AranhaHunter, Arbero, Arcaynn, Archer3, Arebuffe8, Ari21, ArieTrouw, Arif92, Armaced, Arnon Chaffin, Art LaPella, Artrix, Arukun14, Asarch, Asc112, Asclepius, AshSert, Ashenai, Ashmellor, Ashroh, Ashthecat3, Ashton305, Aska, AssassinXel, Assistance, Astonmartini, At45y, Atf487, Atomic23112, Attitude2000, Auger Martel, Aussie Evil, Austboss, AustinZ, Autumn Forrester, Av, Avangion, Aviado, Avicennasis, Avillia, Avtech, Awesomeness95, AySz88, Aydee, Azentium, Azetab, Azurepalm, B DEETS, BCKE11, BHC, BMT, Babakmomen, Babar92, BaconBitzL0lz, Badreligion, Bailey1987, Baker013, Bakerja03, Bakka2, Ball of pain, Bambamfan91, BambinoPrime, BananaDesk, Banej, Banes, Barek, Barneyboo, Barnolde, Barronlroth, Bartboy919, Battle medic, Bavgang123, Bbatsell, Bcorr, Bdavid, Bdhoff, Be300x, Be4r, BeL1EveR, BeanyFans, Beardy10, Beast432, Beautiful so ur, Beeblebrox5000, Ben5ive, BenGriffiths, Bencey, Bencherlite, Bencossette, Bender235, Benea, Beneaththelandslide, BenjaminWillcock, Bennity, Benparfitt, Berkut, Beta, Bevan.coleman, Beyond silence, Beyondevolution, Big Andy, Big Bird, Big Merl, BigBoy52690, BigEgo, Bigger Boss, Bigsnake 19, Biker14, Bill.matthews, Bill25, Billcheese1, Billyjoe99theone, Bizhaoqi, Bizkit moorse, Bizub4, Bjørn, Blackfinale, Blackwatch21, Bladepro91, Bladestorm, Blanchardb, BlightedEye, Blooga245, BlueMint, Bluedude3011, Bluephazon, BlythMeister, Bmicomp, Bngrybt, Boarder8925, Bob A, Bob EggHead, Bobblehead, Bobblewik, Bobdragster, Bobo192, Boffo, Boing! said Zebedee, Bollinger, Bongwarrior, Boninater1, BonzoESC, Boogey mon, Bookoflove, Bopher1, Borameer, BoricuaPR, Boringcarlover, Boris Red, Born2beSlicker, Bornhj, Bovineone, Bowmanjj, Bowserlx, Br4d d, BrOnXbOmBr21, Bradeos Graphon, Bradwart, BranER, Brandom, Brandon10092, Bratsche, Brazil4Linux, BrenRich101, Brent Butler, Brian Ryans, BrianSkelly, Briankel, Brighat, Brighter7, Brittflick 1994, BrokenSegue, Brokendownferrarienzoferrari, Brokenscope, Broomeboy, BrotherEstapol, Browne.dean, Bryan H Bell, BryanG, Bsimmons666, Btball, Btipling, Bttfpromo, BuckeyeFanatic25, Buckeyes1186, Burgessm, Burner0718, Burnte, Burntgreen, Burto88, Butros, Buzzoven767, C0nanPayne, CPO Pieman, Cabanaguy, Cabe6403, Caboose2233, Caesura, Calamity-Ace, Calderra, Caligulathegod, Calmer Waters, Calton, CalusReyma, Cambrant, CambridgeBayWeather, Cameron789787, Cammi666, Can't sleep, clown will eat me, Canadian-Bacon, CanadianCaesar, Canderson7,

CanesOL79, CanisRufus, Canley, Cantara, CanucksRule, Caomdayn, Captain Disdain, Captain Waffles, CaptainAmerica, CaptainHaZ, CaptainVindaloo, Cardboard boxA, Cardew, Cardsplayer4life, CaseyPenk, Cash RingofFire, Casliber, Casper2k3, Castaa, Catbert107, Causa sui, Cavenba, Cb2k76, Ccardali, Cccccc, Cdamian, Cdc, Cdude1034, Ceecookie, Celerityfm, Cenarium, Central2, Centrx, Cfailde, ChAoTiCpInOy89, Chadwickschool, Chairboy, Chairman S., Chambece, Chan Yin Keen, Chanlord, Chanman121, Chanting Fox, ChaosAngelZero, Charliebruce, Charliedelaseux, ChasePuglisi, Chasekorfhage, ChazBeckett, Check two you, Cheesor, Cheezmeister, Chensiyuan, ChicagoRules, Chicito21154, Chill Pill Bill, Chmod007, Chochopk, Chocobogamer, Cholmes75, Chorizo245, Chowbok, Chowells, Chris3145, Chrislk02, Chrismear, Christofer C. Bell, Chroniclev, Chronulator, Chuck369, Chykka207, Ciao 90, Cjcamilla, Cjoyce1980, Ck lostsword, Ckoicedelire, ClapBoy380, Clarince63, Clarkey4boro, Classicalgas, Claudmar, Clayhalliwell, Clearhistory, Cliché Online, Closedmouth, Cloud109, CloudNine, Cloudrider7, Clutzy jamie, Cm8765, Cmc0, Cmdrjameson, Cncxbox, CoJaBo, Cobraman156, Coconutfred73, Cody 88, Coelacan, Col. Hauler, ColdShoulder, Colin Keigher, CombatEvolved, Combination, Cometstyles, Commander Shepard, Commando3200, Commodore 6400, Computer97, Computerdude1032, Computergeek1507, Comradeash, Con 017, Conf, Connor10, Conor65, Consumed Crustacean, Conti, CoolKid1993, Coolcato, Coolg35748, Coredesat, Corporal Punishment, Couchpotato99, Courcelles, Cpc464, CraigD1993, CranberryPub, Crashedata, CrazySlyHawk, Crazycomputers, Crazytales, Credema, Creol, CritterNYC, Crobichaud, Crosbie, CrossHouses, Cruddycakes, Crumb, Cryptanalysis, Cschock, Cubanaso, Cumbiagermen, Cumulus Clouds, Curps, Curtis234, Curtmack, CyberSkull, Cybersonic10, Cynical, Czechnmymail, D prime, D.e.Iverson, D.luo117, D510r9, DBGFrost, DBaba, DKruse, DLand, DaKing, DaL33T, Daa89563, DabMachine, Daemon Lotos, Daft tv, Daghall, Daku0g, Daler, Damian Yerrick, Damien Russell, Dammit, Dan D. Ric, Dan100, Dan2313, Dancter, Dandelions, Danhash, Daniel Vollmer, Daniel.Cardenas, DanielCD, Danno uk, Dannybu2001, Danthomas99, Darealclub, Darin-0, DarkPhoenix32, DarkRain42, Darkcloud4491, Darkcross, Darkhunger, Darkman007e, Darkness0110, Darkquest21, Darkraider9, Darkshark0159, Darranc, DarthShrine, Darthblaze99, Darthbob100, Darthmarth37, Darthnader37, Darwin-rover, DasRakel, Dashwarts, Dat11, Daunrealist, DaveJB, Daveoc64, Davetheman44, Davewho2, DavidBrowne, DavidMann, Davidhorman, Davidyorke, Dboisen, Dbrodbeck, Dcandeto, Dcsutherland, Dctcool, De728631, DeFFiance, Dead Midgets, Deaddude, Deank06, DeathSeeker, Deathawk, Deathtrap3000, Debresser, Deckiller, Decky, Decompiled, DeeKay64, Deepen03, Deeptrivia, Defjamrules, Defkkon, Dekonegawa, Delldot, Delmothop, Delta Elite, Deltabeignet, Deltasquared, Demonslay677, Dengarde, Denzelio, DerHexer, Desertsnowman, Desperadonige, DetroitDan2004, Devilboy1015, Devl2666, Df747jet, Dffgd, Dgcaste, Dibol, Dicconb, Diceman, Dickdogers49, Diegofrgc, Diggitydiggity, Digitalme, Dilip gerba, Dimre01, Dina, Dinohunter999, Dionyseus, Dirt Pecker, Disavian, Discospinster, Dispenser, Diverman, DivideByZero14, DjSyndrome, Djfost, Djlollyb, Dlkessel, Dmarom, Dmitribendoliuv, DmitryKo, Doberman Pharaoh, Doc glasgow, Doctor Hexagon, Doctorinthahouse, Dog Eat Dog World, Dominar Rygel XVI, Dominicanpapi82, Domstang68, Don Joes, DonEd, Donatj, Dondondon, Donevc, Donkey100, Doodledoo, Doodoobutter, Doom127, DoomBringer, Dopeykid, Dorro, Dorsal Axe, Dosner, Dou Gweler, Doughnuthead, Dpcough, Dpw6, Dr doris, Dr. Pizza, Dr. Sunglasses, DrKiernan, DrYak, Dragon32, Dragon76, DragonRider12, Dragonilm, Dragonpulsar, Drakino, Drat, Drendy88, Drewerd, Drknockers, Drmies, Drqshadow, Drseudo, Drw25, Dstancioff, DuBbL3 U To0, DubaiTerminator, Dude1010101, Duel007, Dunnette, Dupz, Dvferret, Dylanwithers, Dynamax, Dynendal, Dysalot, E barroga, E-Magination, EEMIV, ERADEKator, ERW1980, ESkog, EVula, EarthSage14, East718, Echo1, Ecioa, Ecksemmess, EcthelionGenesis, Ed g2s, Eddiet19, Edgar181, Editor18, Edward4321, Edwincruz654, Ee player, Efe, Ehavir, Either way, Ejg930, Ekrub-ntyh, El $corpio, El C, El Capitaine, Elad476, Electricmoose, Electron9, ElementFire, EliasAlucard, Elliotcraen, EllisW360, Elsan, Elvaangohst, Elven6, Elwood00, Emersoni, Emhilradim, Emufarmers, Endangered Human, Endersdouble, Energyfreezer, EnglishArds, Enjoi, EoGuy, Eoyount, Epbr123, Eptin, Eran of Arcadia, Eric1016, Erish, Erzengel, Escape Orbit, Escobar4Life, Espiox, EstebanF, Estel, Estoy Aquí, EtTuBrute, EthanB, Euchiasmus, Eurocave, Eurovictor, Evan-Amos, Evansam, Evercat, Everyday17, Everyking, Evice, Evil Merlin, Evil Monkey, Evil Vin, Evilgohan2, Evilgrug, Evonsky1, Excirial, Exeunt, ExilorX, Exmachina23, Exodus94, Extraordinary Machine, Exukvera, FF2010, FPAtl, Facts707, FaithRaven, Falcon866, Falcon8765, Falcon9x5, Fanboy, Fantasydragon, Farazaraf, FatFox, FatM1ke, Fatjoooooooseph, Favonian, Fb767, Fbchad, Feitclub, FellowWikipedian, Ferrarimanf355, Fethers, Ffgamera, Fid9735, Fighting Zucchini, FinalRockerdude, Finitecesar, Fireaxe888, Firetrap9254, Fish and karate, FishStik, Flame Rising4, Flash512, Flaviohmg, Fleminra, FlieGerFaUstMe262, Flipsterz4life, Flockmeal, Flood Master, Flooey, Floor3013, Florescent, Fnagaton, Fnfd, Foogin, Footballracer7, FootiejeM, Forteblast, Fortenium, Foster-smells, Four24oh5, Four8split, Fourohfour, Foxhunt99, FrYGuY, Fractal3, Francs2000, Frank Bitterhof, Frankbkkcam1, Freakofnurture, Fred J, Fredgamer12, Free Bear, Freestyleking012, Freezydemon, FreplySpang, Friday, Frin, Frodo1511, Frodoandgandalf, Fruit666, Fuhghettaboutit, Full Metal 01, FullMetal Falcon, Fullstop, Fultron89, Fumo7887, Funkydunkleman, Funper, Furrykef, Furrypig, FutureNJGov, Fuzzycasserole, Fvw, G8or8de, GCFreak2, GEO VA, GILLY1089, GRevolution824, GSK, Gaara42, Gabbe, Galuitui, Galwhaa, Gamaliel, Gamem8ker, Gamer2325, Gamer28, Gamer93, Gamester17, Gaming&Computing, Gargaj, Garglebutt, Garion96, Garric, Gatoatigrado, Gault8121, GaussNegotiator, Gavyn Sykes, Geckopulse, Geeko2, Gene Nygaard, General Heed, GeneralAtrocity, GeneralDuke, Genesys 4, Geordanh, GeorgeMoney, Georgieboy212121, Gerardm, Gerbon689, Gfrewq12, GhostMaster3000, Ghostchild23, Gigi Galli, Gilawson, GinaDana, Ginsengbomb, Gitman87, Gizzakk, Glane23, Glen, Gmrtom7, Gmrxbox, Gobonobo, Gogo Dodo, Gold Stur, GoldDragon, Golden Dragoon, Goodnightmush, GorbaLor, GordonFreechman, Gordonfreeman 19, Gothics-j, Gracefool, GraemeL, Grafikm fr, Graham87, Graumail, Gray Spot, GreenUpGreenOut, Greenpuppet, GregAsche, Gregory1981, Gresser, Grey2100, GrimRepr39, Grindking, Grm wnr, GrooveDog, GroovySandwich, Grubish, Grzesik, Gsp, Gspawn, Gsvaughn, Gta95, Guam00, Guinea pig warrior, Guitarman63mm, Gundamforce, Gunmetal Angel, Gunsmithy, Gurch, Gurukid, Guticb, Guyinblack25, Guyjin, Gwendibbley, Gwernol, GwiDan, Gyrferret, HHS SpartanBaller, Haakon, Hackazor, Hadal, Hadlock, Hagerman, Hahnchen, Haipa Doragon, Hairchrm, HalfShadow, Hall Monitor, Halo2vista, Halo3 gleeson, Halofan333, Hamiltha, Handymanjr, Hangid, Hanjos88, HappyVR, Hardy172, Harro5, Harty23690, Haruyasha, Harvey5150, Hasanahmad, Hasek is the best, Haseo9999, Hashmi, Usman, Havok, Hawkypunk, Hayden120, Hdtvinfoeu, HeartofaDog, Hede2000, Hedymiguel, Heia09, Heimstern, Hellboy1975, Hellno2, HelloAnnyong, Helo2, Helohe, Hemightbesoup, Hempeater, Henriok, Henry Flower, HenryTouchdown, Hesperian, Heymid, Hideyuki, Highwitdr, Hijak, Hilts99, Hippo99, Hitmanhackerzero, HoCkEy PUCK, Hobart, Hobofuzz, Hohum, Holek, Holodoctor1, HolyRunts, Hour, How dare you?, Hu, Hubblek4, Hugh G. Rection, Humble Guy, HumphreyW, Huntman92, Hurricane111, Husond, HweeWieng, Hyad, Hyperscope, I already forgot, IAMDUDE70, IAnything21, IDuDaChaCha, IRWeta, ITurtle, Iamagamesadict, Iams, Ian Dalziel, Ian Moody, Ian Pitchford, Ian13, Iapetus, IceDrake523, IceT13TSOSIM, Icedragon57, Icep, Icey, Igglybuff63, Igordebraga, Igotoparker, Ihoudini, Iisryan, Ikanreed, Ike-bana, Ikh, Ildfisk, Illegal Operation, Ilyanep, Imaginary heroes, Imaninjapirate, Imdaking, Ime Kong, Imhungry, Imjustmatthew, Imnotminkus, Imroy, InShaneee, Inerte, Infamous.angel, Infomanager, Insert cool name, Interframe, Interiot, Interrupt feed, Inuboy1000, Irishguy, IronTek, Isaac Rabinovitch, Isaiah1996, Israeljamesbond, Italianboy10, Iuhkjhk87y678, Ixfd64, Ixoria, Izanbardprince, J Di, J Milburn, J. M., J. Nguyen, J.delanoy, J.reed, J00bsrnub3s, JAHRASATAMON, JARHEAD999, JAZAM007, JBsupreme, JCRules, JDoorjam, JForget, JHMM13, JLaTondre, JR98664, JTBX, JZ Sizzle, JaGa, Jaap3, Jabbawocky Inmate, Jacek Kendysz, Jacj, Jack Cox, Jackwiliamsteiner, Jacoplane, Jadedchaos, Jaeger5432, Jagged21, Jaisonline, Jajabinks92, Jake0geek, James pic, James.stanbridge@gmail.com, Jameschen810, Jamesfgreer, Jamesfords, Jamesjp2114, JamieSI, Jamott, Jan otto, Jarenls, JarlaxleArtemis, Jarlz0r, Jasmeet 181, Jason One, Jasonpero, Java13690, Javert, Javier Donoso, Javierito92, Jaws2022, Jaxl, Jay Gatsby, Jay Turner, JayC, JayKeaton, Jaymacdonald, Jdcaust, Jdm64, Jebuonag, Jecates, Jedistud, Jeff G., Jennica, Jerdon13, Jericho1337, Jess Flaps, JesseGarrett, Jesus9752, JettaMann, Jety111, Jfg284, Jgarzafe, Jgnm, Jh51681, Jhsounds, Jigahurtz, Jinxed, Jitterro, Jj;evene, Jjintheuk, Jk31213, Jklin, Jmundo, Jmwollny, Jn2minella, JoanneB, Joaquin008, Joe96, JoeSmack, Joemc72, Joey Joe Joe Junior Shabadoo, Johann Wolfgang, JohannVisagie, John Quincy Adding Machine, John Reid, John a s, John of Reading, John254, JohnWittle, Johnbrownsbody, Johndude1321, Johnjohnston, Johnleemk, Johnporcaro, Johntex, Joker9678, Jon Harald Søby, Jon Rob, Jonamerica, Jonapello22, Jonathan Wash, Jonathan179, JonathanDP81, Jonnycc23, Jonnykool, Jontintinjordan, Jordsta95, Jorge279, Jort227, Josh Parris, Josh18230, Josh3580, JoshRoulston, Joshcvb, JoshuaMD, Joslop500, Josquius, Jossi, Journalist, Joyous!, Jrayk, Jrvegeeta, Jschuur, Jsmlw, JtMinahan, Jtalledo, Jtkiefer, Julius Sahara, Jusdafax, Just James, Justinsane5x, Juventino4ever, Jvandyke, Jwoodger, JzG, K1Bond007, K7 madman, KAMiKAZOW, KAPITALIST88, KCMraz, Kajisol, Kamasutra, Kamekichi, Kanzler31, Karada, Karafias, Karebear 1022, Karmakazy64, Kashman, Kateshortforbob, KazekageTR, Kazeneko, Kazrak, Kbdank71, Kbh3rd, Kbolino, Kedi the tramp, Keira Salia, Kelestar, KenKeller, Kenaldinho10, Kencaesi, Kenimaru, KennedyBaird, Ketiltrout, Kevin 11, Kevin B12, KevinKim, Kevinh456, Kevmoo, Kflester, Khoikhoi, Kianoosh kani, Kibblesworth, KickahaOta, KieferSkunk, Killamike, Killer coon, Killermidget123, Killfire72, Killrhedake, Killswitch Engage, Kilo-Lima, Kim Lee, King of Hearts, KingEmperor24, Kinghajj, Kingofgames2, KingpinE7, Kinneyboy90, KirbyMeister, Kismeth, KitAlexHarrison, Kitch, Kitchov, KittenKiller, Kizor, Kkeller1116, Klimtkunst, Kmg90, Kmoe333, KnowitallWiki, KnowledgeOfSelf, Knuckles Dawson, Knulclunk, Koavf, Kodster, Kokoriri, Kollision, Konman72, Kookro, Koosh, Koratnia, Kostas.karachalios, Koweja, Kozuch, KrazyCaley, Krazykris83, Krich, Krisirk, Kross, Krossedout, Krusty111, Krynnmaliki, Krystyn Dominik, KuSh, Kungfuadam, Kurai, Kurieeto, Kurotsyn, Kuru, Kwiki, Kyle, Kyle1278, Kylewpeterson, Kyoko, L daruwala, L1ghtwave, LAX, LOL, Ladlergo, Lakerraptor, Lalex2, Lamename3000, Lanhoj, Lankiveil, Lanway, Lar, Larakin, Lardo, Larry laptop, Larsinio, Last apoc, Lasthand, Latics, Laug, Lavenderbunny, Lazarus666, Lazychris2000, Leachyboy77, Leadguitar1211, Lebelu2007, Lectonar, LedRush, Leflyman, Legend78, Lehx, Leonffs, Leopyrofox, Leslie222, Lewisgreen, Lewispb, LexLuthorMetropolis!, Lhasapso, LiamUK, Liasos, Liddle esco, Liftarn, Lightdarkness, Lightmouse, Lightsup55, Ligulem, Likelife, Lil AzZa, LilPro22, Lilac Soul, Link 486, Link83, LinkinPark9297, Linkinman199, Little Professor, Llort, Lmm0611, LobStoR, Local hero, Locazreal, Locriani, Loft, Logalot, Logan, Logan GBA, Lomn, LonerXL, Lord Hawk, Lord Psyko Jo, Lord Voldemort, Lord d'Eath, LordViD, Lordmenace, Los76, Lotr faramir, Lou-Dawg, Louis3333, Luigi30, Luigi6138, Luismbs, Luk, Luke Vandelanotte, Luke255, Lukep913, Luna Santin, Luph25, Lupin, M1rth, M2, M412k, MAINEiac4434, MC10, MDM, MER-C, MKD Majstor, MVP5, MZMcBride, Mad onion, Madchester, Madcowmnky, Madskunk, Maeglin Lómion, Mael-Num, Maestro25, Magister Mathematicae, Mahjongg, Majorly, Majorsuave, Makron1n, Malckeor, Malcolm the samurai, Malfacteur, Malinaccier, Mallerd, Malo, Malphonic, Mamyles, Man123123, Maneman, Mapleafsrock, Marc-Olivier Pagé, Mareg, MarioV, Marioman12, MarkS, Markdang, Markgriz, Markthemac, Marquez, Mars729, Martarius, Martin871, Martinicognac, Martinman11, Marudubshinki, Master Deusoma, Master Kim, Master of the Orichalcos, Master784, MasterEditor99, Mastershake329, Mat10169, Matariel, Mathguyscott, Matmoo4, Matster9090, Matt Britt, Matt Deres, Matt Heard, Matt Neuteboom, Mattbash, Mattbr, Mattclarkie, Mauimonica, MaulYoda, Mauls, Maxamegalon2000, Maxim, Maxninjaman, Maxtro, MaxwellIsYumYum, Maythammahdipour, Mazca, Mbg6231, Mboverload, Mbrandall, Mbslrm, McDonaldsGuy, McGeddon, Mcdffdfdaydy, Mcjakeqcool, Mckaysalisbury, Mcman777, Mddyer, Mdhmdh31, Mdnthrvst, Me412, Me676, Measure, Meelar, Megamanfan3, Member, Mentifisto, Meow07, Meph1sto, Mephiles602, Merovingian, MerseyMal, MetaFinal49, Metal moe, Metallica x, Metroidfan23, Mets501, Mewcenary, Mewtu, Mexcellent, Mexican playa12, Mga, Mgiganteus, Mgillespie, Mhking, Micahmn, Michael Devore, Michael Hardy, MichaelBillington, Michaeljoeseph, Micheal Charles, MickWest, Midlem, MightyCortex, MightyJordan, Mike mgoblue, MikeTru, Mikejm07, Mikem77, Mikeycal, Milad Mosapoor, Milki, Millahnna, MindWraith, Mindgiver, Minidude426, Minna Sora no Shita, Mipadi, Mirror Vax, Misocurdo, Mister macphisto, Mistrbily, Misza13, Mitchello, Mjnhbgv, Mjpieters, Mlgshadowdog, Mmmdonuts210, Mnemeson, Mnts, Moe Epsilon, Moeron, Moigle, Molteanu, Moltovivace, Momo Hemo, Momus, Monger, Monkeyking 808, Montecarlocars, Moogle10000, Mookey, MoondyneAWB, Morgul12, Mortense, Morton devonshire, Mossman93, Mousti, Movellon, Mr Adequate, Mr Bartels, Mr Beige, Mr Chuckles, Mr Pillows, Mr. Tacopants, Mr.Do!, Mr.bonus, MrFish, MrNobodyWTB, MrOllie, MrVoluntarist, Mrbadboy3, Mrstickball, Mrwojo, Mrxg4, Msc44, Msikma, Muddy9494, Muffi1005, MuffinFlavored, Muhand, Multiply, Mulvad, MunchableSandwich, Munchrazz, Munkyamador, Munkypk, Munkyxtc, Mushroom, Musical Linguist, Mutchy126, Muzz2006, Mvent2, Mweflen, Mwillard, Mxdxcxnx, MySchizoBuddy, MykReeve, Myke2020, Mynabird, Myscrnnm, MysteriousMystery, Mythdon, N i grunze, N00body, N22YF, NASCARforlife, NERVUN, NES Wii, NN3210, NP Chilla, NYC5Boro, NYkid0709, Naaa, Nakon, Nandesuka, Nandryshak, Narutometal, Naryathegreat, Nascarfans, Natalya, NateDan, NativeForeigner, Navito, Navstar, NawlinWiki, Nayl, Nchester, Ndz275, NeO Latvianracer, Neckername, Nedron12, Neil916, Neilc, Neo Geo, NeoChaosX, Neodarksaver, Neolandes, Nerddawg, NescioNomen, Netallica, Netgem21, Netpilot43556, New Republic Jedi, New guy, Newbie82, Newbiez, Newco, Newsnight Watcher, Nexexumbra, Nick123, NickBush24, NickCatal, Nickelbackrules1518, Nickin, Nickj, Nickle5000, Nicknw909, Nidnid, Nifboy, Nighthawkzx, Nightscream, Nima1024, Nintendo7, Nintendog master 54, Nintendogeek05, Nintendude, Nirutochi, Nishkid64, Nitro378, Nixdorf, Nixeagle, NoBull6, NoPity2, NoSeptember, Noj9, Nolanjurgens, Nolelover, NorCalHistory, Nords, Northside777, Nrobbins, Nrvn93, Nscheffey, Nsigniacorp, Nsisx9x, Nsjoker, Ntg, Nu89, Nufy8, Nuggetboy, Numlockfishy, Nutcrackr, Nxxus, Nymf, Nytemunkey, Nyttend, Nzk, ONEtryout, OSCA7R, OVde, Obelisk1, ObfuscatePenguin, Obli, Oblivious, Obs1dian, Oddball163, Odie5533, Oemenia, Off we go, Off!,

Ohnoitsjamie, Oklonia, Old Guard, Oldag07, Oli Filth, Oli5893, Oliverbeattie, Oliverdl, Olly150, Olorin28, Olpus, Omicronpersei8, OnePt618, Onejaguar, Oni Lukos, Oni Ookami Alfador, Onishenko, Onorem, Oogles, Oppiesomes, Optichan, OranL, Orangemango, Orbzon, Origami kid, OrigamiGuy, Original Name, Oriol003, Orion Minor, Orjanlothe, Oroso, Orphic, Orz, Orzornn, Oscarthecat, Osman leon, Ost316, Osu8907, Otchiman, Outcast316, OuterHeaven, Outlamed, OutlawMcgraw, Outsidethewall, Ovrd, Owen, OwenX, P1234p, P924s88, PDH, PPecken, Pablomartinez, Pacquiao, Paddyperdue, Pagrashtak, Pakokonka, Palfrey, Pandapower.1, Pandini, Pappa26, Parallel or Together?, Parhamr, Parivindi, Particularist11, Pastordavid, PatCheng, Pathoschild, PatrikR, Patrolman89, Patstuart, Paul August, Paul0991, Paulbag, Paulguiang, Pauli133, Pax85, Paxsimius, Pcbene, Pdc004, Pdmckinley, Peeto87, Pemilligan, Peng, Penpen123, Pentagram6, Peppery9, Perceval, PerfectStorm, Peripitus, PerryTachett, Persian Poet Gal, Person333, Peruvianllama, Peter, Peter.C, Petesmiles, Petrarch, Pevarnj, Pgk, Phil Sandifer, PhilHibbs, Philip Trueman, PhilipO, Phillip Bromley, Phily D, Phlegat, Phloog, Phred, PiaCarrot, Pichu1988, Pickleman415, Pie and bob, Pielover87, Pillcrusher, Pillow2011, Pilotguy, Pimpstate, PinchasC, Pinnecco, Pixarian, Pjh3000, Plantmanz, Playstationpsycho, Pleasantville, Pleasetaketicket, Plonk420, Pmsyyz, Poison the Well, Poiuyt Man, Pokeman, Politicalmachine, Ponydepression, Poor Yorick, Porges, Porqin, Portillo, Poweroverwhelming, Prem555, PrestonH, Private Butcher, Professional Gamer, Progamer97, ProjectSavior, Prothean, Prototime, Proud Ho, Ps2fanboy, Ps3link, PseudoKirby, Psi edit, Psp.crazed, Pspking247, PsychoNeo18, Psykechan, Pumapayam, PuppetMaster123, PureLegend, Purplepride1, PyroGamer, QX100, Qdr, Qelery, Quaker*oats, Quasar281, QuasiAbstract, Quentin mcalmott, Quikbit, Quikthoughts, QuillOmega0, Qxz, Qz3, R-21, RB McLeroy, RC Master, RC-0722, RFSamigo, RT100, RTBoyce, RUL3R, Rab V, Racjsingamac53, Rada, RadicalBender, Radicalfaith360, Rafert, Rafichamp, Raftysworld, Raidcrisis, Raijin Z, Ramennx, Random24, RandomP, Randommoocow, RandorXeus, RandySavageFTW, Raoulharris, Raptor007, Rare, Ras29, RasputinAXP, Rastrojo, Ratchetcomand, Rattlesnake, Raymondyo, Razermile, Rcandelori, Rdsmith4, ReCkLeSs X, Realm of Shadows, Rebel shadow, Red Platinum, RedWolfX, Redarmyman, Redlineryan, Redneckblues, Redskins0756, Redsparta, Reedy, Reevnar, Regre7, Regular mario, Remsen122, Remurmur, Remy B, RenegadeSniper7, Renegadeviking, Repetition, Rescbr, Resiak, Retodon8, Rettetast, Revo331, Revth, RexImperium, RexNL, ReyBrujo, Reyn116, Reyqune, Rgsgww, Rhobite, Rhysfoster, Riahc3, Riana, Ribeirowaldo, Rich Farmbrough, Richardcavell, RichiH, Richiekim, Richmd, Ricky81682, Riddles123456, Riff, Riotpirate90, Rjm656s, Rjwilmsi, RkOrton, Rkoking, Rob1n, Robdog 10, RobertG, Robertvan1, Robincognito, Robivy64, Robocoder, Roboy600, Robth, RockMFR, RockOfVictory, Rocket976, Rocko b, Roddie Digital, Roguegeek, Roguemaster83, Rohawn, Rohirok, Roketjack, Rokvok, Rolltide3738, Root value, Rorkster2, Rory096, Rory88uk, Rorytinker2009, Rorzo, RossPatterson, Rouge2, RoyBoy, Royboycrashfan, Rpmart, Rpyle731, Rubberkeith, Russau, Rusty23, RustyBadger, Ruy Pugliesi, Rwalker, RxS, Ryajinor, Ryan Norton, RyanD743, RyanKajito, Ryanbomber, Ryanbutterworth, Ryanmcv, Ryanv316, Ryno23cubbies, Ryosuke100, Ryulong, Ryzoop, S, S3ank1m, SCB '92, SD6-Agent, SONY, SPARTEN II, ST47, STLocutus, Sabu99, Sachinwalia, Saeroswind, Sag969, Sahrin, Saimcheeda, Saint-Paddy, Saiyanofsteel, Salvio giuliano, Sam Burke, Sam Korn, Sam Li, Sam916, Samuel Blanning, Samuel Boisvert, Samuele Rosa, Samuraid, Samuraispy, Samwise801, SanGatiche, SandBoxer, Sango123, Sanius, Sanke88, Sanya, Sapherald, Sarenne, Sarty, Saru786, Satori Son, SaveTheWhales, Saxbryn, Sc7, Scapermoya, Scarian, Scars39, Scepia, Sceptre, Schamps, SchfiftyThree, Schyler, Schzmo, Sciurinæ, Scott Gall, Scott Nash, Scott Richter, Scott3, Sd31415, Sdhunter, Sdornan, Sean WI, SeanMooney, Seano1, Seattlefan68, Sebbe94, Seece, Seedo, Seemee, Segatavis, Seidenstud, Seishirou Sakurazuka, Selmo, Sen91, Senorquigles, Sephcoco, Sephiroth878, SeptemberX1990, Seraosha, SeraphimXI, SergioNext, Seriphyn, Seuj, Sexyginga, Sferrier, Sfnhltb, Sgeureka, Sgt Squiggles, Shadow Hog, Shadowlynk, Shadowmario64fan, Shadowphax, Shady 22, Shahiboomboom, ShakataGaNai, Shanel, Shanes, Sharebandit, Shawnc, SheeEttin, ShelfSkewed, Sherool, Shifty36, Shii, Shimerdron, Shinealight, Shizane, Shredder46, Sidasta, Silentreflex, Silly rabbit, Silver Edge, Silver Sonic Shadow, Silver86, Silverhurley, Silvestre Zabala, Simer190, Simmie, Simon Shek, Simon12, Sin-man, Sir Nicholas de Mimsy-Porpington, Sirgregmac, Sixteen Left, Sjakkalle, Skeejay, SkepticMuhs, Sketchmoose, Skinnyweed, Skl1983, Skorp, Skullcap, SkyWalker, Skybon, Skylar44, Slady, Slam23, Slapshot24, Slark, Slearl, Sleepaholic, Sleigh, Sliderkk, Slo-mo, Slordak, Slyguy06, Small Profit, Smarterdeals, Smatts, Smeagol.i.am, Smithx807, Smokeyfire, Smomo, Smoothjazzlover, Smoothy, Snarius, Snoopdop, Snorbaard, Snowolf, Snowy150, Snowydweather, Soban, Soccergeek43, Socer1104, Socialservice, Sockatume, Soir, Soldier of Unfortune, Soldieroffortune7, Sonance, Sonic0967, SonicNiGHT, Sonicsean89, Sonicwolf, Sony465, Sorcre, Soulseeker86, Soumyasch, SouthernNights, Sp!ke, Sp33dyphil, Spacecataz90, Spacewizard, Spagman, Sparksalot, Sparuuto, Spawn316x, SpeakFree, Speqter, Spii, Spike Wilbury, Splash, Spliffy, Splintax, Splinter hells, Spookcentral, Spoonboy42, Sporeiscool123, Spyrus, SquirrelPhister, Squirrelist, Sreha, Srw985, Sryan2k1, StJarvitude, StargateX1, Starionwolf, StarkyD, Starship Trooper, Starwarsrulez, StationHub, Statsone, Stebucko360, Steel, SteelyDave, Stephen, Stephenb, Stephenc1, SteveSims, Steveketchup, Steven2197, StevenAFC, Stewiegriffin, Stickee, Stickeylabel, StolenIdentity, Stormedelf, Storms15, Strata8, Stratadrake, Strelic, Strifedaniel, Strolls, Strykie-boy, Strykur, Stuart D., Stuart P. Bentley, Stump, Stupid2, SubSeven, Subscriber0100, Subscriber0101, SudoGhost, Sully1311, Sum0, SupaSoldier, Super Coconut, Super2k, SuperBuuBuu, SuperHamster, SuperTails92, Supermaner, Surfyspaz, Surplu, Swaaye, Swanner, Swarm, Swedenboy722, SwordfishII, Sybaris (usurped), Syferus, Sylphie, SynergyBlades, Syrthiss, Syvanen, T3k-Ni9e, TDS, TGilmour, THEN WHO WAS PHONE?, TITROTU, TJ Spyke, TJDay, TKD, Tabletop, Tachikoma22, Tachyon01, Tagasaki, Tails623, Taladar, Tangerineduel, TangibleGhost, Tangotango, Targetter, Tassive Mits, TastyCakes, Tawker, Taxioverdose, Tbhotch, Tbjablin, Tcardone05, Tcncv, Tech2blog, Technoguy123, Ted87, Teddy.Coughlin, Tedlogan123, Tegel, Tehjord, Tehw1k1, Template namespace initialisation script, Tempshill, TenaciousBeelzeboss, Tenchi2, Tenniru, Terence, TerokNor, Terrapin, Terrillja, Terror G, Terrx, TestPilot, TexMurphy, TexasAndroid, Tezero, Tgraham87, Th1rt3en, ThaddeusB, Thanatos465, That Guy, From That Show!, Thatguykalem, Thax, The R, The Anome, The Bandsaw Vigilante, The Brotherhood of NOD, The Captain Returns, The Haunted Angel, The Iceman2288, The Original Juggernautical, The Phantomnaut, The Requiem, The Stick Man, The Thing That Should Not Be, The enthusiast, The jazz musician, The undertow, The unknown22, The wub, The1physicist, The93owner, TheAndrewMayShow, TheArmadillo, TheBilly, TheEnlightened, TheExtruder, TheKMan, TheKoG, TheListener, TheN1Armyguy, TheProject, TheShadowHitMan, TheSpectator, TheVinylHippo, Thebaboon, Thebogusman, Theda, Thedcm, Thedrycold, Thegame4ever, Thegreatestmoever, Thegreene2010, Thegreyanomaly, Thekittenofterra, Themasterofwiki, Themat21III, Themaxviwe, Theoneintraining, Thepatness, Thepwnster, Therealchadwarden, Thewikipedian, Thewizard ofozz, Theyarecomingforyou, Thingg, Thingy28123, This user has left wikipedia, Thiseye, ThomasHarte, Thornstrom, Thorpe, Threeocs, Throup, Thu, Thue, ThunderPeel2001, ThunderPower, Thunderbrand, Thunderfox933, Thunderpenguin, Thygard, TicketMan, Tide rolls, Tigen, TigerShark, Tigerman81, Tigermave, Tigga123, Tiglalawilliam, Tik, Tim curry, Timharwoodx, Timmins92, Timofonic, Timwi, Tiptoety, Titoxd, Tjdodd, Tjoneslo, Tmcw, Tmwerty, Toadthetoad, ToadyX, Toehead2001, TogTogTogTog, Tokugawapants, Tom, Tom Hek, Tom harrison, Tom k&e, Tomberry92, Tome711, Tommo h, Tommycm02, Tomwchow, ToneLa, Tonker83, Tony1, TonySt, Tonyr791, Toomontrangle, Torcute, Toytoy, Tpbradbury, Tregoweth, Trevor Bekolay, Trevor MacInnis, Trey56, Treyvo, Tricon, Triona, Trippz, TroubleBrewing3, Troyoda1990, Truthanado, Tsiehta, Tsmith96, Tsunamishadow, Tsurayu, Tubbimura91, TurboBoom, Turk brown, Tut74749, Tuxthepenguin933, Tvtechhead, Twholt, Twile, TwilightPhoenix, Twinxor, Twlighter, TwoOneTwo, Tyguy92, Tyrftg, Tzarius, Tzion, U$er, USMA, UberScienceNerd, Ugotownedo, Ugur Basak, Ukexpat, Ulquiorra, Ultra megatron, Ultraflame, Uncle Dick, Unclesallybone, Unheard Of, Unregistered.coward, UnrivaledShogun, Unyolo, Urahara, Urbndude, Urhixidur, Us3r--I33T, Useight, User27091, Username550, Utcursch, Uttaddmb, UzEE, V3ctorx, Vader99, Valoem, Valravn, Van helsing, VanFowler, Vanderdecken, Vanruels, Variant, Varsindarkin, Vary, Vash63, Vastu, Vbgamer45, VegaDark, Vegaswikian, Veinor, Venge, Venomaru 2.0, Vespristiano, Viakenny, VibroAxe, Vic sweet, VicVega123, Viciouspretzel, Viewtiful joe, Vildricianus, Vilerage, Vin789, Violetriga, ViperSnake151, Virchu, Virek, Viriditas, Vitz-RS, Vlad Dracula, Voidxor, Volcanux, Voltarios, Vortexor, Vrinan, Vrixmaster, Vulcanstar6, Vulian, VzjrZ, W00tm4n101, WATP, WJBscribe, Wageslave, Waggers, Waisybabu, Wanders1, Wangmike, WarPig66, Warlock2, Warren, Warren12, Wasdead917, Wasell, Watsona.2, Wavelength, Wayne317, Web comic, Webmaestro, Weejocky, Werdguy, Weregeek, Werididiot, Wernher, Wes!, Whatchitfoool, Where, WhiteMinority, Whitepaw, Who.was.phone, Whodunit, Whomp, Wicked595, Wiggster, WiiLord, WiiWii25, WiiWillieWiki, Wiidude, Wiimonkey9, Wikeel, Wiki alf, Wiki fanatic, WikiBone, WikiZorro, Wikifan21century, Wikilord, Wikipedia brown, WikipedianMarlith, Wikipug, Wikiuser387, Wikiwikimania, WildKard, Wildnox, Wilester, WillAndrews, Willhsmit, William Avery, Williamstormymauldin, Williamtaft, Williamwade, Windy906, Wisesabre, Wisq, Wizelf94, Wizmo, Woip, Wolfofwar, Wonchop, Woodyer99, Woohookitty, Wowowwiki, Wrightbus, Wtatour, X1987x, X201, X42bn6, XMillar, XXDucky21Xx, Xadith12, Xander Deathstalker, Xander756, Xaosflux, Xbxg32000, Xdiabolicalx, XeWoN, Xelliz, Xeno, Xenoslayer, Xezbeth, Xfirefire, Xgreenyukx, Xigaxu, Xino, Xizer, Xmoogle, Xnacional, Xplosneer, Xpyrofuryx, XreDuex, XxXDUNERIDERXxX, Xxhopingtearsxx, Y2j420, Y2kcrazyjoker4, YQTWRE, Yahadreas, Yamamoto Ichiro, Yamla, Yankeeboy1865, Yanksox, Yansa, Ybbor, Ydam, Yeanold Viskersenn, Yekrats, Yelyos, Yerpo, Yincrash, Yiwen017, Ynhockey, Yogi de, Yoosq, Yoshinatsu, Yourboychris69, Yubal, Yuefairchild, Yuhong, Yuri Elite, Zabiba, Zachary, Zachschwartz, Zainboy, Zanter, Zazumage, Zcomuto, Zedarkmarine, Zeppelin462, Zer0faults, ZeroOne, ZeroUm, Zeromaru, Zerueu, Zhou Yu, Zichalo, Zidus, Ziggymank, Zimbabweed, Zippanova, Zoboa, Zocky, Zombie007, Zr2d2, Zserghei, Zsinj, Zword97, ^demon, Žmogus, ЯЯЯ, on ✱!, 5538 anonymous edits

Artificial_intelligence *Source*: http://en.wikipedia.org/w/index.php?title=Artificial_intelligence *Contributors*: 100110100, 132.204.25.xxx, 172.141.188.xxx, 17Drew, 1dragon, 200.191.188.xxx, A157247, AAAAA, APH, APL, ARC Gritt, AVBAI, AVGavrilov, Abdullahazzam, Abeg92, Academic Challenger, Acerperi, Acroterion, AdSR, Adam Roush, Adamreiswig, AdjustShift, AdultSwim, AgadaUrbanit, Agemoi, AgnosticPreachersKid, Ahoerstemeier, Ai24081983, Aitias, AlGreen00, Alan Rockefeller, Alansohn, Aldux, Alex naish, Alexf, AlistairMcMillan, Allstarecho, Aloisdimpflmoser, Alsandro, Amareshjoshi, AmiDaniel, Anandcv, Andre Engels, AndrewHZ, Andreworkney, Andy Dingley, Andy Marchbanks, Angr, Aniu, Ankank, Anniepoo, Annonnimus, Antandrus, Anthony Appleyard, Anville, ArielGold, Arne Heise, Arthena, Arthur Rubin, Arthur Smart, ArthurWeasley, ArtificioSapiens, Arvindn, Asbestos, Ashish krazzy, Astrobase36, Atamyrat, Avenged Eightfold, Avery.mabojie, AxelBoldt, Axl, Axon, Aykantspel, AzaToth, Azlan Iqbal, B3virq3b, Baby16, Banes, Banus, Barnaby dawson, BarryList, Basploeger, Bbewsdirector, Beland, Ben Ben, Bender235, Benpryde, Bensin, BertSeghers, Beta Trom, Betacommand, Betterusername, Bforte, Bill52270, Bissinger, Bitsmart, Bjklein, Blackshadow153, Bo Jacoby, Bobblewik, Bobby D. Bryant, Bobianite, Bobo192, Boing! said Zebedee, Boneheadmx, Bongwarrior, Bootedcat, Bootstoots, Borislav, Bornslippy, Bovlb, Bqdemon, Bradgib, Brentdax, Brianjfox, BrightBlackHeaven, Brion VIBBER, BrokenSegue, Bryan Derksen, Bryan Seecrets, Bubba2323, Bugone, Burto88, Burzmali, Bustter, CO, COMPFUNK2, CRGreathouse, CWenger, Cabiria, Cacuija, Caiaffa, Can't sleep, clown will eat me, CanadianLinuxUser, Capricorn42, CarlHewitt, Casia wyq, Casper2k3, Cassianp, Centrx, Ceyockey, ChangChienFu, Chaosdruid, ChaoticLogic, Charles Gaudette, CharlesC, CharlesGillingham, Chessmaniac, Chicarelli, Chocolateboy, Chopchopwhitey, Chris Roy, Chris.urs-o, Chuckino, Chucks91, Chun-hian, Chuq, Ck lostsword, ClaretAsh, Clarityfiend, Coasterlover1994, Cobratom, Coder Dan, CoderGnome, Coffee2theorems, ColinMcMillen, CommodiCast, ConceptExp, Condem, Conskeptical, Conversion script, Coolcatfish, Corpx, CosineKitty, Cremepuff222, Crunchy Numbers, Csörföly D, Curious1i, Curly Turkey, Cvdwalt, CygnusPius, D, D Monack, DARTH SIDIOUS 2, DHN, DVD R W, Damiens.rf, DanMS, Daniel, DanielDemaret, Danim, Daphne A, Dar-Ape, DarkGhost08, Darker Dreams, Darth Chyrsaor, Darth Panda, Darthali, Davespice, Davewild, David D., David Martland, David.Monniaux, DavidBourguignon, DavidCary, Ddxc, DeStilaDo, Dee Jay Randall, Dejitarob, DerHexer, Derek Ross, Dessimoz, Devantheryv, Devonmann101, Dheer7c, Dicklyon, Digsdirt, Diogeneselcinico42, Dionyseus, Dirkbb, Disavian, Discordant, Discospinster, Dispenser, Diza, Dlloader, Dlohcierekim, Dmcq, Dmsar, Docu, DonSiano, DonutGuy, DouglasGreen, Dougofborg, DougsTech, Dpbsmith, Dpupek, Dream land2080, Drift chambers, Drmies, Dublinclontarf, Dudesleeper, Dv82matt, Dysprosia, ELApro, ELDRAS, ENeville, EPIC MASTER, ERcheck, ESkog, Eastlaw, EdH, Edgar181, Edivorce, Edward, Eggman64, EisenKnoechel, El C, El aprendelenguas, Elassint, Ellmist, Ellywa, Elsendero, Eluchil, Emarus, Emir Arven, Emperorbma, Emptymountains, Emurph, Emuzesto, Eoghan, Epbr123, Erayman64, Eric Mathews Technology, Eroark, Escientist, Eubulides, Eumolpo, EvanProdromou, Evanherk, Evercat, Everyking, Excirial, Eyrian, Eyu100, Ezavarei, Falcon8765, FalconL, Farmer21, Farquaadhnchmn, Fartherred, Fillepa, Finetooth, Finn-Zoltan, Flewis, Floquenbeam, Foryourinfo, Francob, Francs2000, FrankCostanza, Frans Fowler, Frans-w1, Fraziergeorge122, Frecklefoot, Freek Verkerk, Freemarket, Fribbler, Fricanod, Funandtrvl, Furrykef, Fuzheado, Fvw, GTof, Gabbe, Gadfium, Gaius Cornelius, Galoubet, Gamkiller, Garas, Gary King, Gdm, Gene s, Gengiskanhg, Geoinline, GerryWolff, Giftlite, Gilliam, Gimmetrow, Glacialfox, Glane23, Glen Pepicelli, Glenfarclas, Gmaxwell, Gnome de plume, Gogarburn, Gogo Dodo, Goodnightmush, GraemeL, Grafen, Graham87, Graymornings, GreenReaper, Greensburger, Gregbard, Gregman2, GreyCat, Gridlinked, Grokmoo, Grunt, Guanaco, Guoguo12, Gurch, Gustavo1255, Gwernol, Gyrofrog, Gzabers, Hadal, Halmstad, HamburgerRadio, Hamster Sandwich,,

Harderm, Harnil, Harriv, Harryboyles, Haseo9999, Haymaker, Healeyb, Herbee, HereToHelp, Heron, Herschelkrustofsky, Hervegirod, Hezarfenn, Hfastedge, Hgamboa, Hi878, Hike395, Horris, Hpdl, Hthth, Hu, Hu12, Hulagutten, Humanrobo, Hydrargyrum, Hydrogen Iodide, I already forgot, IRP, IYY, IanElmore, Iceworks, Ignatzmice, Igoldste, Ihavenolife, Ilcmuchas, ImGz, Implements, Imran, Indyfitz, Inego, Insomniak, IntellectToday, Interested, Intrealm, Introgressive, Inwind, Iohannes Animosus, Ioverka, Isheden, ItsZippy, Ivant, Ixfd64, Iyerakshay, J. Spencer, J.delanoy, J04n, JRR Trollkien, Jackol, Jagdeepyadav, Jagginess, Jaibe, Jake11, James childs, Jamyskis, Jan eissfeldt, Jared555, Jason Palpatine, Java7837, Javidjamae, Javierito92, Jayjg, Jdzarlino, Jebba, Jeeny, JesseHogan, Jimothytrotter, Jj137, Jkaplan, JoeSmack, Joeblakesley, Jogers, Johannes Simon, John Newbury, John Vandenberg, John of Reading, JohnOwens, Johnkoza1992, Johnuniq, Jojit fb, Jon Awbrey, Jonathan Métillon, Jonrgrover, Jordan123, Jorfer, Joseph Solis in Australia, Josephorourke, Joydurgin, Jpbowen, Jpgordon, Jroudh, Jrtayloriv, Jsun027, Judson, Julian Mendez, JuniorMonkey, Jvoegele, Jwoodger, K.Nevelsteen, KF, Kaldosh, Kane5187, Karada, KarenEdda, Karol Langner, KellyCoinGuy, Kevin143, Kevyn, Khalid, Khaydarian, Kimveale, King of Hearts, Kingboyk, Kjellmikal, Klafubra, Kneiphof, Koavf, Kostisl, Koyaanis Qatsi, Kramlovesmovies, Kri, Kukini, Kurowoofwoof111, Kzollman, L Kensington, LFaraone, Lacatosias, Larry laptop, LaserBeams, LauriO, Laurusnobilis, Lectonar, LedgendGamer, Lee Daniel Crocker, Leire Sánchez, Lenehey, LeoNomis, Leoneris, Leontolstoy2, Leszek Jańczuk, Levineps, Lewblack, Lexor, Libcub, Liftarn, Liger, Lightmouse, Lilac Soul, Lilinjing cas, Lindmere, Linweizhen, Livajo, Logperson, Loki en, Looxix, Loremaster, Loxley, Lumos3, Luna Santin, Lylodo, Lynn Wilbur, M, MCTales, MD87, MER-C, MHLU, MIT Trekkie, MJBurrage, MTuffield, Mac, Mac-steele, MacMed, Macduffman, Macilnar, MadSurgeon, Madd4Max, Maghnus, Mailseth, MakeRocketGoNow, Malignedtruth, MangoWong, Mani1, Marc Girod, Marc Venot, Marek69, MarkAb, Markus.Waibel, Martha2000, Martinp, Marudubshinki, Marysunshine, Master Jaken, MasterOfHisOwnDomain, MattBan, Matteh, Matthew Stannard, Maurreen, Mausy5043, Mav, Maver1ck, Maximus Rex, Mayfly may fly, McCart42, McGeddon, Mdd, Mdebets, Me pras, MeekMark, Melody, Memming, Mendel, Mercury, Metamagician3000, MiNombreDeGuerra, Micahmn, Michael Fourman, Michael Hardy, Michal Jurosz, MichalJurosz, Mikael Häggström, Mike Schwartz, MikeCapone, MikeLynch, Mild Bill Hiccup, MilesMi, Mindmatrix, Minesweeper, Mintguy, Mmernex, Mmxx, Mneser, Monkee13, Moralis, Mostlymark, Moxon, Mr.Sisgay, MrJones, MrOllie, Mrsolutions, Mrvasin, Mschel, Mschures, Mudd1, Mullhawk, Muness, Myanw, Myasuda, Myncknm, NCurse, Nabarry, Nabeth, Nach0king, Nanshu, Naohiro19, NapoleonB, Nappy1020, Nayanraut, Ndenison, Neckelmann, NeilN, Neilc, Neko-chan, Netesq, Neverquick, Newkidd11, Nick, Nickj, Nightscream, Nishkid64, Nistra, Nixdorf, Noctibus, Nondescript, Normxxx, NorrYtt, Novum, Nposs, Nuclear Treason, NuclearWarfare, Nufy8, Nunocordeiro, Nx7000, Obscuranym, Olathe, Oldhamlet, Oleg Alexandrov, Olethros, Olinga, Oliver Pereira, Olivier, Onorem, Opelio, Orangemike, Ost316, Ott, OwenX, Owlbuster, Oxwil, Oxymoron83, Pacas, Palace of the Purple Emperor, Paradoctor, Pawyilee, Pegua, Peterdjones, PetroKonashevich, Peyre, Pgr94, Phileas, Philip Trueman, Piet Delport, Pigsonthewing, Pilotguy, Pinethicket, Piotrus, Pkirlin, Pluke, PluniAlmoni, Poccil, Poiman, Pokrajac, Polocrunch, Poor Yorick, Pooya.babahajyani, Populus, Prakashavinash, Pravdaverita, Prohlep, Prolog, PrometheusX303, Psb777, Pspoulin, Psychonaut, Public Menace, Qartis, Qswitch426, Qwertyus, R. S. Shaw, RCPayne, RJASE1, RK, RabidDeity, Rafael.perez.rosario, Rahuldharmani, Rajah, Ralf Klinkenberg, Rangoon11, Ranjithsutari, Rasmus Faber, Ratiocinate, Reckiz, Recognizance, Reconsider the static, RedHillian, RedWolf, Redaktor, Redvers, Reidgsmith, Remember the dot, Remy B, Renaissancee, Renamed user 3, Rene halle, Rettetast, RexNL, Rgarvage, Rhobite, Rholton, Riccardopoli, Rich Farmbrough, Richard001, Richardcavell, Rick Block, Ricvelozo, Rinea, Ripe, Risk one, Rjwilmsi, Rlw, Rnb, Robert Merkel, Robin klein, Roesslerj, Ronz, Rootxploit, Rpmorrow, Rror, Rumping, Ruslik0, Ruud Koot, RyanParis, Ryguasu, Rzwitserloot, S0aasdf2sf, SDC, SWAdair, Sagaciousuk, Saizai, Salamurai, Sam Hocevar, Sam Korn, Sander123, Sango123, Sarnholm, Sbennett4, Sceptre, Schaefer, Schlegel, SchreyP, Science History, Scorpion451, Sdornan, SeanMack, Seaphoto, Searchmaven, Searchme, Seb az86556, Sebaldusadam, Sebastian scha., SecretStory, Selket, Sgunteratparamus.k12.nj.us, Shadow1, Shadowjams, Shaggorama, Shanes, Shannonbonannon, Shanoman, Shenme, Shirik, Shpiget, Siddharthsk2000, Sillybilly, Simoneau, Sirkad, Smallclone2, SnoopY, Snoutholder, Snoyes, SocratesJedi, Sodium, Sole Soul, SouthernNights, Sparkygravity, Special-T, Spinningobo, Spitfire, Srinivasasha, Staberind, Stardust8212, Stefanomione, SteinbDJ, Stephane.magnenat, Stephen G Graham, StephenReed, Stevage, Steve Quinn, Stevertigo, Stoni, Stormie, Stroppolo, Supten, Susvolans, Svick, Sweetness46, Symane, Syncategoremata, Szeldich, TAnthony, THEN WHO WAS PHONE?, Tabortiger, Taemyr, Taggart Transcontinental, Tailpig, Talon Artaine, Tarotcards, Tassedethe, Taw, Taxisfolder, Taylormas229, Tazmaniacs, Tdewey, Technopat, Tedickey, Teo64x, ThG, The Anome, The Cunctator, The Divine Fluffalizer, The Evil IP address, The Magnificent Clean-keeper, The Thing That Should Not Be, The Transhumanist (AWB), The world deserves the truth, The wub, TheDoober, TheRanger, TheSix, Thecheesykid, Thede, Themfromspace, Theo Pardilla, Theyer, Thiseye, Thomas Kist, Thumperward, Thunderhippo, Thüringer, Tiddly Tom, Tideflat, Tillander, Time for action, Timwi, Titus III, Tkinias, Tkorrovi, Tobias Bergemann, Tobias Kuhn, Toby Bartels, Tokek, Tommy Herbert, Tone, Tony1, Took, Touisiau, Tpbradbury, Tremilux, Trevor MacInnis, Truthnlove, Turk oğlan, TurntableX, TutterMouse, Twexcom, TwoOneTwo, Tzartzam, Ugen64, Una Smith, Uncle Dick, Unknown, Unreal7, Urod, Useight, Utcursch, VSimonian, Velho, Vesal, Vespine, Vfrias, Vidstige, Vilerage, Virtualerian, Viskonsas, Vivohobson, Voices cray, Vrenator, Vrossi1, Vsmith, WBTtheFROG, WMod-NS, Wadeduck, Wafulz, Walter Fritz, Wasell, WaveRunner85, Wavelength, Wdfarmer, WereSpielChequers, Wesleymurphy, Westwood25, Wfructose, WhatisFeelings?, Wiki alf, WikiNing, Wikicolleen, Wikiklrsc, Wikireviewer99, Wikiwikifast, Wile E. Heresiarch, William conway bcc, Witbrock, Wizardist, Wmahan, WojPob, Wookiee cheese, Wyckster, Wyklety, Xklsv, Xobritbabeox10, Yakota21, Yankee-Bravo, Yardcock, Ykalyan, YoungFreud, Yoyosocool, Ypetrachenko, Yuckfoo, Yungjibbz, ZENDELBACH, Zaklog, Zapvet, ZeroOne, Zntrip, Zsinj, Zubenelgenubi, Zundark, Zzthex, Zzuuzz, ²¹², Ásgeir IV., Δ, Иь Лю Ха, 1965 anonymous edits

Fighting_game *Source*: http://en.wikipedia.org/w/index.php?title=Fighting_game *Contributors*: *NormalGamer*, *drew, 1wngdngl, A Merchant, Acs4b, Addit, After Midnight, Akinara, Alexmcmorris, Alphathon, AmDrag21, AnY FOUR!, Andrevan, Andrzejbanas, Anticipation of a New Lover's Arrival, The, Antonio Lopez, AntonioMartin, Arede, Arrowned, Arthena, Artichoker, Austriacus, BBTEAK, BRG, Barek, Bbobjs, Beowulph, BlueAzure, Bridies, Bubbachuck, BusSDriver, Caltas, CardinalDan, Carlosguitar, Cherkash, Chigurgh, Chinju, Chris the speller, Citicat, Colin Kimbrell, Combination, Conversion script, Crazysane, Dabcio, Dabomb87, Dangerous-Boy, Darkpower, Darkside 88, Davichito, David Schaich, Depressio, Discopriest, Dispenser, Draicone, Drbovell, Drwarpmind, Ed Poor, Eekerz, Elamaton, Elcalen, Elm-39, Endlessdan, Enigmaman, EoGuy, Everyking, Feitclub, Firsfron, Freakofnurture, Frecklefoot, FriedMilk, Frogacuda, Gamer93, GamerPro64, Geniac, GlassCobra, Goodraise, Grandpafootsoldier, Ground, Gtrmp, Gurch, Guyinblack25, Gökhan, Haakon, Hairy Dude, Haruth, ICAPTCHA, Iceberg3k, Introgressive, Iorirox, J 1982, J.delanoy, JFD, JHFTC, Jacoplane, Jagged 85, Jason One, Jason Quinn, Jbmurray, Jeff Silvers, Jeff3000, Jer0en 1988NL, Jhsounds, Johandomonji, John of Reading, JohnnyMrNinja, Johnnyfog, Jonny2x4, JuJube, Kbdank71, KelvSYC, Ken Gallager, Kevin, LGagnon, Larsinio, Leileilol, LilHelpa, Loona, Lowellian, Luna Santin, Luvcraft, Mackeriv, Macy, Magicus69, Majed, MajinHurricane, Manberzil, Marasmusine, Martarius, Master Bigode, Master Deusoma, Master Thief Garrett, Mattg82, Maueze, Meami, Merotoker1, Michael j. iwai, Mikademus, MissingNOOO, Mitaphane, Modulatum, Momomomo111, MoonSoo, Mrmister001, MuZemike, N3m6, NHRHS2010, NIS agent, NawlinWiki, Ndenison, Neeklamy, NeoChaosX, NeoDoubleGames, New Age Retro Hippie, Nick R, Nifboy, Nightscream, Nilmerg, Nintendax, Notheruser, Nyttend, Oberiko, Oscarthecat, Ost316, PJ, PabloG, Parrothead1983, Patriotic dissent, Pengo, Penguinboyroy, PhilHibbs, Pixelface, Prawn, Prime Blue, ProjectPlatinum, PyroGamer, R'n'B, Rameriz, Randomran, Redketchup67, Remurmur, Rettetast, Revth, Rhe br, Richiekim, Rjwilmsi, Robertheadley, RockMFR, Rossumcapek, RussBlau, Rustyfence, SAMAS, SBGamesCone, Selket, Sephiroth BCR, Seryass, Sesshomaru, SevereTireDamage, Sfan00 IMG, Shadoman, SharkD, Sikyanakotik, Silentaria, Sin-man, Sir Lothar, Sjakkalle, SkyWalker, Someguy1221, Someone another, Sparky the Seventh Chaos, Sss333, Stabby Joe, Straatmeester, Subzerosmokerain, SuperMarius, SuperSonicTH, Surge79uwf, Sykoguy, Symplectic Map, Tearlock, Tekkenfanbrendan, Template namespace initialisation script, Tenks, TheLoverofLove, Thunderbrand, Timo Honkasalo, Tintor2, Tjansen, TowerDefender, Triskaideka, Tyan23, Unyoyega, Vianello, Violetriga, Visor, Wakipudeo, Wik, WikiFew, Wikieditor1988, WikipedianMarlith, Wmahan, Wonchop, Woohookitty, X201, Xiong Chiamiov, Xubelox, Xueguo, Y2kcrazyjoker4, Yar Kramer, YemeniteCamel, Yngvarr, Yoshi, You Can't See Me!, ZS, Zanetu, ZeroOne, ZimZalaBim, Zumbo, 701 anonymous edits

Single-player_video_game *Source*: http://en.wikipedia.org/w/index.php?title=Single-player_video_game *Contributors*: 041744, Aiken drum, Al3xil, AlistairMcMillan, Angelbo, Bilbo571, Black Falcon, BluesD, Cadsuane Melaidhrin, Careydaniels, Chaos5023, Cherkash, Dadude3320, Dark Pulse, Dbiel, Deuxhero, Diego Moya, Disavian, Falcon8765, Fish and karate, FriedMilk, Gtrmp, Holek, IGeMiNix, INkubusse, JForget, JaGa, Jagged 85, Jeffrey Mall, Jezhotwells, JimmyBlackwing, K1Bond007, Kariteh, Kchishol1970, LOL, LedgendGamer, LiDaobing, Lightsup55, Marasmusine, Mauler90, Mechasheherezada, Meelar, Mitaphane, Mrwojo, Necoplay, OboeCrack, Optim, Ost316, Panser Born, Panther991, Piecemealcranky, Poetic Decay, Quoth, RAlafriz, Ras, Razorflame, Remurmur, Resound, Retodon8, Rhe br, Rigadoun, RoyBoy, S@bre, SMcCandlish, SharkD, Shrumster, Sitearm, Slike, Smurrayinchester, Snowolf, Soetermans, Sopemting, SuperMarius, Surachit, TheCatalyst31, Theblackgecko, Theresa knott, Vendettax, Video game fan11, Violinmaster4321, Vishnava, Zacatecnik, 91 anonymous edits

Multiplayer_video_game *Source*: http://en.wikipedia.org/w/index.php?title=Multiplayer_video_game *Contributors*: 2005, ASpafford, Ace Faolchu, Addshore, Aervanath, Afasmit, Aitias, Alai, Alansohn, Ale jrb, Alexkin, AlistairMcMillan, Amiteli, Amren, Andres Velasco y Coll, Andrevan, Andrew Maiman, Andypandy.UK, Anirazarina, Anurmi, Atroposmodel, Attitude2000, B7T, Barnabypage, Beyond silence, Biasoli, Bilbo571, Bissinger, Black Kite, Blackbox77, Bloodshedder, Bobo192, Bongwarrior, Bulbous, CarlKenner, Cfryar, Chaos5023, CharlotteWebb, ChristopherCaldwell, Chromaticity, Cocoaguy, Coll7, Colonel Warden, CounterValue, Crossmr, CryptoDerk, Cyberpower678, D.brodale, DMather, Damian Yerrick, Dancter, Dawnseeker2000, Delldot, DiscipleOfKnowledge, Divine Auror, Dogposter, Dooms, Download, Drable, Draconus, Drat, Dreadstar, Duanearnett, Dust Filter, Edward, Eik Corell, Eike, Einhanderkiller, Elassint, Elfguy, Elmoswrold, Emodinol, Eric-Wester, Faramir1138, Fayenatic london, Felecita, Fredrik, FreplySpang, FriedMilk, Gakon5, Gardar Rurak, Giggy, Gscshoyru, Holek, Hornpipe2, INkubusse, Ianman2001, Idleguy, Immaheadtoday, J.delanoy, JNW, JackSparrow Ninja, Jackson McArthur, Jagged 85, Jake7789, Jhsounds, Jls33fsls, John smith 99, John254, Johneyjay, Josh900, Joylock, Jéské Couriano, Ke5crz, Kimchi.sg, Kingaschi, Kubigula, Kyz, Leroy573, LittleSmall, Locationx3, Lunarbunny, MLauba, Majed, Marasmusine, Martarius, Matthew Yeager, Matthäus Wander, Mcbball55, Mechasheherezada, Merlinsorca, Mgiganteus1, Mike Rosoft, Minghong, Montecarlocars, MrSomeone, Mrwojo, Msikma, Muchness, N5iln, NaBUru38, Nathaneddy, NawlinWiki, Nazi 2007, Nitesshadow, NoFear2009, NorwegianBlue, OinkOink, Oliver202, Olpus, Oni Lukos, Ost316, Panser Born, Paper Luigi, Pharaoh of the Wizards, Piano non troppo, Pilotguy, Porqin, Preeeemo, Qxz, R.123, R3PULSE, RTC, RainbowOfLight, Rankiri, Reaper Eternal, RedWolf, RememberingLife, Riana, Richard W.M. Jones, Richi, Rotor DB, Rupertslander, SMcCandlish, Saberwyn, Scepia, Sdornan, Seaphoto, Serpent's Choice, Sherool, Singularity, SkyWalker, Skysmith, Slike, Sm8900, Smcc, Sockatume, Spankological, SpeedyGonsales, Spinnick597, Stefan, Stormie, Sylph, TJ Spyke, TKD, Tempshill, Th1rt3en, TheKMan, TheLinKing, Thingg, Thonan, Thunderbrand, Tide rolls, Tordek, Triwbe, Tslocum, Tustin2121, Tyler, Ugen64, Unint, Uogl, Utcursch, Vanderdecken, Vendettax, Video game fan11, Voidvector, Vsmith, Wassupmygs, Wikier9, Wimt, Xensyria, Xplogen, Xtremeblur, Yashtulsyan, Yerpo, Young david, Zhatt, ZimZalaBim, Zirkov, Zzuuzz, 303 anonymous edits

Multiplayer_online_game *Source*: http://en.wikipedia.org/w/index.php?title=Multiplayer_online_game *Contributors*: 1ForTheMoney, Akking34, Biasoli, Bilbo571, Boing! said Zebedee, Cadsuane Melaidhrin, Chaos5023, Curps, DarqFlare, Erickjonathanradja, Evolve2k, Faramir1138, Freshmaniac, Fyyer, J.delanoy, JHunterJ, JLaTondre, Jagged 85, Jununhere, Keith D, Madame Sosostris, McGeddon, Oldag07, Phantomsteve, Randomran, Sertion, Skater, Sm8900, Smalljim, Sophie, Sylph, Tabbitha Trash, Tommy2010, Toussaint, Wedb, Writelabor, Wyatt Riot, Xeno, Xenoveritas, 37 anonymous edits

Image Sources, Licenses and Contributors

File:Ultimate Tenkaichi Cover.PNG *Source*: http://en.wikipedia.org/w/index.php?title=File:Ultimate_Tenkaichi_Cover.PNG *License*: unknown *Contributors*: Sarujo

File:Hero Mode Menu.PNG *Source*: http://en.wikipedia.org/w/index.php?title=File:Hero_Mode_Menu.PNG *License*: unknown *Contributors*: Mika1h, Sarujo

File:DB Tankōbon.png *Source*: http://en.wikipedia.org/w/index.php?title=File:DB_Tankōbon.png *License*: unknown *Contributors*: AnmaFinotera, Delldot, Hot Dog Wolf, Linkdude20002001, Mayur, Ninja neko, Philip Trueman, Sesshomaru, TheFarix, 8 anonymous edits

File:Flag of Australia.svg *Source*: http://en.wikipedia.org/w/index.php?title=File:Flag_of_Australia.svg *License*: unknown *Contributors*: Anomie, Mifter

File:Flag of New Zealand.svg *Source*: http://en.wikipedia.org/w/index.php?title=File:Flag_of_New_Zealand.svg *License*: unknown *Contributors*: Adambro, Arria Belli, Avenue, Bawolff, Bjankuloski06en, ButterStick, Denelson83, Donk, Duduziq, EugeneZelenko, Fred J, Fry1989, Hugh Jass, Ibagli, Jusjih, Klemen Kocjancic, Mamndassan, Mattes, Nightstallion, O, Peeperman, Poromiami, Reisio, Rfc1394, Shizhao, Tabasco, Transparent Blue, Väsk, Xufanc, Zscout370, 35 anonymous edits

File:Flag of Canada.svg *Source*: http://en.wikipedia.org/w/index.php?title=File:Flag_of_Canada.svg *License*: unknown *Contributors*: Anomie

File:Flag of the United States.svg *Source*: http://en.wikipedia.org/w/index.php?title=File:Flag_of_the_United_States.svg *License*: unknown *Contributors*: Anomie

File:Flag of the United Kingdom.svg *Source*: http://en.wikipedia.org/w/index.php?title=File:Flag_of_the_United_Kingdom.svg *License*: unknown *Contributors*: Anomie, Mifter

File:Flag of Greece.svg *Source*: http://en.wikipedia.org/w/index.php?title=File:Flag_of_Greece.svg *License*: unknown *Contributors*: (of code) (talk)

File:Flag of the Philippines.svg *Source*: http://en.wikipedia.org/w/index.php?title=File:Flag_of_the_Philippines.svg *License*: unknown *Contributors*: Aira Cutamora

File:Flag of Russia.svg *Source*: http://en.wikipedia.org/w/index.php?title=File:Flag_of_Russia.svg *License*: unknown *Contributors*: Anomie

File:Wiki DragonBall Earth.png *Source*: http://en.wikipedia.org/w/index.php?title=File:Wiki_DragonBall_Earth.png *License*: unknown *Contributors*: AnmaFinotera, Explicit, Jj98, Sesshomaru, Son Goharotto, Sreejithk2000, TTN, 5 anonymous edits

File:Db TCI.jpg *Source*: http://en.wikipedia.org/w/index.php?title=File:Db_TCI.jpg *License*: unknown *Contributors*: Akira Toriyama

File:Manga in Jp.svg *Source*: http://en.wikipedia.org/w/index.php?title=File:Manga_in_Jp.svg *License*: unknown *Contributors*: User:TheOtherJesse

File:Sazae-san kamishibai.jpg *Source*: http://en.wikipedia.org/w/index.php?title=File:Sazae-san_kamishibai.jpg *License*: unknown *Contributors*: Adaobi, Bkell, Peregrine Fisher, Radishes, SchuminWeb, Sfan00 IMG, 4 anonymous edits

File:Eshibun Nipponchi.jpg *Source*: http://en.wikipedia.org/w/index.php?title=File:Eshibun_Nipponchi.jpg *License*: unknown *Contributors*: Jump Guru

File:Manga reading direction.svg *Source*: http://en.wikipedia.org/w/index.php?title=File:Manga_reading_direction.svg *License*: unknown *Contributors*: User:TheOtherJesse

File:Young boy reading manga.jpg *Source*: http://en.wikipedia.org/w/index.php?title=File:Young_boy_reading_manga.jpg *License*: unknown *Contributors*: User:Ragesoss

File:Speaker Icon.svg *Source*: http://en.wikipedia.org/w/index.php?title=File:Speaker_Icon.svg *License*: unknown *Contributors*: Blast, G.Hagedorn, Mobius, 2 anonymous edits

File:Anime cell 1917.jpg *Source*: http://en.wikipedia.org/w/index.php?title=File:Anime_cell_1917.jpg *License*: unknown *Contributors*: Kyaa the Catlord, Tenmei, 1 anonymous edits

File:Momotaro's Divine Sea Warriors-screeny.JPG *Source*: http://en.wikipedia.org/w/index.php?title=File:Momotaro's_Divine_Sea_Warriors-screeny.JPG *License*: unknown *Contributors*: Shochiku / directed by Mitsuyo Seo

File:Modernanime.jpg *Source*: http://en.wikipedia.org/w/index.php?title=File:Modernanime.jpg *License*: unknown *Contributors*: Addict 2006, Dinoguy1000, Fireaza, 4 anonymous edits

File:Spike com logo.gif *Source*: http://en.wikipedia.org/w/index.php?title=File:Spike_com_logo.gif *License*: unknown *Contributors*: Banej

Image:Namco Bandai Games Logo.png *Source*: http://en.wikipedia.org/w/index.php?title=File:Namco_Bandai_Games_Logo.png *License*: unknown *Contributors*: Bovineboy2008, N. Harmonik, Nintendogeek05, SWFlash

image:NBGI Headquarters.JPG *Source*: http://en.wikipedia.org/w/index.php?title=File:NBGI_Headquarters.JPG *License*: unknown *Contributors*: Mapio

Image:PLAYSTATION 3 logo.svg *Source*: http://en.wikipedia.org/w/index.php?title=File:PLAYSTATION_3_logo.svg *License*: unknown *Contributors*: Sony

File:PlayStation 3 Logo neu.svg *Source*: http://en.wikipedia.org/w/index.php?title=File:PlayStation_3_Logo_neu.svg *License*: unknown *Contributors*: MisterSanderson, Mizunoryu, My76Strat, Oxguy3, StG1990

Image:PS3Versions.png *Source*: http://en.wikipedia.org/w/index.php?title=File:PS3Versions.png *License*: unknown *Contributors*: User:Evan-Amos

File:PS3s and controllers at E3 2006.jpg *Source*: http://en.wikipedia.org/w/index.php?title=File:PS3s_and_controllers_at_E3_2006.jpg *License*: unknown *Contributors*: Anetode, Dancter, Ed g2s, Evian Pepper, Ffgamera

File:PS3-Fat-Console-Vert.png *Source*: http://en.wikipedia.org/w/index.php?title=File:PS3-Fat-Console-Vert.png *License*: unknown *Contributors*: User:Evan-Amos

File:PS3-Slim-Console-Vert.png *Source*: http://en.wikipedia.org/w/index.php?title=File:PS3-Slim-Console-Vert.png *License*: unknown *Contributors*: User:Evan-Amos

File:DualShock3-in-Hand.jpg *Source*: http://en.wikipedia.org/w/index.php?title=File:DualShock3-in-Hand.jpg *License*: unknown *Contributors*: User:Evan-Amos

File:Yellow Light of Death- PlayStation 3.png *Source*: http://en.wikipedia.org/w/index.php?title=File:Yellow_Light_of_Death-_PlayStation_3.png *License*: unknown *Contributors*: Ffgamera, using the image made by Ssolbergj (:File:Playstation3vector.svg). Original uploader was Ffgamera at en.wikipedia

File:PhotoGalleryPS3.png *Source*: http://en.wikipedia.org/w/index.php?title=File:PhotoGalleryPS3.png *License*: unknown *Contributors*: ChimpanzeeUK, Ffgamera

File:PSN logo color trans.png *Source*: http://en.wikipedia.org/w/index.php?title=File:PSN_logo_color_trans.png *License*: unknown *Contributors*: ChimpanzeeUK, Frosted14, KiasuKiasiMan, LordDarkPhantom, Opertinicy, Thingg, WikiCopter, 2 anonymous edits

File:PlayStationStore-Screenshot.png *Source*: http://en.wikipedia.org/w/index.php?title=File:PlayStationStore-Screenshot.png *License*: unknown *Contributors*: ChimpanzeeUK, Jj98, KiasuKiasiMan

File:Whats-new.png *Source*: http://en.wikipedia.org/w/index.php?title=File:Whats-new.png *License*: unknown *Contributors*: ChimpanzeeUK, Ffgamera

File:LifeWithPlayStation Weather.jpg *Source*: http://en.wikipedia.org/w/index.php?title=File:LifeWithPlayStation_Weather.jpg *License*: unknown *Contributors*: Chester Markel, ChimpanzeeUK, Ffgamera, KiasuKiasiMan, N. Harmonik, 1 anonymous edits

File:XBOX 360 LOGO.svg *Source*: http://en.wikipedia.org/w/index.php?title=File:XBOX_360_LOGO.svg *License*: unknown *Contributors*: JCRules

File:Xbox-360-Consoles-Infobox.png *Source*: http://en.wikipedia.org/w/index.php?title=File:Xbox-360-Consoles-Infobox.png *License*: unknown *Contributors*: User:Evan-Amos

File:Xbox-360S-Console-Set.jpg *Source*: http://en.wikipedia.org/w/index.php?title=File:Xbox-360S-Console-Set.jpg *License*: unknown *Contributors*: User:Evan-Amos

File:Xbox-360-arcade.jpg *Source*: http://en.wikipedia.org/w/index.php?title=File:Xbox-360-arcade.jpg *License*: unknown *Contributors*: User:Evan-amos

File:Xbox-360-Elite-Console-Set.png *Source*: http://en.wikipedia.org/w/index.php?title=File:Xbox-360-Elite-Console-Set.png *License*: unknown *Contributors*: User:Evan-Amos

File:Xbox 360 Halo 3 Special Edition.jpg *Source*: http://en.wikipedia.org/w/index.php?title=File:Xbox_360_Halo_3_Special_Edition.jpg *License*: unknown *Contributors*: w?odi from Szczecin, Poland

File:specialeditionlaunchteamconsolexbox360.jpg *Source*: http://en.wikipedia.org/w/index.php?title=File:Specialeditionlaunchteamconsolexbox360.jpg *License*: unknown *Contributors*: User:Killrhedake

File:Xbox-360-white-wireless-controller.jpg *Source*: http://en.wikipedia.org/w/index.php?title=File:Xbox-360-white-wireless-controller.jpg *License*: unknown *Contributors*: User:Evan-amos

File:Xbox-360-Kinect-Standalone.png *Source*: http://en.wikipedia.org/w/index.php?title=File:Xbox-360-Kinect-Standalone.png *License*: unknown *Contributors*: User:Evan-Amos

File:1home30rock0531.JPG *Source*: http://en.wikipedia.org/w/index.php?title=File:1home30rock0531.JPG *License*: unknown *Contributors*: Jericho1337, N. Harmonik, 4 anonymous edits

File:Gameschannel6slots0531.JPG *Source*: http://en.wikipedia.org/w/index.php?title=File:Gameschannel6slots0531.JPG *License*: unknown *Contributors*: Jericho1337

File:Xbox360-ringofdeath.jpg *Source*: http://en.wikipedia.org/w/index.php?title=File:Xbox360-ringofdeath.jpg *License*: unknown *Contributors*: Original uploader was Droobey at en.wikipedia

File:TOPIO 3.0.jpg *Source*: http://en.wikipedia.org/w/index.php?title=File:TOPIO_3.0.jpg *License*: unknown *Contributors*: User:Humanrobo

Image:GFO taxonomy tree.png *Source*: http://en.wikipedia.org/w/index.php?title=File:GFO_taxonomy_tree.png *License*: unknown *Contributors*: Original uploader was Leechuck at en.wikipedia

Image:Hierarchical-control-system.svg *Source*: http://en.wikipedia.org/w/index.php?title=File:Hierarchical-control-system.svg *License*: unknown *Contributors*: user:Pbroks13

File:ParseTree.svg *Source*: http://en.wikipedia.org/w/index.php?title=File:ParseTree.svg *License*: unknown *Contributors*: User:Stannered

File:Kismet robot at MIT Museum.jpg *Source*: http://en.wikipedia.org/w/index.php?title=File:Kismet_robot_at_MIT_Museum.jpg *License*: unknown *Contributors*: User:Tsca.bot, User:Waldir, user:Polimerek

File:ArtificialFictionBrain.png *Source*: http://en.wikipedia.org/w/index.php?title=File:ArtificialFictionBrain.png *License*: unknown *Contributors*: Gengiskanhg, Jtneill, Scientist94, 1 anonymous edits

File:Artificial neural network.svg *Source*: http://en.wikipedia.org/w/index.php?title=File:Artificial_neural_network.svg *License*: unknown *Contributors*: Cburnett, JokerXtreme, Mdd, 2 anonymous edits

File:Automated online assistant.png *Source*: http://en.wikipedia.org/w/index.php?title=File:Automated_online_assistant.png *License*: unknown *Contributors*: User:Mikael Häggström

File:Street Fighter II.png *Source*: http://en.wikipedia.org/w/index.php?title=File:Street_Fighter_II.png *License*: unknown *Contributors*: MuZemike, N. Harmonik, Randomran, X201

File:Virtua Fighter.png *Source*: http://en.wikipedia.org/w/index.php?title=File:Virtua_Fighter.png *License*: unknown *Contributors*: User:Tyan23

File:Fatality.png *Source*: http://en.wikipedia.org/w/index.php?title=File:Fatality.png *License*: unknown *Contributors*: Chrysaor, Crt, EVula, Fabricationary, Jeff Silvers, Master Bigode, MuZemike, N. Harmonik, Randomran, Tintor2, 2 anonymous edits

File:Kyoto arcade.jpg *Source*: http://en.wikipedia.org/w/index.php?title=File:Kyoto_arcade.jpg *License*: unknown *Contributors*: Ethan Hein

CPSIA information can be obtained at www.ICGtesting.com
Printed in the USA
LVOW101633151112

307507LV00006B/53/P